SURROGATE MODELING FOR HIGH-FREQUENCY DESIGN RECENT ADVANCES

Other World Scientific Titles by the Author

Multi-Objective Design of Antennas Using Surrogate Models
by Slawomir Koziel and Adrian Bekasiewicz
ISBN: 978-1-78634-147-1

Simulation-Based Optimization of Antenna Arrays
by Slawomir Koziel and Stanislav Ogurtsov
ISBN: 978-1-78634-598-1

Simulation-Driven Aerodynamic Design Using Variable-Fidelity Models
by Leifur Leifsson and Slawomir Koziel
ISBN: 978-1-78326-628-9

Simulation-Driven Design Optimization and Modeling for Microwave Engineering
edited by Slawomir Koziel, Xin-She Yang and Qi-Jun Zhang
ISBN: 978-1-84816-916-6

SURROGATE MODELING FOR HIGH-FREQUENCY DESIGN RECENT ADVANCES

Editors

Slawomir Koziel
Reykjavik University, Iceland

Anna Pietrenko-Dabrowska
Gdansk University of Technology, Poland

NEW JERSEY • LONDON • SINGAPORE • BEIJING • SHANGHAI • HONG KONG • TAIPEI • CHENNAI • TOKYO

Published by

World Scientific Publishing Europe Ltd.
57 Shelton Street, Covent Garden, London WC2H 9HE
Head office: 5 Toh Tuck Link, Singapore 596224
USA office: 27 Warren Street, Suite 401-402, Hackensack, NJ 07601

Library of Congress Cataloging-in-Publication Data
Names: Koziel, Slawomir, editors. | Pietrenko-Dabrowska, Anna, editors.
Title: Surrogate modeling for high-frequency design : recent advances /
    editors, Slawomir Koziel, Reykjavik University, Iceland,
    Anna Pietrenko-Dabrowska, Gdansk University of Technology, Poland.
Description: London ; Singapore ; Hackensack, NJ : World Scientific, [2022] |
    Includes bibliographical references and index.
Identifiers: LCCN 2021028027 (print) | LCCN 2021028028 (ebook) |
    ISBN 9781800610743 (hardcover) | ISBN 9781800610750 (ebook) |
    ISBN 9781800610767 (ebook other)
Subjects: LCSH: Microwave devices--Mathematical models. | Radio frequency integrated circuits--
    Mathematical models. | Microwaves--Mathematical models.
Classification: LCC TK7876 .S78 2022 (print) | LCC TK7876 (ebook) | DDC 621.381/3--dc23
LC record available at https://lccn.loc.gov/2021028027
LC ebook record available at https://lccn.loc.gov/2021028028

British Library Cataloguing-in-Publication Data
A catalogue record for this book is available from the British Library.

Copyright © 2022 by World Scientific Publishing Europe Ltd.

All rights reserved. This book, or parts thereof, may not be reproduced in any form or by any means, electronic or mechanical, including photocopying, recording or any information storage and retrieval system now known or to be invented, without written permission from the Publisher.

For photocopying of material in this volume, please pay a copying fee through the Copyright Clearance Center, Inc., 222 Rosewood Drive, Danvers, MA 01923, USA. In this case permission to photocopy is not required from the publisher.

For any available supplementary material, please visit
https://www.worldscientific.com/worldscibooks/10.1142/Q0317#t=suppl

Desk Editors: Jayanthi Muthuswamy/Michael Beale/Shi Ying Koe

Typeset by Stallion Press
Email: enquiries@stallionpress.com

© 2022 World Scientific Publishing Europe Ltd.
https://doi.org/10.1142/9781800610750_fmatter

Preface

Contemporary high-frequency engineering design heavily relies on full-wave electromagnetic (EM) analysis. This is primarily due to its versatility and the ability to account for phenomena that are important from the point of view of system performance but cannot be adequately represented using traditional means, mostly analytical descriptions or equivalent networks. Notwithstanding, versatility comes at a price, which is a high computational cost of accurate system evaluation. This issue has been only partially mitigated by the tremendous progress in computer hardware and software. For modern real-world applications, it is counterbalanced by the growing demands concerning simulation accuracy, the increasing complexity of the systems, as well as the necessity of enlarging the computational domains, e.g., by including various environmental components such as connectors, installation fixtures, or radomes in the case of antennas and antenna arrays. Consequently, utilization of simulation models in the design processes, although highly desirable, remains challenging.

Surrogate modeling offers practical workaround the aforementioned problems by providing cheaper representations of the systems, which enable massive evaluations without incurring prohibitive computational expenses. By far, the most popular class of surrogates are data-driven (or approximation) models, where the surrogate is constructed by approximating sampled data from the original (or high-fidelity) simulation model. A large variety of techniques have been developed, including a range of response surface approximation models (e.g., polynomial regression but also more sophisticated techniques such as orthogonal matching pursuit or least angle regression), radial basis function interpolation, kriging, support vector regression, artificial neural networks, and Gaussian process regression. Unfortunately, majority of these techniques are affected by the curse of dimensionality, which limits the number of independent parameters of the system the surrogate can be constructed for, but also the parameter ranges. This is especially pronounced in high-frequency electronics (microwave engineering, antenna design, photonics), where typical system responses are highly nonlinear. Increasing the parameter space dimensionality affects the number of training data points required to obtain sufficient model accuracy in a significant manner. At the same time, practically useful (i.e., design ready) models need to cover wide ranges of parameters and operating conditions. This is a considerable challenge as indicated in the literature and revealed through the analysis of reported modeling methods and application cases studies. The latter — in the areas relevant to the scope of this book, i.e., high-frequency electronics — are typically limited to a few parameters and narrow parameter ranges.

The second major class of surrogates, physics-based models (e.g., space mapping, or various response correction methodologies) offer improved generalization capability at the expense of reduced versatility. Physics-based surrogates are normally constructed using underlying low-fidelity models (e.g., coarse-mesh EM simulations in the case of antenna structures, or network equivalents in the case of microwave components). Appropriate selection of such models, specifically, the balance between speed and accuracy, are the critical factors deciding about reliability of the resulting surrogates.

This volume presents a selection of works representing the recent advancements in surrogate modeling (both forward and inverse) and their applications to high-frequency design. The book begins with the two chapters that address the fundamentals of data-driven and physics-based modeling. Further chapters, provide a review of specific topics such as neural network modeling of microwave components, inverse modeling for radio-frequency integrated circuits, or performance-driven modeling of antennas in constrained domains, whereas others describe the recent attempts to improve over the existing modeling methodologies, whether to enhance their computational efficiency, reliability, or scalability (e.g., with respect to the parameter space dimension). Furthermore, the book features numerous applications of surrogate modeling methodologies to design optimization and uncertainty quantification of antenna, microwave, and analog RF circuits.

Slawomir Koziel and Anna Pietrenko-Dabrowska
December 2021

© 2022 World Scientific Publishing Europe Ltd.
https://doi.org/10.1142/9781800610750_fmatter

About the Editors

Slawomir Koziel received the M.Sc. and Ph.D. degrees in electronic engineering from Gdansk University of Technology, Poland, in 1995 and 2000, respectively. He also received the M.Sc. degrees in theoretical physics and in mathematics, in 2000 and 2002, respectively, as well as the Ph.D. in mathematics in 2003, from the University of Gdansk, Poland. He is currently a Professor with the Department of Engineering, Reykjavik University, Iceland. His research interests include CAD and modeling of microwave and antenna structures, simulation-driven design, surrogate-based optimization, space mapping, circuit theory, analog signal processing, evolutionary computation and numerical analysis.

Anna Pietrenko-Dabrowska received the M.Sc. and Ph.D. degrees in electronic engineering from Gdansk University of Technology, Poland, in 1998 and 2007, respectively. Currently, she is an Associate Professor with Gdansk University of Technology, Poland. Her research interests include simulation-driven design, design optimization, control theory, modeling of microwave and antenna structures, numerical analysis.

© 2022 World Scientific Publishing Europe Ltd.
https://doi.org/10.1142/9781800610750_fmatter

List of Contributors

Mobayode O. Akinsolu
Faculty of Arts, Science and Technology, Wrexham Glyndŵr University, Mold Rd, Wrexham, Wales, United Kingdom
mobayode.akinsolu@glyndwr.ac.uk

Qingsha S. Cheng
Department of Electrical and Electronic Engineering, Southern University of Science and Technology, 1088 Xueyuan Ave., Nanshan, Shenzhen, Guangdong Province, China
chengqs@sustc.edu.cn

Ahmed E. Hammad H. Elqenawy
Engineering Mathematics and Physics Department, Faculty of Engineering, Cairo University, Giza, Egypt
aessam_2008@yahoo.com

Peter Excell
Faculty of Arts, Science and Technology, Wrexham Glyndŵr University, Mold Rd., Wrexham, Wales, United Kingdom
p.excell@glyndwr.ac.uk

Feng Feng
School of Microelectronics, Tianjin University, Tianjin, China
ff@tju.edu.cn

Francesco Ferranti
Microwave Department, IMT Atlantique, CNRS UMR 6285 Lab-STICC, Brest CEDEX 3, France
francesco.ferranti@imt-atlantique.fr

Paul Franzon
North Carolina State University, Raleigh, NC, USA
paulf@ncsu.edu

Abdel-Karim S.O. Hassan
Engineering Mathematics and Physics Department, Faculty of Engineering, Cairo University, Giza, Egypt
asho_hassan@yahoo.com

Jing Jin
School of Microelectronics, Tianjin University, Tianjin, China
jingjin5@cmail.carleton.ca

Slawomir Koziel
Reykjavik University, Menntavegur 1, Reykjavik, Iceland
koziel@ru.is

Leifur Leifsson
Department of Aerospace Engineering, Iowa State University, 537 Bissel Rd., Ames, IA, USA
leifur@iastate.edu

Bo Liu
James Watt School of Engineering, University of Glasgow, Scotland, United Kingdom
bo.liu@glasgow.ac.uk

Weicong Na
Faculty of Information Technology, Beijing University of Technology, Beijing, China
weicongna@bjut.edu.cn

Jethro Nagawkar
Department of Aerospace Engineering, Iowa State University, 537 Bissel Rd., Ames, IA, USA
jethro@iastate.edu

Anna Pietrenko-Dabrowska
Gdansk University of Technology, Narutowicza 11/12, Gdansk, Poland
anna.dabrowska@pg.edu.pl

Francisco E. Rangel-Patiño
Intel Corporation, Zapopan, Mexico
francisco.rangel@intel.com

José E. Rayas-Sánchez
ITESO — The Jesuit University of Guadalajara, Tlaquepaque, Mexico
erayas@iteso.mx

Yves Rolain
Department of Fundamental Electricity and Instrumentation, Vrije Universiteit Brussel, Boulevard de la Plaine 2, Ixelles, Brussels, Belgium
yves.rolain@vub.ac.be

Yi Wang
North Carolina State University, Raleigh, NC, USA
ywang73@ncsu.edu

Xhesila Xhafa
Istanbul Technical University, Faculty of Electrical and Electronics Engineering, Electronics and Communications Engineering Department, Istanbul, Turkey
xhafa15@itu.edu.tr

Mustafa Berke Yelten
Istanbul Technical University, Faculty of Electrical and Electronics Engineering, Electronics and Communications Engineering Department, Istanbul, Turkey
yeltenm@itu.edu.tr

Jianan Zhang
Department of Electronics, Carleton University, Ottawa, ON, Canada
jiananzhang@doe.carleton.ca

Qi-Jun Zhang
Department of Electronics, Carleton University, Ottawa, ON, Canada
QiJunZhang@cunet.carleton.ca

Zhen Zhang
Department of Electrical and Electronic Engineering, Southern
University of Science and Technology, Shenzhen, China; Harbin
Institute of Technology, Harbin, China
11849553@mail.sustech.edu.cn

Maral Zyari
Department of Fundamental Electricity and Instrumentation, Vrije
Universiteit Brussel, Boulevard de la Plaine 2, Ixelles, Brussels,
Belgium
mzyari@vub.ac.be

© 2022 World Scientific Publishing Europe Ltd.
https://doi.org/10.1142/9781800610750_fmatter

Acknowledgments

We would like to thank all the contributing authors for their contributions and their help during the peer-review process. The Editors would also like to thank the Production Editor Michael Beale, and staff at World Scientific for their help and professionalism.

© 2022 World Scientific Publishing Europe Ltd.
https://doi.org/10.1142/9781800610750_fmatter

Contents

Preface	v
About the Editors	ix
List of Contributors	xi
Acknowledgments	xv

1. Fundamentals of Data-Driven Surrogate Modeling   1
   Slawomir Koziel and Anna Pietrenko-Dabrowska

   1.1 Data-Driven Surrogates: Overview 2
   1.2 Design of Experiments 4
       1.2.1 Factorial designs 5
       1.2.2 Space-filling designs 6
       1.2.3 Sequential sampling 7
   1.3 Modeling Methods 9
       1.3.1 Polynomial regression 10
       1.3.2 Radial basis functions 11
       1.3.3 Kriging 12
       1.3.4 Polynomial chaos expansion 14
       1.3.5 Support vector regression 18
       1.3.6 Other methods 20
   1.4 Surrogate Model Validation 24
   Acknowledgments 28
   References 28

2. Fundamentals of Physics-Based Surrogate Modeling   39
   Anna Pietrenko-Dabrowska and Slawomir Koziel

   2.1 Physics-Based Surrogates: Overview 39

2.2 Low-Fidelity Models ... 42
2.2.1 Overview ... 42
2.2.2 Variable-resolution and variable-accuracy models ... 44
2.2.3 Variable-fidelity physics models ... 48
2.2.4 Selecting low-fidelity models ... 49
2.3 Physics-Based Surrogates — Basic Techniques ... 50
2.4 Response Correction Techniques ... 57
2.4.1 Quasi-global modeling using multi-point space mapping ... 57
2.4.2 Space mapping with a function approximation layer ... 58
2.4.3 Multi-point output space mapping ... 59
2.4.4 Generalized shape-preserving response prediction ... 61
2.4.5 Feature-based modeling ... 64
2.5 Physics-Based Surrogates for Design Optimization ... 69
Acknowledgments ... 72
References ... 73

3. Parametric Modeling of Microwave Components Using Combined Neural Network and Transfer Function ... 81

Feng Feng, Jianan Zhang, Weicong Na, Jing Jin, and Qi-Jun Zhang

3.1 Introduction ... 82
3.2 Parametric Modeling Using Neuro-Transfer Function in Rational Format ... 86
3.2.1 Formulation of the neuro-TF model in rational format ... 86
3.2.2 Two-stage training process of the neuro-TF model ... 87
3.3 Parametric Modeling Using Neuro-Transfer Function in Pole/Zero Format ... 89
3.3.1 Formulation of the neuro-TF model in pole/zero format ... 89
3.3.2 Pole/zero-matching algorithm for addressing the issue of mismatch of poles and zeros ... 90

	3.4	Parametric Modeling Using Neuro-TF Model in Pole/Residue Format	93
		3.4.1 Formulation of the neuro-TF model in pole/residue format	93
		3.4.2 Vector fitting for parameter extraction	93
		3.4.3 Pole–residue tracking technique for order-changing	94
	3.5	Sensitivity-Analysis-Based Neuro-TF Modeling Technique	98
		3.5.1 Structure of the sensitivity-analysis-based neuro-TF model	98
		3.5.2 Preliminary training process for the sensitivity-analysis-based neuro-TF model	99
		3.5.3 Refinement training process for the sensitivity-analysis-based neuro-TF model	100
	3.6	Neuro-TF Modeling Examples and RF/Microwave Applications	101
		3.6.1 Neuro-TF modeling in pole/zero format of a three-pole H-plane filter	101
		3.6.2 Neuro-TF modeling in pole/residue format of a microwave junction	105
		3.6.3 Sensitivity-analysis-based neuro-TF modeling of a diplexer	108
	3.7	Discussion	111
	3.8	Conclusion	115
	References	115
4.	Surrogate Model-Assisted Global Optimization for Antenna Design	123

Mobayode O. Akinsolu, Peter Excell, and Bo Liu

4.1	Introduction	124
4.2	Overview of the SADEA Algorithm Family	128
4.3	The PSADEA Method	129
	4.3.1 GP surrogate modeling	129
	4.3.2 The DE algorithm	131
	4.3.3 Implementation of PSADEA	133
4.4	Case Study	134
	4.4.1 Example one	135
	4.4.2 Example two	141

	4.5	Conclusions	147
	References		148

5.	Surrogate-Based Modeling and Design Optimization Techniques for Signal Integrity in High-Performance Computer Platforms	153

Francisco E. Rangel-Patiño and José E. Rayas-Sánchez

	5.1	Introduction			154
	5.2	Post-Silicon Validation			155
		5.2.1	Post-silicon tuning		157
		5.2.2	System margining		158
		5.2.3	Jitter tolerance testing		160
	5.3	PHY Surrogate Modeling			162
		5.3.1	Design of experiments		163
			5.3.1.1	Box Behnken	163
			5.3.1.2	Orthogonal arrays	164
			5.3.1.3	Sobol sequence	164
		5.3.2	Polynomial-based surrogate modeling		165
		5.3.3	Generalized regression neural networks		166
		5.3.4	Supported vector machines		166
		5.3.5	Kriging		166
		5.3.6	Artificial neural networks		169
			5.3.6.1	ANN topology	170
			5.3.6.2	ANN modeling and training	170
		5.3.7	PHY surrogate modeling results and comparisons		172
			5.3.7.1	SATA Gen3 HSIO link PHY modeling	176
			5.3.7.2	USB3.1 Gen1 HSIO link PHY modeling	180
	5.4	Receiver Equalization Surrogate-Based Optimization			183
		5.4.1	Objective function for system margining		183
		5.4.2	Surrogate-based optimization for system margining		186
		5.4.3	Objective function for system margining and jitter tolerance		187

		5.4.4	Surrogate-based optimization for system margining and jitter tolerance 189
			5.4.4.1 Test case 1: USB3 189
			5.4.4.2 Test case 2: SATA3 192
			5.4.4.3 Test case 3: PCIe 194
	5.5	Space Mapping Optimization for PHY Tuning 196
		5.5.1	Broyden-based input space mapping 196
		5.5.2	Fine model 197
		5.5.3	Coarse model 198
		5.5.4	ASM optimization 199
		5.5.5	Optimization results 201
	5.6	Discussion and Conclusion 202
	References 204

6. Performance-Driven Inverse/Forward Modeling of Antennas in Variable-Thickness Domains   213

Slawomir Koziel and Anna Pietrenko-Dabrowska

	6.1	Introduction 214
	6.2	Antenna Modeling Using Nested Kriging 217
		6.2.1 Design space objective space: First-level (inverse) surrogate 217
		6.2.2 Surrogate model domain: Second-level surrogate 220
	6.3	Variable-Thickness Domain 222
		6.3.1 Domain thickness: Model accuracy vs. utility trade-offs 222
		6.3.2 Variable-thickness domain: Definition and properties 223
		6.3.3 Nested kriging with variable-thickness domain 225
	6.4	Demonstration Case Studies 228
		6.4.1 Case I: Dual-band microstrip dipole antenna 228
		6.4.2 Case II: Broadband patch antenna 231
	6.5	Summary and Discussion 237
	Acknowledgments 238
	References 238

7. Sampling Methods for Surrogate Modeling and
   Optimization   245
   Qingsha S. Cheng and Zhen Zhang

   7.1 Introduction 246
   7.2 Conventional Sampling Methods 247
        7.2.1 Full factorial sampling method 248
        7.2.2 Monte Carlo sampling method 248
        7.2.3 Latin hypercube sampling method 249
        7.2.4 Space-infill sampling method 250
   7.3 A Hybrid Sampling Method for Surrogate
        Modeling and Optimization 251
        7.3.1 Local sampling 251
        7.3.2 Global sampling 253
        7.3.3 Microwave verification example 253
               7.3.3.1 Modeling performance 255
               7.3.3.2 Optimization performance 256
   7.4 Adaptive Sampling Region Updating for
        Surrogate-Assisted Optimization 257
        7.4.1 Adaptive sampling region updating
               strategy 258
        7.4.2 Surrogate refinement using fine model 260
        7.4.3 Antenna verification examples 261
   7.5 Conclusions 265
   Acknowledgments 266
   References 266

8. Statistical Design Centering of Microwave Systems
   via Space Mapping Technology and Modified Trust
   Region Algorithm   271
   *Abdel-Karim S.O. Hassan and Ahmed E. Hammad
   H. Elqenawy*

   8.1 Introduction 272
        8.1.1 Design centering problem 272
        8.1.2 Microwave design centering 276
   8.2 New Statistical Design Centering Technique for
        Microwave Systems 277
        8.2.1 Modified trust region algorithm 278

		8.2.1.1 Steps for the modified TR algorithm 281
	8.2.2	Generalized space mapping technique 282
	8.2.3	The statistical design centering algorithm . . . 283
8.3	Practical Examples 284	
	8.3.1	Bandstop microstrip filter with open stubs . . . 284
	8.3.2	Ultra-wideband multiple-input–multiple-output antenna 285
8.4	Conclusion 288	
Acknowledgments 289		
References 289		

9. Expedited Yield-Driven Design of High-Frequency Structures by Kriging Surrogates in Confined Domains 293

    Anna Pietrenko-Dabrowska and Slawomir Koziel

    9.1 Introduction 294
    9.2 Yield Optimization Problem and Benchmark Algorithms 298
        9.2.1 Yield optimization problem 298
        9.2.2 Surrogate-based yield optimization — benchmark Algorithm 1: One-shot optimization 301
        9.2.3 Surrogate-based yield optimization — benchmark Algorithm 2: Sequential approximate optimization 301
    9.3 Surrogate-Based Yield Optimization with Domain Confinement 302
        9.3.1 Yield optimization of multi-band antennas . . . 303
        9.3.2 Yield optimization of microwave couplers 305
    9.4 Demonstration Case Studies 307
        9.4.1 Case I: Ring-slot antenna 308
        9.4.2 Case II: Dual-band uniplanar dipole antenna . . 311
        9.4.3 Case III: Triple-band uniplanar dipole antenna 313
        9.4.4 Case IV: Compact microstrip rat-race coupler . 317
    9.5 Summary and Discussion 320
    Acknowledgments 321
    References 321

10. Solving the Inverse Problem Through
    Optimization — Applications to Analog/RF IC Design   325
    Yi Wang and Paul Franzon

    10.1 Introduction 325
    10.2 Overview of Proposed Design Flow 326
         10.2.1 Design space analysis 326
         10.2.2 Surrogate modeling method 327
         10.2.3 Adaptive sample strategy 327
    10.3 Bayesian Optimization Framework 328
         10.3.1 Overview of Bayesian optimization 328
              10.3.1.1 Acquisition function 328
    10.4 Candidate Point Search 330
    10.5 Design Analysis: Weight Setting 332
         10.5.1 Weight setting and optimization 334
    10.6 Example of Optimization and Design Reuse with
         Bayesian Optimization 335
    10.7 Surrogate Model Extension in Physical Design: Multi-
         Fidelity Optimization for Electromagnetic Simulation
         Acceleration 337
         10.7.1 Space mapping 338
         10.7.2 Overview of EM simulation acceleration ... 340
              10.7.2.1 Overview of multi-fidelity surrogate-
                       based optimization with candidate
                       search 340
         10.7.3 Multi-fidelity surrogate-based optimization
                with candidate search flow 341
              10.7.3.1 Design exploration 341
              10.7.3.2 Statistical surrogate model 344
              10.7.3.3 Adaptive sampling with dropout ... 346
              10.7.3.4 Adaptive samples filtering 350
              10.7.3.5 Low-fidelity dataset update 350
              10.7.3.6 Sample generation for model rebuild . 350
    10.8 Experimental Results 351
         10.8.1 Inductor design 351
              10.8.1.1 Optimization result comparison for
                       inductor 352
              10.8.1.2 IP redesign example — VCO 354

10.9 Conclusions	357
Acknowledgments	357
References	357

11. **An Automated and Adaptive Calibration of Passive Tuners Using an Advanced Modeling Technique** — 361

Maral Zyari, Francesco Ferranti, and Yves Rolain

11.1 Introduction	361
11.2 Behavior of a Passive Mechanical Tuner	364
11.3 The Proposed Algorithm	365
11.3.1 Adaptive sampling technique	366
11.3.2 Final model generation	369
11.3.3 Validation and error estimation	370
11.4 Numerical Results Based on Measurements	371
11.5 Conclusion	378
References	378

12. **Surrogate Modeling of High-Frequency Electronic Circuits** — 381

Xhesila Xhafa and Mustafa Berke Yelten

12.1 Introduction	381
12.2 Surrogate Modeling as a Circuit Optimization Tool	385
12.3 Variability Analysis of an LNA Using Surrogate Modeling	389
12.3.1 Case study I: Surrogate modeling of LNA performance parameters based on bondwire inductances	391
12.3.1.1 Surrogate modeling of S_{21} and S_{22}	393
12.3.1.2 Surrogate modeling of NF and S_{11}	396
12.3.2 Case study II: Surrogate model of the LNA gain variability	397
12.4 Concluding Remarks	400
References	401

13. Sensitivity Analysis and Optimal Design with
    PC-co-kriging   405
    Leifur Leifsson and Jethro Nagawkar

    13.1 Introduction 405
    13.2 Methods 407
         13.2.1 Surrogate modeling and analysis workflow . . . 407
         13.2.2 Sampling plan 408
         13.2.3 Constructing the surrogate model 409
         13.2.4 Validation 410
         13.2.5 Surrogate-based sensitivity analysis 410
         13.2.6 Surrogate-based optimal design 411
    13.3 Application Examples 412
         13.3.1 Surrogate modeling of the borehole
                function 412
         13.3.2 Model-based sensitivity analysis of
                ultrasonic testing 413
         13.3.3 Optimal design of transonic airfoil
                shapes 416
    13.4 Conclusion 421
    Acknowledgments 422
    References 422

Index   427

© 2022 World Scientific Publishing Europe Ltd.
https://doi.org/10.1142/9781800610750_0001

Chapter 1

Fundamentals of Data-Driven Surrogate Modeling

Slawomir Koziel and Anna Pietrenko-Dabrowska

Abstract

The primary topic of the book is surrogate modeling and surrogate-based design of high-frequency structures. The purpose of the first two chapters is to provide the reader with an overview of the two most important classes of modeling methods, data-driven (or approximation), as well as physics-based ones. These are covered in Chapters 1 and 2, respectively. The remaining parts of the book give an exposition of the specific aspects of particular modeling methodologies and their applications to solving various simulation-driven design tasks such as parametric optimization or uncertainty quantification.

Data-driven models are by far the most popular types of surrogates. This is due to several reasons, including versatility, low evaluation cost, a large variety of matured methods, and — important from the point of view of practical utility — widespread availability through third-party toolboxes implemented in programming environments such as Matlab. This chapter covers the fundamentals of approximation-based modeling. We discuss the surrogate modeling flow, design of experiments, selected modeling methods (e.g., kriging, radial basis functions, support vector regression, or polynomial chaos expansion), as well as discuss model validation approaches. The presented material is intended to provide the readers who are new to the subject with the basics necessary to understand the remaining parts of the book. On the other hand, it is by no means exhaustive, and the readers interested in a more detailed exposition can refer to a rich literature of the subject (e.g., Queipo *et al.*, 2005; Forrester and Keane, 2009; Biegler *et al.*, 2014; Chugh *et al.*, 2019; Jin, 2005; Santana-Quintero *et al.*, 2010; Gorissen *et al.*, 2009).

Keywords: Surrogate modeling, data-driven modeling, design of experiments, model validation

1.1 Data-Driven Surrogates: Overview

Approximation surrogates belong to the most popular class of metamodels. They are widely used in numerous disciplines of engineering and science. The literature offers a large variety of specific techniques in many variations (e.g., Simpson *et al.*, 2001; Søndergaard, 2003; Forrester and Keane, 2009; Couckuyt, 2013; Wang and Shan, 2006; Chen *et al.*, 2005). Furthermore, most of these techniques are readily accessible through various toolboxes (e.g., Lophaven *et al.*, 2002; Gorissen *et al.*, 2010; Marelli and Sudret, 2014), implemented in high-level programming environments such as Matlab. The attractiveness of data-driven models stems from the following features:

- Approximation models are based on the data acquired from the system of interest (in the case of high-frequency structures, typically, through full-wave EM simulation), thus, no a priori knowledge about the system of interest needs to be involved.
- Approximation models are generic, and can be easily transferred between various application domains.
- The models are computationally cheap to evaluate due to being based on explicit analytical formulations (e.g., linear combinations of basis functions, Montegranario, 2014).
- Availability of various (usually Matlab) toolboxes makes them easy to handle even by non-experts.

Despite these advantages, data-driven surrogates are affected by a fundamental issue, i.e., a normally large amount of training data is required to ensure reasonable predictive power. The number of necessary training samples grows quickly with the dimensionality of the design space (the problem referred to as the curse of dimensionality), and, even more importantly, with the ranges of the system parameters (e.g., geometry dimensions in the case of most high-frequency structures such as antennas or microstrip circuits). This is a fundamental problem from the point of view of design utility of the surrogates. The issue is more pronounced whenever the responses of the components and systems are highly nonlinear, e.g., in antenna and microwave engineering (Rayas-Sanchez *et al.*, 2017; Goudos, 2017; Rossi and Rizzo, 2009; Hausmair *et al.*, 2017). For cases like these, data-driven modeling is often limited to low-dimensional spaces (up to four or five parameters) and narrow parameter ranges

(Wu et al., 2019; Koziel and Leifsson, 2016). Techniques such as high-dimensional model representation (HDMR; Ma and Zabaras, 2010), model-order reduction (MOR; Baur et al., 2014), or variable-fidelity modeling (Fernández-Godino et al., 2019) might alleviate these difficulties to a certain extent.

The process of constructing data-driven surrogate normally consists of the following stages:

- *Design of experiments* (DoEs), i.e., allocating the training samples within the design space pertinent to the problem at hand. The sampling strategy depends on the source of the training data (e.g., space-filling designs are usually employed for computer-generated data; Simpson et al., 2001) and available computational budget.
- *Training data acquisition*. In most cases, the training data is nowadays obtained from computer simulation models.
- *Model identification*. The model parameters are found by solving a suitably formulated minimization problem, e.g., kriging (Kleijnen, 2009) or neural networks (Rayas-Sanchez, 2004), or analytically, by solving an appropriate linear regression problem (polynomial approximation, Queipo et al. (2005)).
- *Model validation*. At this step, the predictive power of the model is verified, including its approximation capability (i.e., fitting the training data set) and generalization capability (i.e., fitting the designs not seen during the identification stage). A proper balance between model approximation and generalization needs to be found in practice (Arlot and Celisse, 2010; James et al., 2013).

Figure 1.1 illustrates the surrogate modeling flowchart. The process may be iterative with the data acquisition, model identification and validation repeated upon extending the training set using additional (infill) samples. The procedure continues until the accuracy goals are met or the computational budget is exceeded.

The relationship between the surrogate model bias (the expected quality of approximating the training data samples) and variance (sensitivity of the surrogate output to particular data sets) should be emphasized. Generally, the bias can be decreased by using more complex (flexible) models, which would normally lead to the increase of the model variance. On the other hand, reducing the variance by means of regularization (e.g., smoothing achieved by penalizing the model complexity, Tikhonov and Arsenin (1977)) would

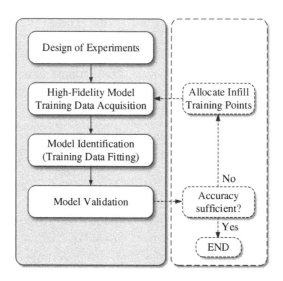

Fig. 1.1. A graphical illustration of the surrogate modeling process. An optional iterative procedure involving generation of additional training data (using a suitable infill strategy) is shown on the right-hand-side of the picture.

increase the bias error. Reduction of both the bias and the variance requires the enlargement of the training data set. This, however, may not be possible due to a limited computational budget. Furthermore, the effects of the additional data on the modeling error is limited especially for higher-dimensional problems due to poor scaling of point-to-point distance as a function of the data set size.

1.2 Design of Experiments

Design of experiments (DoEs) is the strategy for distributing the training data samples in the design space (Giunta *et al.*, 2003; Santner *et al.*, 2003; Koehler and Owen, 1996; Kleijnen, 2018; Santner *et al.*, 2018). DoE is an important stage of the modeling process as it determines the way of gathering the knowledge about the system of interest. Experimental design is followed by data acquisition from the high-fidelity simulation model. Clearly, the amount of information about the system depends on the training set size.

Notwithstanding, in most cases, available computational budget limits the number of samples that can be assigned.

This section outlines the three classes of DoE procedures: factorial designs (Section 1.2.1), space-filling designs (Section 1.2.2), and sequential sampling methods (Section 1.2.3). The last group has been attracting attention because feeding back information about the current sample distribution or the model performance allows us to allocate the data points in a more efficient manner (Couckuyt, 2013; Devabhaktuni *et al.*, 2001; Park *et al.*, 2018; Woods and Lewis, 2015).

1.2.1 *Factorial designs*

Factorial DoEs are traditional strategies (Giunta *et al.*, 2003) that allow for estimating the main and cooperative effects of design variables without using excessive numbers of samples. The points are typically allocated in the corners, edges and/or faces of the design space. At the same time, some of the factorial designs are rather "economical" in terms of the number of samples, which is important if the computational budget for data acquiring is limited. Figure 1.2 shows some representative examples of factorial designs.

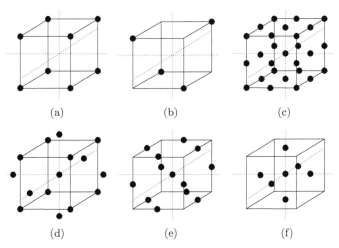

Fig. 1.2. Examples of factorial DoEs: (a) full factorial design, (b) fractional factorial design, (c) block design, (d) central composite design, (e) box-Behnken design, and (f) star distribution.

1.2.2 Space-filling designs

Nowadays, the data about the system of interest comes from deterministic computer simulations and spreading out the samples to reduce the effects of random errors that result from imperfect measurements is not necessary. Consequently, the majority of contemporary DoE algorithms are space-filling designs attempting to allocate the training points uniformly within the design space (Queipo et al., 2005). This is especially useful for constructing an initial surrogate model when the knowledge about the system is limited.

There are many space-filling DoEs available, including rudimentary approaches such as pseudo-random sampling (Giunta et al., 2003; Fig. 1.3(a)) and uniform grid sampling (Fig. 1.3(b)). Unfortunately, random sample allocation is associated with poor uniformity, whereas, distributing samples on the rectangular grid is only practical for low-dimensional spaces as the number of samples is restricted to $N_1 \cdot N_2 \cdots N_n$, where N_j is the number of samples along jth axis of the design space.

One of the most popular space-filling DoEs is Latin Hypercube Sampling (LHS) (McKay et al., 1979). To allocate p samples with LHS, the range for each parameter is divided into p bins, which for n design variables, yields a total number of p^n bins in the design space. The samples are distributed into the bins using the following two rules: (i) each sample is randomly placed inside a bin, and (ii) for all 1D projections of the p samples and bins, there is exactly one sample in each bin. Figure 1.3(c) provides an illustration example.

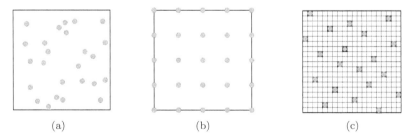

Fig. 1.3. Elementary space-filling DoEs: (a) pseudo-random sampling, (b) grid sampling, (c) Latin hypercube sampling (LHS) with a typical allocation of $p = 20$ samples in a 2D space.

Because the above rules are insufficient to ensure sample allocation uniformity, various improvements have been developed (e.g., Beachkofski and Grandhi, 2002; Leary et al., 2003; Ye, 1998; Palmer and Tsui, 2001).

Other popular space-filling DoEs include orthogonal array (OA) sampling (Queipo et al., 2005), quasi-Monte Carlo sampling (Giunta et al., 2003), or Hammersley sampling (Giunta et al., 2003). Space-filling DoE can also be realized as an optimization task by minimizing a suitably defined non-uniformity measure, e.g., $\sum_{i=1,\ldots,p} \sum_{j=i+1,\ldots,p} d_{ij}^{-2}$ (Leary et al., 2003), where d_{ij} is the Euclidean distance between samples i and j. Solving such problems may be challenging due to a large number of parameters that have to be optimized (equal to $p \cdot n$).

1.2.3 *Sequential sampling*

Sequential DoE (also known as adaptive sampling; Lehmensiek et al. (2002), or active learning; Sugiyama (2006)) is an iterative process, where the data obtained from the previous iterations (both the surrogate model and samples) are analyzed in order to allocate the new (infill) samples, primarily in the regions that are more difficult to approximate. The benefit is a more efficient distribution of samples as compared to traditional DoEs (Sasena et al., 2002).

One of the crucial aspects of sequential DoEs is to maintain the balance between exploration and exploitation (Crombecq et al., 2011). Exploration aims at identifying the design space regions that contain discontinuities, optima, etc. In practice, it boils down to filling up the domain in a uniform manner. The system response is not involved in the process. Yet, the locations of the infill samples in each iteration depend on the samples already distributed in the design space, which is the major difference between the sequential and one-shot DoEs. Exploitation focuses on the regions that have been already identified and tries to allocate infill samples therein. The goal is to obtain a better representation of the specific parts of the space, e.g., vicinities of the optima. Exploitation involves the system outputs, evaluated at the previously allocated points (Crombecq et al., 2011).

The development of experimental design strategies needs to take into account several criteria, including granularity (e.g., fine-grained DoEs select a small number of infill samples in each iteration, preferably one, Qian, 2009), space-filling (quantified using suitable metrics, e.g., the Manhattan; van Dam et al. (2007), or the Maximin; Joseph and Hung (2008)), as well as the projective properties. The details are omitted for the sake of brevity. The interested reader may refer to the rich literature of the subject (van Dam et al., 2007; Viana et al., 2009; Xiong et al., 2009).

A large variety of sequential DoE strategies have been developed. The most popular ones include sequential LHS (Liu et al., 2016; Wang, 2003; Tong, 2006), low-discrepancy sequences such as Halton sequence or Sobol sequence (Giunta et al., 2003; Chi et al., 2005), techniques based on Delaunay triangulation (Davis and Ierapetritou, 2010, see Fig. 1.4), as well as techniques involving a Voronoi tessellation (Crombecq et al., 2011). These are all exploratory methods.

For sequential designs involving exploitation, an important consideration is the surrogate model purpose. In particular, choosing appropriate infill criteria is an important aspect of surrogate-assisted optimization (Forrester and Keane, 2009). The two main goals for selecting the infill points include reduction of the objective function value and improvement of the global accuracy of the surrogate (Couckuyt, 2013). The simplest infill strategy is to allocate a single

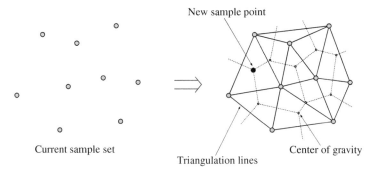

Fig. 1.4. Sequential sampling using Delaunay triangulation. The current experimental design is marked using large gray circles. Solid lines represent its Delaunay triangulation, whereas dashed lines determine the gravity centers of the corresponding simplexes. The new sample (marked using a large black circle) is allocated in the center of the largest-volume simplex. The procedure is iterated until allocating a required number of points.

sample at the surrogate model optimum (Alexandrov et al., 1998). In general, allocation of the new training point may be oriented towards global search or constructing a globally accurate surrogate. In this context, information about the expected model error is necessary (Kleijnen, 2018; Jones et al., 1998; Gorissen et al., 2010). The popular infill criteria include:

1. Maximization of the expected improvement, i.e., the improvement one expects to achieve at an untried point x (Jones et al., 1998).
2. Minimization of the predicted objective function $\hat{y}(x)$, i.e., surrogate optimization already mentioned earlier (Liu et al., 2012).
3. Minimization of the statistical lower bound, i.e., $LB(x) = \hat{y}(x) - As(x)$ (Forrester and Keane, 2009), where $\hat{y}(x)$ is the surrogate model prediction and $s^2(x)$ is the variance; A is a user-defined coefficient.
4. Maximization of the probability of improvement, i.e., identifying locations that give the highest chance of improving the objective function value (Forrester and Keane, 2009).
5. Maximization of the mean square error, i.e., finding locations where the predicted mean square error is the highest (Liu et al., 2012).

Note that identifying the new samples according to the above infill criteria requires global optimization (Couckuyt, 2013). Putting more focus on exploitation usually leads to a reduced computational cost. Exploration results in higher cost but also global search capability (Forrester and Keane, 2009). Yet, global exploration is often impractical, especially for expensive functions. Maintaining balance between exploitation and exploration is particularly important in the optimization context. Minimization of the statistical lower bound is an example of achieving such a balance controlled by the constant A (from pure exploitation, i.e., $LB(x) \to \hat{y}(x)$, for $A \to 0$ to pure exploration for $A \to \infty$) (Forrester and Keane, 2009).

1.3 Modeling Methods

There is quite a few data-driven modeling techniques that are well established and widely used, including polynomial regression

(Queipo et al., 2005), radial-basis functions (Wild et al., 2008), kriging (Forrester and Keane, 2009), neural networks (Haykin, 1998), support vector regression (Gunn, 1998), Gaussian Process Regression (Rasmussen and Williams, 2006), polynomial chaos expansion (Blatman and Sudret, 2010). For the convenience of the reader, this chapter outlines the selected approaches. More details can be found in the literature (e.g., Simpson et al., 2001; Jin, 2005; Chen et al., 2005; Wang and Shan, 2006; Goel et al., 2007).

In the following, the training samples will be denoted as $\{\boldsymbol{x}\}^{(i)}, i = 1, \ldots, p$, whereas the corresponding high-fidelity model evaluations as $f(\boldsymbol{x}^{(i)})$. The surrogate model is constructed by approximating the data pairs $\{\boldsymbol{x}^{(i)}, f(\boldsymbol{x}^{(i)})\}$.

1.3.1 *Polynomial regression*

Approximation by low-order polynomials is a simple yet efficient way of constructing the surrogates especially when the region of interest is small or the functional landscape of the high-fidelity model is regular. The surrogate model is defined as (Queipo et al., 2005)

$$\boldsymbol{s}(\boldsymbol{x}) = \sum_{j=1}^{K} \beta_j v_j(\boldsymbol{x}), \quad (1.1)$$

where β_j are unknown coefficients and v_j are the (polynomial) basis functions. The model parameters can be found as a least-square solution to the linear system

$$\boldsymbol{f} = \boldsymbol{\psi}\boldsymbol{\beta}, \quad (1.2)$$

where $\boldsymbol{f} = [f(\boldsymbol{x}^{(1)})\ f(\boldsymbol{x}^{(2)}) \cdots f(\boldsymbol{x}^{(p)})]^T$, ψ is a $p \times K$ matrix containing the basis functions evaluated at the sample points, and $\boldsymbol{\beta} = [\beta_1 \beta_2 \cdots \beta_K]^T$. The number of sample points p should be consistent with the number of basis functions considered K (typically $p \geq K$). If the sample points and basis functions are taken arbitrarily, some columns of ψ can be linearly dependent. If $p \geq K$ and rank $(\psi) = K$, a solution to (1.2) in the least-squares sense can be computed through ψ^+, the pseudoinverse of ψ (Golub and Van Loan, 1996), i.e., $\boldsymbol{\beta} = \boldsymbol{\psi}^+\boldsymbol{f} = (\boldsymbol{\psi}^T\boldsymbol{\psi})^{-1}\boldsymbol{\psi}^T$.

One of the simplest yet useful examples of a regression model is a second-order polynomial one defined as

$$s(x) = s\left([x_1\ x_2\ \cdots\ x_n]^T\right) = \beta_0 + \sum_{j=1}^{n} \beta_j x_j + \sum_{i=1}^{n} \sum_{j \leq i}^{n} \beta_{ij} x_i x_j, \quad (1.3)$$

with the basis functions being monomials: 1, x_j, and $x_i x_j$.

1.3.2 Radial basis functions

Radial basis function interpolation/approximation surrogates (Forrester and Keane, 2009; Wild et al., 2008) are regression models employing combinations of K radially symmetric functions ϕ

$$s(x) = \sum_{j=1}^{K} \lambda_j \phi\left(||x - c^{(j)}||\right), \quad (1.4)$$

$\lambda = [\lambda_1 \lambda_2 \cdots \lambda_K]^T$ denotes the vector of model parameters, and $c^{(j)}, j = 1, \ldots, K$, are the basis function centers. The model parameters can be calculated as $\lambda = \Phi^+ f = (\Phi^T \Phi)^{-1} \Phi^T f$, where $f = [f(x^{(1)}) f(x^{(2)}) \cdots f(x^{(p)})]^T$, and the $p \times K$ matrix $\Phi = [\Phi_{kl}]_{k=1,\ldots,p;l=1,\ldots,K}$, with the entries defined as

$$\Phi_{kl} = \phi\left(||x^{(k)} - c^{(l)}||\right). \quad (1.5)$$

If $p = K$, the centers of the basis functions coincide with the data points, and are all different, Φ is a regular square matrix. Then, $\lambda = \Phi^{-1} f$. However, finding the model coefficients by directly solving the system $\Phi \lambda = f$ is not practical when the number of training points is large (e.g., a few thousand) because matrix inversion requires computational time of the order of p^3 and the storage space $\sim p^2$. Also, sparse methods cannot be used as Φ is normally non-sparse. Alternative methods of dealing with large numbers of samples include multilevel (Liu, 2004) and multipole methods (Lschenbrand, 2016). A popular choice of the basis function is a Gaussian, $\phi(r) = \exp(-cr^2)$, where c is the scaling parameter, typically adjusted using cross-validation (Fasshauer and McCourt, 2012). Other types of basis functions are listed in Table 1.1.

Table 1.1. Commonly used radial basis functions.

Name	Formulation	Parameters
Gaussian	$\phi(r) = e^{-cr^2}$	$r \geq 0, c > 0$
Multiquadric	$\phi(r) = (r^2 + c^2)^{1/2}$	$r \geq 0, c > 0$
Inverse multiquadric	$\phi(r) = (r^2 + c^2)^{-1/2}$	$r \geq 0, c > 0$
Thin place spline	$\phi(r) = r^2 \log r$	$r \geq 0$
Polyharmonic spline	$\phi(r) = r^k$	$r \geq 0, k = 1, 3, 5, \ldots$

1.3.3 Kriging

Kriging is one of the most popular techniques for interpolating deterministic noise-free data (Journel and Huijbregts, 1981; Simpson *et al.*, 2001; Kleijnen, 2009; O'Hagan, 1978). In its basic formulation, kriging (Journel and Huijbregts, 1981) assumes that the function of interest is of the following form:

$$f(x) = g(x)^T \beta + Z(x), \tag{1.6}$$

where $g(x) = [g_1(x) \, g_2(x) \cdots g_K(x)]^T$ are known (e.g., constant) functions, $\beta = [\beta_1 \beta_2 \cdots \beta_K]^T$ are the unknown model parameters (hyperparameters), and $Z(x)$ is a realization of a normally distributed Gaussian random process with zero mean and variance σ^2. The regression part $g(x)^T \beta$ is a trend function for f, and $Z(x)$ takes into account localized variations. The covariance matrix of $Z(x)$ is given as

$$Cov\left[Z(x^{(i)}) Z(x^{(j)})\right] = \sigma^2 R\left([R(x^{(i)}, x^{(j)})]\right), \tag{1.7}$$

where R is a $p \times p$ correlation matrix with $R_{ij} = R(x^{(i)}, x^{(j)})$. Here, $R(x^{(i)}, x^{(j)})$ is the correlation function between sampled data points $x^{(i)}$ and $x^{(j)}$. A widely used correlation function is the Gaussian one

$$R(x, y) = \exp\left[-\sum_{k=1}^{n} \theta_k |x_k - y_k|^2\right], \tag{1.8}$$

where θ_k are the unknown correlation parameters (hyperparameters), and x_k and y_k are the kth components of the vectors $\boldsymbol{x}$ and $\boldsymbol{y}$, respectively.

The kriging predictor (Simpson et al., 2001; Journel and Huijbregts, 1981) is defined as

$$s(\boldsymbol{x}) = \boldsymbol{g}(\boldsymbol{x})^T \boldsymbol{\beta} + \boldsymbol{r}^T(\boldsymbol{x}) \boldsymbol{R}^{-1}(\boldsymbol{f} - \boldsymbol{G}\boldsymbol{\beta}), \quad (1.9)$$

where

$$\boldsymbol{r}(\boldsymbol{x}) = \left[R(\boldsymbol{x}, \boldsymbol{x}^{(1)}) \cdots R(\boldsymbol{x}, \boldsymbol{x}^{(p)}) \right], \quad (1.10)$$

$$\boldsymbol{f} = \left[f(\boldsymbol{x}^{(1)}) f(\boldsymbol{x}^{(2)}) \cdots f(\boldsymbol{x}^{(p)}) \right]^T, \quad (1.11)$$

and $\boldsymbol{G}$ is a $p \times K$ matrix with $G_{ij} = g_j(\boldsymbol{x}^{(i)})$. The vector of model parameters β can be computed as

$$\boldsymbol{\beta} = (\boldsymbol{G}^T \boldsymbol{R}^{-1} \boldsymbol{G})^{-1} \boldsymbol{G}^T \boldsymbol{R}^{-1} \boldsymbol{f}. \quad (1.12)$$

Model fitting is accomplished by maximum likelihood for θ_k (Journel and Huijbregts, 1981), i.e., by maximizing $-[p \ln(\sigma^2) + \ln |\boldsymbol{R}|]/2$, in which both σ^2 and $\boldsymbol{R}$ are functions of θ_k.

An important property of kriging is that the random process $Z(\boldsymbol{x})$ provides information on the approximation error that can be used for implementing various sequential sampling strategies (Forrester and Keane, 2009; Journel and Huijbregts, 1981), see also Section 1.2.3. This feature is also utilized in various global optimization methods (see Couckuyt (2013), and references therein). Figure 1.5 illustrates the kriging models for an exemplary function of two parameters.

A generalization of kriging that allows for incorporation of variable-fidelity training data is co-kriging (Forrester et al., 2007; Toal and Keane, 2011). Co-kriging works by constructing a regular kriging model using densely-sampled low-fidelity (and cheap to acquire) data points; subsequently, a separate kriging model is generated on the residuals of the high- and low-fidelity samples. The number of the former is much smaller than for the low-fidelity model. Co-kriging is a rather recent method with relatively few applications in engineering (Toal and Keane, 2011; Huang and Gao, 2012; Laurenceau and Sagaut, 2008; Koziel et al., 2013).

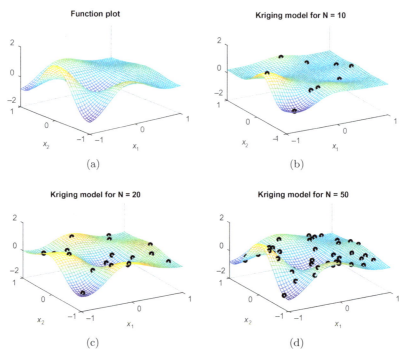

Fig. 1.5. Exemplary function of two variables and its kriging surrogate obtained for various number of training samples: (a) function plot, (b)–(d) kriging models with 10, 20, and 50 samples, respectively. Design of experiments: Latin hypercube sampling.

1.3.4 *Polynomial chaos expansion*

Polynomial chaos expansion (PCE) attempts to build a model of stochastic variations of the system of interest (Xiu and Karniadakis, 2002). The model inputs are probability distributions of the considered random parameters (e.g., manufacturing tolerances, uncertainties concerning operating conditions, etc., Kaintura et al., 2018). The technique has gained considerable popularity over the last 20 years or so owing to some recent developments (Kim et al., 2013; Du and Roblin, 2017; Manfredi et al., 2017). PCE allows for direct estimation of the statistical moments of the output probability distributions without the necessity of conducting Monte Carlo simulations (Gong et al., 2012). This section provides a brief outline of the method, including basic properties, truncation schemes, and methods for calculating the expansion coefficients.

We denote by $\boldsymbol{X} \in R^n$ a random vector described by the joint probability density function $f_{\boldsymbol{X}}$. The output Y of the system of interest is described by a map $Y = M(\boldsymbol{X})$ such that the second-order moments of Y are finite. The PCE of $M(\boldsymbol{X})$ is defined as (Blatman and Sudret, 2010)

$$Y = M(\boldsymbol{X}) = \sum_{\alpha \in N^n} a_\alpha \Psi_\alpha(\boldsymbol{X}), \quad (1.13)$$

where $\Psi_\alpha(\boldsymbol{X})$ are multivariate polynomials (orthonormal with respect to $f_{\boldsymbol{X}}$), $\alpha \in N^n$ are multi-indices identifying polynomial components, whereas $a_\alpha \in R$ are the expansion coefficients (Kaintura et al., 2018). In practice, the infinite sum (1.13) has to be truncated so that

$$M(\boldsymbol{X}) \approx M^{PC}(\boldsymbol{X}) = \sum_{\alpha \in A} a_\alpha \Psi_\alpha(\boldsymbol{X}), \quad (1.14)$$

where A is the finite set of multi-indices.

The polynomial basis construction starts from univariate orthonormal polynomials $\phi_k^{(i)}(x_i)$ that fulfill the conditions

$$\left\langle \phi_j^{(i)}(x_i), \phi_k^{(i)}(x_i) \right\rangle = \int_{D_{X_i}} \phi_j^{(i)}(x_i) \phi_k^{(i)}(x_i) f_{X_i}(x_i) dx_i = \delta_{jk}, \quad (1.15)$$

where i is the index of the input variable, j and k are the polynomial degrees, $f_{X_i}(x_i)$ is the ith marginal distribution, whereas δ_{jk} is the Kronecker symbol (Sudret, 2008). The multivariate polynomials are then

$$\Psi_\alpha(\boldsymbol{x}) = \prod_{i=1}^{n} \phi_{\alpha_i}^{(i)}(x_i). \quad (1.16)$$

Table 1.2 shows the most popular families of univariate polynomials and the distributions with respect to which these polynomials are orthonormal.

Using PCE, the stochastic moments of the system output can be conveniently obtained from the expansion coefficients. More

Table 1.2. Univariate orthogonal polynomials for PCE applications.

Distribution	Polynomials	PDF	Hilbertian basis $\psi_k(x)$	Support range
Gaussian	Hermite $He_k(x)$	$\dfrac{1}{\sqrt{2\pi}}e^{-x^2/2}$	$He_k(x)/\sqrt{k!}$	$[-\infty, \infty]$
Uniform	Legendre $P_k(x)$	$\dfrac{1}{2}$	$P_k(x)/\sqrt{\dfrac{1}{2k+1}}$	$[-1, 1]$
Gamma	Laguerre $L_k^a(x)$	$x^a e^{-x}$	$L_k^a(x)/\sqrt{\dfrac{\Gamma(k+a+1)}{k!}}$	$[0, \infty]$
Beta	Jacobi $J_k^{a,b}(x)$	$\dfrac{(1-x)^a(1+x)^b}{B(a)B(b)}$	$J_k^{a,b}(x)/J_{a,b,k}$	$[-1, 1]$

$$J_{a,b,k}^2 = \frac{2^{a+b+1}}{2k+a+b+1}\frac{\Gamma(k+a+1)\Gamma(k+b+1)}{\Gamma(k+a+b+1)\Gamma(k+1)}$$

specifically, the mean and the variance can be calculated as

$$\mu^{PC} = E[M^{PC}(\boldsymbol{X})] = a_0, \tag{1.17}$$

$$(\sigma^2)^{PC} = E\left[(M^{PC}(\boldsymbol{X}) - \mu^{PC})^2\right] = \sum_{\substack{\alpha \in A \\ \alpha \neq 0}} a_\alpha^2. \tag{1.18}$$

Practical PCE can only use a finite number of basis functions. The most straightforward (standard) truncation scheme uses all polynomials of the n input variables of the total degree less than or equal to p, i.e., $A^{n,p} = \{\alpha \in N^n : |\alpha| \leq p\}$. The number of basis functions (and, therefore, expansion coefficients) is then

$$\text{card } A^{n,p} = P = \binom{n+p}{p}, \tag{1.19}$$

and grows quickly with both n and p. Other schemes are often used such as the maximum interaction scheme involving αs that have at most r non-zero elements (Blatman, 2009), or hyperbolic truncation, where $A^{n,p,q} = \{\boldsymbol{\alpha} \in A^{n,p} : \|\boldsymbol{\alpha}\|_q \leq p\}$, where the q-norm is defined as

$$\|\boldsymbol{\alpha}\| = \left(\sum_{i=1}^n \alpha_i^q\right)^{1/q}. \tag{1.20}$$

For $q = 1$, the hyperbolic truncation is the same as the standard one.

The polynomial chaos coefficients can be found using several methods. Here, only the non-intrusive methods are outlined, i.e., those that determine the coefficients based on post-processing of a set of system evaluations obtained through sampling the input random variables. In the projection method, directly following the PCE definition (1.13), we have

$$a_{\boldsymbol{\alpha}} = E[\Psi_{\boldsymbol{\alpha}}(\boldsymbol{X}) \cdot M(\boldsymbol{X})], \tag{1.21}$$

(1.21) can be cast into numerical integration problem solved by quadrature methods, e.g., the Gaussian quadrature (Gander and Gautschi, 2000).

We have

$$a_\alpha = \int_{\Omega_X} M(\boldsymbol{x})\Psi_\alpha(\boldsymbol{x})f_X(\boldsymbol{x})d\boldsymbol{x} \approx \sum_{i=1}^{N} w^{(i)} M\left(\boldsymbol{x}^{(i)}\right)\Psi_\alpha\left(\boldsymbol{x}^{(i)}\right), \quad (1.22)$$

where the weights $w^{(i)}$ and the quadrature points $\boldsymbol{x}^{(i)}$ come from Lagrange polynomial interpolation to guarantee evaluation exactness of the integral of functions of polynomial complexity (Gander and Gautschi, 2000).

Another approach to PCE coefficient determination is least-square regression. More specifically, one can write

$$Y = M(\boldsymbol{X}) = \sum_{j=0}^{P-1} a_j \Psi_j(\boldsymbol{X}) + \varepsilon_P = \boldsymbol{a}^T \Psi(\boldsymbol{X}) + \varepsilon_P, \quad (1.23)$$

where $P = \text{card } A^{n,p}$, ε_P stands for the truncation error, $\boldsymbol{a} = [a_0 \cdots a_{P-1}]^T$ is the coefficient vector, and $\Psi(\boldsymbol{x}) = [\Psi_0(\boldsymbol{x}) \cdots \Psi_{P-1}(\boldsymbol{x})]^T$ is the matrix of all orthonormal polynomials in $\boldsymbol{X}$ (Berveiller et al., 2006). The least-square problem is then defined as

$$\boldsymbol{a}^* = \arg\min E[(\boldsymbol{a}^T \Psi(\boldsymbol{X}) - M(\boldsymbol{X}))^2], \quad (1.24)$$

and can be solved using ordinary least-squares (OSL) (Berveiller et al., 2006).

In practical problems, low-order variable interactions are typically dominant, therefore, low-rank truncation schemes are preferred. An alternative strategy is penalizing the least-square problem (1.24) using a regularization term that favors low-rank solutions (Blatman, 2009; Blatman and Sudret, 2010; Kaintura et al., 2018). One of popular realizations of this idea is the LAR (least-angle regression) algorithm (Ahadi et al., 2016), where the basis functions are iteratively added to the so-called active set based on their correlation with the current residual (i.e., the difference between the data set and the model constructed using the basis functions selected so far).

1.3.5 *Support vector regression*

Support vector regression (SVR) (Gunn, 1998) is another popular modeling technique that found applications in many areas, including high-frequency electronics (Ceperic and Baric, 2004; Rojo-Alvarez et al., 2005; Yang et al., 2005; Meng and Xia, 2007; Xia et al., 2007;

Andrés et al., 2012; Zhang and Han, 2013). SVR exhibits good generalization capability (Angiulli et al., 2007) and easy training by means of quadratic programming (Smola and Schölkopf, 2004).

SVR uses the structural risk minimization (SRM) principle, which has been shown to be superior (Gunn, 1998) to traditional empirical risk minimization (ERM) principle, employed by, e.g., neural networks.

In this section, SVR is formulated for a vector-valued function f. Let $\boldsymbol{f}^k = f(\boldsymbol{x}^k) = [f_1^k f_2^k \cdots f_m^k]^T, k = 1, 2, \ldots, N$, denote the sampled high-fidelity model responses. The objective is to use SVR to approximate $\boldsymbol{f}^k$ at the base points $\boldsymbol{x}^k$, $k = 1, 2, \ldots, N$. For linear regression, we aim at approximating a training data set, the pairs $D_j = \{(\boldsymbol{x}^1, f_j^1), \ldots, (\boldsymbol{x}^N, f_j^N)\}, j = 1, 2, \ldots, m$, by a linear function $f_j(\boldsymbol{x}) = \boldsymbol{w}_j^T \boldsymbol{x} + b_j$. The optimal regression function is given by minimizing the functional

$$\Phi_j(\boldsymbol{w}, \xi) = \frac{1}{2}\|\boldsymbol{w}_j\|^2 + C_j \sum_{i=1}^{N} (\xi_{j.i}^+ + \xi_{j.i}^-). \qquad (1.25)$$

In (1.25), C_j is a user-defined value, whereas $\xi_{j.i}^+$ and $\xi_{j.i}^-$ are the slack variables representing upper and lower constraints on the output of the system. The typical cost function used in SVR is an ε-insensitive loss function defined as

$$L_\varepsilon(y) = \begin{cases} 0 & \text{for } |f_j(\boldsymbol{x}) - y| < \varepsilon \\ |f_j(\boldsymbol{x}) - y| & \text{otherwise.} \end{cases} \qquad (1.26)$$

The value of C_j determines the trade-off between the flatness of f_j and the amount up to which deviations larger than ε are tolerated (Gunn, 1998).

Here, we describe nonlinear regression employing the kernel approach, in which the linear function $\boldsymbol{w}_j^T \boldsymbol{x} + b_j$ is replaced by the nonlinear function $\Sigma_i \gamma_{j.i} K(\boldsymbol{x}^k, \boldsymbol{x}) + b_j$, where K is a kernel function. Thus, the SVR surrogate model is defined as

$$\boldsymbol{s}(\boldsymbol{x}) = \begin{bmatrix} \sum_{i=1}^{N} \gamma_{1.i} K(\boldsymbol{x}^i, \boldsymbol{x}) + b_1 \\ \vdots \\ \sum_{i=1}^{N} \gamma_{m.i} K(\boldsymbol{x}^i, \boldsymbol{x}) + b_m \end{bmatrix}, \qquad (1.27)$$

with parameters $\gamma_{j.i}$ and b_j, $j = 1, \ldots, m, i = 1, \ldots, N$ obtained according to a general SVR methodology. In particular, Gaussian kernels of the form $K(\boldsymbol{x}, \boldsymbol{y}) = \exp(-0.5 \cdot \|\boldsymbol{x} - \boldsymbol{y}\|^2 / c^2)$ with $c > 0$ can be used, where c is the scaling parameter. Both c, C_j and ε can be adjusted to minimize the generalization error calculated using crossvalidation (Queipo et al., 2005).

1.3.6 Other methods

Sections 1.3.1–1.3.5 outlined selected data-driven modeling techniques. Clearly, the number of approximation methods is considerably larger. Some of the popular ones are briefly characterized below.

Moving least squares (MLS) (Levin, 1998; Breitkopf et al., 2002) allows for reconstructing continuous functions from a set of sample points through the calculation of a weighted least-square measure biased towards the vicinity of the point at which the reconstruction is requested. The surrogate model is defined as

$$s(\boldsymbol{x}) = \sum_{j=1}^{K} \beta_j(\boldsymbol{x}) v_j(\boldsymbol{x}). \qquad (1.28)$$

In contrast to conventional regression surrogates (cf. Section 1.4.1), the coefficients are functions of $\boldsymbol{x}$. The local objective at $\boldsymbol{x}$ is to minimize

$$\sum_{i=1}^{p} \omega_i \left(\boldsymbol{x}^{(i)} - \boldsymbol{x} \right) \left(\sum_{j=1}^{K} \beta_j \left(\boldsymbol{x}^{(i)} \right) v_j(\boldsymbol{x}) - f\left(\boldsymbol{x}^{(i)} \right) \right)^2, \qquad (1.29)$$

where the weighting factors ω_i are monotonic functions of the distance between $\boldsymbol{x}$ and $\boldsymbol{x}^{(i)}$. A typical choice for the weights is

$$\omega_i \left(\|\boldsymbol{x} - \boldsymbol{x}^{(i)}\| \right) = \exp\left(-\|\boldsymbol{x} - \boldsymbol{x}^{(i)}\|^2 \right). \qquad (1.30)$$

MLS improves flexibility of the surrogate at the expense of the increased computational complexity because finding approximation for each $\boldsymbol{x}$ requires solving a new optimization problem.

Gaussian process regression (GPR) (Rasmussen and Williams, 2006) is another method that — similarly as kriging — addresses the approximation problem from a stochastic point view. Under certain conditions, GPR models can be shown to be equivalent to large neural networks (NN) (Rasmussen and Williams, 2006) while requiring much less regression parameters than NNs. A basic formulation of GPR has been given below.

A Gaussian process (GP) describes a distribution over functions. It can be written as $f(\boldsymbol{x}) \sim GP(m(\boldsymbol{x}), k(\boldsymbol{x}, \boldsymbol{x}'))$, with $\boldsymbol{x}, \boldsymbol{x}' \in R^n$, and $m(\boldsymbol{x})$ and $k(\boldsymbol{x}, \boldsymbol{x}')$ being the mean and covariance functions respectively (Rasmussen and Williams, 2006). The GP encapsulates all possible functions in the space of functions that subscribe to $m(\boldsymbol{x})$ and $k(\boldsymbol{x}, \boldsymbol{x}')$. The GPR model is semi-parametric in the sense that any sample function is not specified in terms of a finite number of parameters (e.g., weights of a linear model), but directly in the space of functions. For a finite (practical) training data set of n observations, $D = \{(\boldsymbol{x}_i, y_i) | i = 1, \ldots, p\}$, where y_i are scalars, the corresponding Gaussian process $f(\boldsymbol{x})$ would be implemented as the collection of random variables $f_i = f(\boldsymbol{x}_i)$, with any n-dimensional point under their jointly Gaussian distribution representing n values of a sample function with index set the set of inputs $\{\boldsymbol{x}_i\}$. The only parameterization that takes place is the specification of hyperparameters, which determine the properties of the mean and covariance functions. A popular function for calculating the covariance between the output random variables $f(\boldsymbol{x})$ and $f(\boldsymbol{x}')$ is the squared-exponential (SE) covariance function with automatic relevance determination (ARD) (MacKay, 1993),

$$\boldsymbol{k}_{SE}(\boldsymbol{x}_i, \boldsymbol{x}_j) = \sigma_f^2 \exp\left(-\frac{1}{2} \sum_{k=1}^{p} \frac{(x_{i.k} - x_{j.k})^2}{\tau_k{}^2}\right). \quad (1.31)$$

In (1.31), $x_{i,k}$ is the kth component of $\boldsymbol{x}_i$, τ_k is the (positive) characteristic length-scale parameter corresponding to the kth components of the input vectors, and $\sigma_f{}^2$ is the signal variance; $\sigma_f{}^2$ and τ_k are the hyperparameters of the covariance function. The hyperparameters are found to minimize the negative log marginal likelihood, which, in the noise-free case is given by (Belyaev et al., 2015)

$$\log P(\boldsymbol{y}|X) = -\frac{1}{2}\boldsymbol{y}^T \boldsymbol{K}^{-1} \boldsymbol{y} - \frac{1}{2} \log |\boldsymbol{K}| - \frac{p}{2} \log 2\pi, \quad (1.32)$$

where $\boldsymbol{K} = (X, X)$ is the $p \times p$ matrix of covariances evaluated between all possible pairs of p training outputs using the covariance function, X is the $n \times p$ matrix of training input (column) vectors $\boldsymbol{x}_i$,

$|\boldsymbol{K}|$ is the determinant of $\boldsymbol{K}$, and $\boldsymbol{y}$ is the training target (column) vector. The GPR predictions are made by assuming a jointly Gaussian (normal) distribution of zero mean over the p random variables representing the training outputs and contained in a column vector $\boldsymbol{f}$, and the n_* random variables representing the test outputs contained in $\boldsymbol{f}_*$. This is the prior distribution:

$$\begin{bmatrix} \boldsymbol{f} \\ \boldsymbol{f}_* \end{bmatrix} \sim N\left(\boldsymbol{0}, \begin{bmatrix} \boldsymbol{K}(X,X) & \boldsymbol{K}(X,X_*) \\ \boldsymbol{K}(X_*,X) & \boldsymbol{K}(X_*,X_*) \end{bmatrix}\right), \qquad (1.33)$$

in which $\boldsymbol{K}(X,X_*)$ is the $p \times p_*$ matrix of covariances evaluated between all pairs of p training and n_* test outputs, with X_* being a matrix containing the test input vectors (similar definitions hold for other matrices).

The distribution of the test outputs conditioned on the known training outputs $\boldsymbol{y}$, or the posterior distribution, can then be expressed as $\boldsymbol{f}_*|X_*,X,\boldsymbol{y} \sim N(\boldsymbol{m}, \sum)$ (Belyaev et al., 2015) with the mean $\boldsymbol{m}$ and covariance matrix $\sum$ given by $\boldsymbol{m} = \boldsymbol{K}(X_*,X)\boldsymbol{K}(X,X)^{-1}\boldsymbol{y}$, and $\Sigma = \boldsymbol{K}(X_*,X_*) - \boldsymbol{K}(X_*,X)\boldsymbol{K}(X,X)^{-1}\boldsymbol{K}(X,X_*)$. The predictive mean $\boldsymbol{m}$ contains the most likely values of the test outputs associated with the test input vectors in X_*, whereas the diagonal of the covariance matrix $\sum$ gives the corresponding predictive variances.

Although artificial neural networks (ANNs) (Haykin, 1998) find their primary application in pattern classification (Ou and Murphey, 2007, prediction and financial analysis (Takahashi et al., 2019), as well as control and optimization (Yan and Wang, 2015), they can be considered as yet another data-driven modeling technique. Neural networks have been popular choices in modeling electromagnetic systems (Christodoulou and Georgiopoulos, 2001; Mishra, 2001): for modeling of antennas (Rawat et al., 2012; Mishra et al., 2015), passive microwave components (Zhang et al., 2003; Rayas-Sanchez, 2004), as well as active devices such as RF power amplifiers (Fang et al., 2000; Xu et al., 2002).

The important component of a neural network (Haykin, 1998; Minsky and Papert, 1969) is the neuron (or single-unit perceptron). A neuron realizes a nonlinear operation illustrated in Fig. 1.6(a), where w_1 through w_n are regression coefficients, β is the bias value of the neuron, and T is a user-defined slope parameter. The most

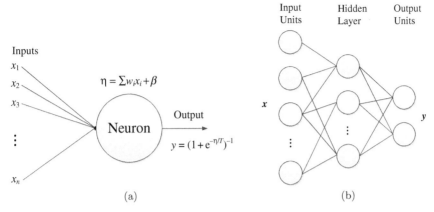

Fig. 1.6. Basic concepts of artificial neural networks: (a) structure of a neuron, (b) two-layer feed-forward neural network architecture.

common neural network architecture is the multi-layer feed-forward network, cf. Fig. 1.6(b).

The construction of a neural network model involves selection of its architecture and the assignment of the values to the regression parameters (referred to as training). Choosing the network architecture is a non-trivial task (Zhang et al., 2003), which can be automated using dedicated software frameworks (Neuromodeler; Zhang and Gupta, 2000).

The network training can be formulated as a nonlinear least-squares regression problem. A popular technique for solving this regression problem is the error back-propagation algorithm (Simpson et al., 2001; Haykin, 1998). Assuming the training data to be $\{(\boldsymbol{x}^{(1)}, y_1), (\boldsymbol{x}^{(2)}, y_2), \ldots, (\boldsymbol{x}^{(p)}, y_p)\}$, the proper values of the network weights are determined to minimize the approximation error

$$E = \sum_j (y_j - \hat{y}_j)^2, \qquad (1.34)$$

where $\hat{y}_j$ is the output given the input $\boldsymbol{x}^{(j)}$. The weights are adjusted based on $(\partial E/\partial y)/(\partial y/\partial w_{ij})$. Training of large networks with complex architectures may require global optimization (Alba and Marti, 2006).

We conclude this section by mentioning some of the techniques developed to handle problems affecting conventional approximation methods. Here, it is placed in the context of response surface modeling (RSM; Khuri and Mukhopadhyay, 2010) that plays an important role in relieving computational cost of simulation-driven design tasks including design optimization (Dorica and Giannacopoulos, 2006; Liu and Fu, 2016), worst-case analysis (Sengupta et al., 2005; Dharchoudhury and Kang, 1995) or parametric yield optimization (Bandler et al., 1993; Li et al., 2007).

In certain areas such as design of analog and mixed-signal (AMS; Rutenbar et al., 2007) circuits, conventional RSM using least-square regression is very challenging due to extremely high-dimensional parameter spaces (e.g., thousands or even tens of thousands of variables necessary to model AMS systems consisting of multiple transistors; Wang et al., 2016) as well as expensive circuit simulation. Although a large number of basis functions are necessary to span the high-dimensional variable space, majority of these functions are of little importance, i.e., many RSM model coefficients are close to zero (Wang et al., 2016). This renders a sparse structure which may be explored in order to reduce the computational cost of surrogate model construction. In particular, a large number of model coefficients (say, from 10^5 up to 10^6) can be identified from a small set (e.g., 10^2 to 10^3) samples without overfitting by means of Orthogonal Matching Pursuit (OMP) or L_1-regularization (Tao et al., 2016; Li, 2010). An alternative is Bayesian Model Fusion (BMF), where reduction of the cost is obtained by re-using the data obtained at the level of a simpler model, utilized at the early stages of the design process, when fitting a late-stage performance model (Wang et al., 2016).

1.4 Surrogate Model Validation

Validation of the surrogate is carried out to assess the model quality, including its predictive power. The two important components of the process are a selection of the error function and approach to estimating the approximation and generalization capability. Both are briefly discussed in the remaining part of this section.

Selection of the error function but also the target accuracy of the surrogate are generally problem dependent. Normally, certain knowledge about the structure of the system response is required along with sufficient understanding of the meaning of a particular generalization estimator to be used (Ling and Mahadevan, 2013). There are two categories of error functions: absolute and relative. Absolute errors are not unit-free and depend on the particular prediction value of the response. Although these features are undesirable, absolute errors are quite popular. On the other hand, relative error measures are generally preferred because of being more context independent. Some of the popular error functions along with their characteristics have been gathered in Tables 1.3 (absolute error functions) and 1.4 (relative error functions).

Having selected the error function, the model generalization capability can be estimated using several approaches. Clearly, evaluating the model quality merely based on its performance on the training set is not a good idea (e.g., interpolative models exhibit zero error therein by definition). Some of the techniques described earlier identify a surrogate model together with some estimation of the attendant approximation error (e.g., kriging or GPR). Alternatively, there are procedures that can be used in a stand-alone manner to validate the prediction capability of a given model beyond the set of training points. Probably the most popular technique is the split-sample method (Queipo et al., 2005), also referred to as a validation set methods, where part of the available data set (the training set) is used to construct the surrogate, whereas the second part (the test set) serves purely for model validation. However, the error estimated by a split-sample method depends strongly on how the set of data samples is partitioned. Also, it may give extremely biased results if only a few testing points are available.

Cross-validation (Queipo et al., 2005; Geisser, 1993) offers a more accurate estimation of the model generalization error. The method works by dividing the data set into K subsets, and sequentially using each of these subsets as testing set for a surrogate constructed on the other $K-1$ subsets.

The prediction error can be estimated with all the K error measures obtained in this process (e.g., as an average value). Its extreme version, the Leave-One-Out error (Vehtari et al., 2017) only uses a single point at a time for error estimation. Cross-validation provides an error estimation that is less biased than with the split-sample method. The disadvantage of this method is that the surrogate has to be constructed more than once.

Table 1.3. Popular absolute error functions.

Name	Formula[a]	Features
Root mean square error (AEE) (Goel et al., 2009)	$\text{RMSE}(y, \tilde{y}) = \sqrt{\frac{1}{N} \sum_{i=1}^{N} (y_i - \tilde{y}_i)^2}$	• Penalizes large errors • Ignores small errors (overly pessimistic) • Unintuitive to interpret
Average Euclidean error (AEE) (Stuart and Ord, 1994)	$\text{AEE}(y, \tilde{y}) = \frac{1}{N} \sum_{i=1}^{N} \sqrt{(y_i - \tilde{y}_i)^2}$	• Also pessimistic but not to the extent of RMSE
Geometric average error (GAE) (Couckuyt, 2013)	$\text{GAE}(y, \tilde{y}) = \left(\prod_{i=1}^{N} \sqrt{(y_i - \tilde{y}_i)^2} \right)^{1/N}$	• Optimistic • Dominated by small error terms
Harmonic average error (GAE) (Mitchell, 2004)	$\text{HAE}(y, \tilde{y}) = \left(\frac{1}{N} \sum_{i=1}^{N} \frac{1}{\sqrt{(y_i - \tilde{y}_i)^2}} \right)^{-1}$	• Optimistic • Dominated by small error terms

Note: [a] $y_i, \tilde{y}_i$ stand for the actual and surrogate-predicted system responses, respectively.

Table 1.4. Popular relative error functions.

Name	Formula[a]	Features
Average relative error (ARE) (Gorissen et al., 2009)	$\mathrm{ARE}(y, \tilde{y}) = \dfrac{1}{N} \sum_{i=1}^{N} \dfrac{|y_i - \tilde{y}_i|}{|y_i|}$	• Intuitively clear • Problematic when the true system outputs are close to zero
Root relative square error (Couckuyt, 2013)	$\mathrm{RRSE}(y, \tilde{y}) = \sqrt{\dfrac{\sum_{i=1}^{N} (y_i - \tilde{y}_i)^2}{\sum_{i=1}^{N} (y_i - \overline{y}_i)^2}}$	• Improves over ARE w.r.t. small outputs • Intuitively attractive (measures improvement over the mean)
Bayesian estimation error quotient (Gorissen et al., 2009)	$\mathrm{BEEQ}(y, \tilde{y}) = \left(\prod_{i=1}^{N} \dfrac{\sum_{i=1}^{N} |y_i - \tilde{y}_i|}{\sum_{i=1}^{N} |y_i - \overline{y}_i|} \right)^{1/N}$	• Improves over both ARE and RRSE by being less sensitive to large errors

Note: [a] $y_i, \tilde{y}_i$ stand for the actual and surrogate-predicted system responses, respectively.

Yet another approach is bootstrapping (Hall, 1986), which is also an iterative procedure but different from cross-validation because the testing set is randomly selected from the available data at each iteration. It has been shown to work better than cross-validation in many cases (Efron and Tibshirani, 1993).

As mentioned before, selection of the error function, the method for estimating the generalization capability, and the target accuracy, are all non-trivial tasks. In practice, some initial guesses concerning these factors are made, and the model is constructed upon acquiring the training data. The decisive stage is typically visual inspection of the model responses and its agreement with the system outputs. Upon this assessment, the modeling procedure may need to be repeated with the changed setup (Couckuyt, 2013).

The surrogate modeling flow, i.e., the procedure of allocating samples, acquiring data, model identification and validation can be repeated until the prescribed surrogate accuracy level is reached. In each repetition, a new set of training samples is added to the existing ones. Some of the strategies of allocating the new samples, especially the exploitative ones (Forrester and Keane, 2009) usually aim at improving the global accuracy of the model, i.e., inserting new samples at the locations where the estimated modeling error is the highest. The details concerning these strategies have been discussed in Section 1.2.3.

Acknowledgments

This chapter was supported in part by the Icelandic Centre for Research (RANNIS) Grant 206606 and by National Science Centre of Poland Grant 2018/31/B/ST7/02369.

References

Ahadi, M., Prasad, A.K., Roy, S. (2016). Hyperbolic polynomial chaos expansion (HPCE) and its application to statistical analysis of nonlinear circuits, in *Proc. IEEE 20th Workshop Sign. Power Integrity (SPI)*, Turin, Italy, pp. 1–4.

Alexandrov, N.M., Dennis, J.E., Lewis, R.M., Torczon, V. (1998). A trust-region framework for managing the use of approximation models in optimization, *Struct. Optimi.*, vol. 15, no. 1, pp. 16–23.

Andrés, E., Salcedo-Sanz, S., Monge, F., Pérez-Bellido, A.M. (2012). Efficient aerodynamic design through evolutionary programming and support vector regression algorithms, *Int. J. Expert Systems with Applications*, vol. 39, pp. 10700–10708.

Angiulli, G., Cacciola, M., Versaci, M. (2007). Microwave devices and antennas modelling by support vector regression machines, *IEEE Trans. Magnetics*, vol. 43, no.4, pp. 1589–1592.

Arlot, S., Celisse, A. (2010). A survey of cross-validation procedures for model selection, *Statist. Surv.*, vol. 4, pp. 40–79.

Bandler, J.W., Biernacki, R.M., Chen, S.H., Grobelny, P.A., Ye, S. (1993). Yield-driven electromagnetic optimization via multilevel multidimensional models, *IEEE Trans. Microw. Theory Tech.*, vol. 41, no. 12, pp. 2269–2278.

Baur, U., Benner, P., Feng, L. (2014). Model order reduction for linear and nonlinear systems: A system-theoretic perspective, *Arch. Computat. Methods. Eng.*, vol. 21, no. 4, pp. 331–358.

Beachkofski, B., Grandhi, R. (2002). Improved distributed hypercube sampling, *American Institute of Aeronautics and Astronautics*, Paper AIAA, 2002-1274.

Belyaev, M., Burnaev, E., Kapushev, Y. (2015). Gaussian process regression for structured data sets, in Gammerman, A., Vovk, V., Papadopoulos, H. (eds.), *Statistical Learning and Data Sciences*, Lecture Notes in Computer Science, vol. 9047, Springer, Cham.

Berveiller, M., Sudret, B., Lemaire, M. (2006). Stochastic finite elements: A non intrusive approach by regression, *Eur. J. Comput. Mech.*, vol. 15, no. 1–3, pp. 81–92.

Biegler, L.T., Lang, Y., Lin, W. (2014). Multi-scale optimization for process systems engineering, *Computers & Chemical Engineering*, vol. 60, no. 10, pp. 17–30.

Blatman, G. (2009). Adaptive sparse polynomial chaos expansions for uncertainty propagation and sensitivity analysis, PhD Thesis, Universite Blaise Pascal, Clermont-Ferrand, France.

Blatman, G., Sudret, B. (2010). An adaptive algorithm to build up sparse polynomial chaos expansions for stochastic finite element analysis, *Probab. Eng. Mech.*, vol. 25, no. 2, pp. 183–197.

Breitkopf, P., Rassineux, A., Villon, P. (2002). An introduction to moving least squares meshfree methods, *Revue Europ.Elements Finis*, vol. 11, no. 7–8, pp. 825–867.

Ceperic, V., Baric, A. (2004). Modeling of analog circuits by using support vector regression machines, *Proc. 11th Int. Conf. Electronics, Circuits, Syst.*, Tel-Aviv, Israel, pp. 391–394.

Chen, V.C.P., Tsui, K.-L., Barton, R.R., Meckesheimer, M. (2005). A review on design, modeling and applications of computer experiments, *IIE Transactions*, vol. 38, no. 4, pp. 273–291.

Chi, H., Mascagni, M., Warnock, T. (2005). On the optimal Halton sequence, *Math. Comput. Simul.*, vol. 70, pp. 9–21.

Christodoulou, C., Georgiopoulos, M. (2001). *Applications of Neural Networks in Electromagnetics*, Norwood, MA, USA, Artech House.

Chugh, T., Sindhya, K., Hakanen, J. Miettinen, K. (2019). A survey on handling computationally expensive multiobjective optimization problems with evolutionary algorithms, *Soft. Comput.*, vol. 23, no. 9, pp. 3137–3166.

Couckuyt, I. (2013). *Forward and inverse surrogate modeling of computationally expensive problems*, PhD Thesis, Ghent University.

Crombecq, K., Gorissen, D., Tommasi, L.D., Dhaene, T. (2009). A novel sequential design strategy for global surrogate modeling, *Proc. 41st Winter Simulation Conf.*, pp. 731–742.

Crombecq, K., Laermans, E., Dhaene, T. (2011). Efficient space-filling and non-collapsing sequential design strategies for simulation-based modeling, *European J. Operational Research*, vol. 214, no. 3, pp. 683–696.

Davis, E., Ierapetritou, M. (2010). A centroid-based sampling strategy for kriging global modeling and optimization, *AIChE J.*, vol. 56, no. 1, pp. 220–240.

Devabhaktuni, V.K., Yagoub, M.C.E., Zhang, Q.J. (2001). A robust algorithm for automatic development of neural-network models for microwave applications, *IEEE Trans. Microwave Theory Tech.*, vol. 49, no. 12, pp. 2282–2291.

Dharchoudhury, A., Kang, S.M. (1995). Worst-case analysis and optimization of VLSI circuit performances, *IEEE Trans. Comput.-Aided Design Integr. Circuits Syst.*, vol. 14, no. 4, pp. 481–492.

Dorica, M., Giannacopoulos, D.D. (2006). Response surface space mapping for electromagnetic optimization, *IEEE Trans. Magn.*, vol. 42, no. 4, pp. 1123–1126.

Du, J., Roblin, C. (2017). Statistical modeling of disturbed antennas based on the polynomial chaos expansion, *IEEE Ant. Wireless Prop. Lett.*, vol. 16, pp. 1843–1846.

Efron, B., Tibshirani, R. (1993). *Introduction to the Bootstrap*, New York, Chapman & Hall.

Fang, Y.H., Yagoub, M.C.E., Wang, F., Zhang, Q.J. (2000). A new macromodeling approach for nonlinear microwave circuits based on recurrent neural networks, *IEEE Trans. Microwave Theory Tech.*, vol. 48, no. 12, pp. 2335–2344.

Fasshauer, G.E., McCourt, M.J. (2012). Stable evaluation of Gaussian radial basis function interpolants, *SIAM J. Sci. Comput.*, vol. 34, no. 2, pp. A737–A762.

Fernández, M.G., Park, C., Kim, N.H., Haftka, R.T. (2019). Issues in deciding whether to use multifidelity surrogates, *AIAA J.*, vol. 57, no. 5, pp. 2039–2054.

Forrester, A.I.J., Keane, A.J. (2009). Recent advances in surrogate-based optimization, *Progress in Aerospace Sciences*, vol. 45, no. 1–3, pp. 50–79.

Gander, W., Gautschi, W. (2000). Adaptive quadrature revisited, *BIT Numerical Mathematics*, vol. 40, no. 1, pp. 84–101.

Geisser, S. (1993). *Predictive Inference*, New York, London, Chapman and Hall.

Giunta, A.A., Wojtkiewicz, S.F., Eldred, M.S. (2003). Overview of modern design of experiments methods for computational simulations, *Paper AIAA*, pp. 2003–0649.

Goel, T., Haftka, R.T., Shyy, W., Queipo, N.V. (2007). Ensemble of surrogates, *Struct. Multidisc. Optim.*, vol. 33, no. 3., pp. 199–216.

Goel, T., Haftka, R.T., Shyy, W. (2009). Comparing error estimation measures for polynomial and kriging approximation of noise-free functions, *Struct. Multidisc. Optim.*, vol. 38, no. 5, pp. 429–442.

Golub, G.H., Van Loan, Ch.F. (1996). *Matrix Computations*, 3rd edn., Baltimore, MD, USA, Johns Hopkins University Press.

Gong, F., Liu, X., Yu, H., Tan, S.X.D., Ren, J., He, L. (2012). A fast non-Monte-Carlo yield analysis and optimization by stochastic orthogonal polynomials, *ACM Trans. Des. Autom. Electron. Syst.*, vol. 17, no. 10, pp. 1–23.

Gorissen, D., Dhaene T., De Turck, F. (2009). Evolutionary model type selection for global surrogate modeling, *J. Mach. Learning Res.*, vol. 10, pp. 2039–2078.

Gorissen, D., Crombecq, K., Couckuyt, I., Dhaene, T., Demeester, P. (2010). A surrogate modeling and adaptive sampling toolbox for computer based design, *J. Machine Learning Research*, vol. 11, pp. 2051–2055.

Goudos, S. (ed.) (2017). *Microwave Systems and Applications*, London, UK, IntechOpen.

Gunn, S.R. (1998). Support vector machines for classification and regression, *Technical Report, School of Electronics and Computer Science*, University of Southampton.

Hall, P. (1986). On the bootstrap and confidence intervals, *Ann. Stat.*, vol. 14, pp. 1432–1452.

Hausmair, K., Gustafsson, S., Sanchez Perez, C., Landin, P.N., Gustavsson, U., Eriksson, T., Fager, C. (2017). Prediction of nonlinear distortion in wideband active antenna arrays, *IEEE Trans. Microw. Theory Tech.*, vol. 65, no. 11, pp. 4550–4563.

Haykin, S. (1998). *Neural Networks: A Comprehensive Foundation*, 2nd edn., Upper Saddle River, NJ, USA, Prentice Hall.

Huang, L., Gao, Z. (2012). Wing-body optimization based on multi-fidelity surrogate model, in *28th Int. Congress of the Aeronautical Sciences*, Brisbane, Australia.

James, G., Witten, D., Hastie, T., Tibshirani, R. (2013). *An Introduction to Statistical Learning*, New York, USA, Springer.

Jin, Y. (2005). A comprehensive survey of fitness approximation in evolutionary computation, *Soft. Computing.*, vol. 9, no. 1, pp. 3–12.

Jones, D., Schonlau, M., Welch, W. (1998). Efficient global optimization of expensive black-box functions, *J. Global Optim.*, vol. 13, pp. 455–492.

Joseph, V.R., Hung, Y. (2008). Orthogonal-maximin latin hypercube designs, *Statistica Sinica*, vol. 18, pp. 171–186.

Journel, A.G., Huijbregts, Ch.J. (1981). *Mining Geostatistics*, London, Academic Press.

Kaintura, A., Dhaene, T., Spina, D. (2018). Review of polynomial chaos-based methods for uncertainty quantification in modern integrated circuits, *Electronics*, vol. 7, no. 30, pp. 1–21.

Khuri, A.I., Mukhopadhyay S. (2010). Response surface methodology: Advanced review, *Computational Statistics*, vol. 2, no. 2, pp. 128–149.

Kim, K.K., Shen, D.E., Nagy, Z.K., Braatz, R.D. (2013). Wiener's polynomial chaos for the analysis and control of nonlinear dynamical systems with probabilistic uncertainties [Historical perspectives], *IEEE Control Syst. Mag.*, vol. 33, no. 5, pp. 58–67.

Kleijnen, J.P.C. (2009). Kriging metamodeling in simulation: A review, *European J. Operational Research*, vol. 192, no. 3, pp. 707–716.

Kleijnen, J.P.C. (2018). Design and analysis of simulation experiments, in Pilz, J., Rasch, D., Melas, V., Moder, K. (eds), *Statistics and Simulation. IWS 2015*, Springer Proceedings in Mathematics & Statistics, vol. 231, Springer, Cham.

Koehler, J.R., Owen, A.B. (1996). Computer experiments, in Ghosh, S., Rao, C.R. (eds.), *Handbook of Statistics*, Elsevier Science B.V. 13, pp. 261–308.

Koziel, S., Ogurtsov, S., Couckuyt, I., Dhaene, T. (2013). Variable-fidelity electromagnetic simulations and co-kriging for accurate modeling of antennas, *IEEE Trans. Antennas Prop.*, vol. 61, no. 3, pp. 1301–1308.

Koziel, S., Leifsson, L. (2016). *Simulation-driven Design by Knowledge-based Response Correction Techniques*, Cham, Springer.

Laurenceau, J., Sagaut, P. (2008). Building efficient response surfaces of aerodynamic functions with kriging and cokriging, *AIAA J.*, vol. 46, pp. 498–507.

Leary, S., Bhaskar, A., Keane, A. (2003). Optimal orthogonal-array-based latin hypercubes, *J. Appl. Statist.*, vol. 30, pp. 585–598.

Lehmensiek, R., Meyer, P., Muller, M. (2002). Adaptive sampling applied to multivariate, multiple output rational interpolation models with application to microwave circuits, *Int. J. RF Microwave CAE*, vol. 12, no. 4, pp. 332–340.

Levin, D. (1998). The approximation power of moving least-squares, *Math. Comput.*, vol. 67, pp. 1517–1531.

Li, X. (2010). Finding deterministic solution from underdetermined equation: largescale performance modeling of analog/RF circuits, *IEEE Trans. CAD Int. Circ. Syst.*, vol. 29, no. 11, pp. 1661–1668.

Li, X., Le, J., Gopalakrishnan, P., Pileggi, L. (2007). Asymptotic probability extraction for nonnormal performance distributions, *IEEE Trans. CAD Integr. Circuits Syst.*, vol. 26, no. 1, pp. 16–37.

Ling, Y., Mahadevan, S. (2013). Quantitative model validation techniques: New insights, *Reliability Eng. System Safety*, vol. 111, pp. 217–231.

Liu, J., Han, Z., Song, W. (2012). Comparison of infill sampling criteria in kriging-based aerodynamic optimization, *28th Int. Congress of the Aeronautical Sciences*, Brisbane, Australia.

Liu, J.S. (2004). Multilevel sampling and optimization methods, in *Monte Carlo strategies in scientific computing*, Springer Series in Statistics, Springer, pp. 205–244.

Liu, X., Fu, W.N. (2016). A dynamic dual-response-surface methodology for optimal design of a permanent-magnet motor using finite-element method, *IEEE Trans. Magn.*, vol. 52, no. 3, pp. 1–4.

Liu, Z., Yang, M., Li, W. (2016). A sequential Latin hypercube sampling method for metamodeling, in Zhang, L., Song, X., Wu, Y. (eds.), *Theory, Methodology, Tools and Applications for Modeling and Simulation of Complex Systems*, AsiaSim 2016, Comm. Comp. Information Science, vol. 643, pp. 176–185.

Lophaven, S.N., Nielsen, H.B., Sndergaard, J. (2002). *DACE: A Matlab kriging toolbox*, Technical University of Denmark.

Lschenbrand, D., Mecklenbrauker, C. (2016). Fast antenna characterization via a sparse spherical multipole expansion, *4th Int. Workshop Comp. Sensing Theory App. Radar, Sonar Remote Sens.*, Aachen, pp. 212–216.

Ma, X., Zabaras, N. (2010). An adaptive high-dimensional stochastic model representation technique for the solution of stochastic partial differential equations, *J. Comput. Phys.*, vol. 229, pp. 3884–3915.

MacKay, D.J.C. (1993). Bayesian methods for backpropagation networks, in van Hemmen, J.L., Domany, E., Schulten, K. (eds.), *Models of Neural Networks II*, Springer.

Manfredi, P., Ginste, D.V., Stievano, I.S., De Zutter, D., Canavero, F.G. (2017). Stochastic transmission line analysis via polynomial chaos methods: An overview, *IEEE Electromagn. Compat. Magaz*, vol. 6, no. 3, pp. 77–84, Third Quarter 2017.

Marelli, S., Sudret, B. (2014). UQLab: A framework for uncertainty quantification in Matlab, in The 2nd Int. Conf. on Vulnerability and Risk Analysis and Management (ICVRAM 2014), University of London, UK, July 13–15, pp. 2554–2563.

McKay, M., Conover, W., Beckman, R. (1979). A comparison of three methods for selecting values of input variables in the analysis of output from a computer code, *Technometrics*, vol. 21, pp. 239–245.

Minsky, M.I., Papert, S.A. (1969). *Perceptrons: An Introduction to Computational Geometry*, Cambridge, MA, USA, The MIT Press.

Mishra, S., Yadav, R.N., Singh, R.P. (2015). Directivity estimations for short dipole antenna arrays using radial basis function neural networks, *IEEE Ant. Wireless Propag. Lett.*, vol. 14, pp. 1219–1222.

Mitchell, D.W. (2004). More on spreads and non-arithmetic means, *Math. Gazette*, vol. 88, pp. 142–144.

Montegranario, H., Espinosa, J. (2014). Radial Basis Functions, in *Variational Regularization of 3D Data*, Springer Briefs in Computer Science, New York, Springer.

O'Hagan, A. (1978). Curve fitting and optimal design for predictions, *J. Royal Statistical Society B*, vol. 40, pp. 1–42.

Ou, G., Murphey, Y.L. (2007). Multi-class pattern classification using neural networks, *Pattern Recognition*, vol. 40, no. 1, pp. 4–18.

Palmer, K., Tsui, K.-L. (2001). A minimum bias latin hypercube design, *IIE Transactions*, vol. 33, pp. 793–808.

Park, D., Chung, I.B., Choi, D.H. (2018). Surrogate based global optimization using adaptive switching infill sampling criterion, in Schumacher, A., Vietor, T., Fiebig, S., Bletzinger, K.U., Maute, K. (eds.), *Advances in Structural and multidisciplinary Optimization*, WCSMO 2017, Cham, Springer, pp. 692–699.

Qian, P.Z.G. (2009). Nested Latin hypercube designs, *Biometrika*, vol. 96, no. 4, pp. 957–970.

Queipo, N.V., Haftka, R.T., Shyy, W., Goel, T., Vaidynathan, R., Tucker, P.K. (2005). Surrogate based analysis and optimization, *Prog. Aerospace Sci.*, vol. 41, no. 1, pp. 1–28.

Rasmussen, C.E., Williams, C.K.I. (2006). *Gaussian Processes for Machine Learning*, Cambridge, MA, USA, MIT Press.

Rayas-Sanchez, J.E. (2004). EM-based optimization of microwave circuits using artificial neural networks: The state-of-the-art, *IEEE Trans. Microwave Theory Tech.*, vol. 52, no. 1, pp. 420–435.

Rayas-Sanchez, J.E., Chávez-Hurtado, J.L., Brito-Brito, Z. (2017). Optimization of full-wave EM models by low-order low-dimension polynomial surrogate functionals, *Int. J. Num. Model.: Electr. Devices Fields*, vol. 30, no. 3–4, pp. e2094.

Rojo-Alvarez, J.L., Camps-Valls, G., Martinez-Ramon, M., Soria-Olivas, E., Navia-Vazquez, A., Figueiras-Vidal, A.R. (2005). Support vector machines framework for linear signal processing, *Signal Process*, vol. 85, pp. 2316–2326.

Rossi, J.O., Rizzo, P.N. (2009). Study of hybrid nonlinear transmission lines for high power RF generation, in *2009 IEEE Pulsed Power Conference*, Washington, DC, 2009, pp. 46–50.

Rutenbar, R., Gielen, G., Roychowdhury, J. (2007). Hierarchical modeling, optimization, and synthesis for system-level analog and RF designs, *Proceedings of the IEEE*, vol. 95, no. 3, pp. 640–669.

Santana-Quintero, L.V., Montaño, A.A., Coello, C.A.C. (2010). A review of techniques for handling expensive functions in evolutionary multi-objective optimization, in Tenne, Y., Goh, CK. (eds.), *Computational Intelligence in Expensive Optimization Problems. Adaptation Learning and Optimization*, vol. 2, Berlin, Heidelberg, Springer.

Santner, T.J., Williams, B., Notz, W. (2003). *The Design and Analysis of Computer Experiments*, New York, USA, Springer-Verlag.

Santner, T.J., Williams, B.J., Notz, W.I. (2018). Space-filling designs for computer experiments, in *The design and analysis of computer experiments*, Springer Series in Statistics. New York, NY, USA, Springer.

Sasena, M., Parkinson, M., Goovaerts, P. (2002). Adaptive experimental design applied to an ergonomics testing procedure, *ASME 2002 Design Eng. Techn. Conf. and Computer and Information in Eng. Conf.*, Montreal, Canada, 2002.

Sengupta, M., Saxena, S., Daldoss, L., Kramer, G., Minehane, S., Cheng, J. (2005). Application-specific worst case corners using response surfaces and statistical models, *IEEE Trans. Comput.-Aided Design Integr. Circuits Syst.*, vol. 24, no. 9, pp. 1372–1380.

Simpson, T.W., Peplinski, J., Koch, P.N., Allen, J.K. (2001). Metamodels for computer-based engineering design: survey and recommendations, *Eng. Computers*, vol. 17, pp. 129–150.

Smola, A.J., Schölkopf, B. (2004). A tutorial on support vector regression, *Statistics and Computing*, vol. 14, pp. 199–222.

Søndergaard, J. (2003). *Optimization using surrogate models – by the space mapping technique*, Ph.D. Thesis, Informatics and Mathematical Modelling, Technical University of Denmark, Lyngby.

Stuart, A., Ord, K. (1994). *Kendall's Advanced Theory of Statistics, Vol. 1: Distribution Theory*, London, Arnold.

Sudret, B. (2008). Global sensitivity analysis using polynomial chaos expansions, *Reliability Eng. System Safety*, vol. 93, no. 7, pp. 964–979.

Sugiyama, M. (2006). Active learning in approximately linear regression based on conditional expectation of generalization error, *J. Machine Learning Research*, vol. 7, pp. 141–166.

Takahashi, S., Chen, Y., Tanaka-Ishii, K. (2019). Modeling financial time-series with generative adversarial networks, *Physica A: Statistical Mechanics and its Applications*, vol. 527, pp. 1–12.

Tao, J., Liao, C., Zeng, X., Li, X. (2016). Harvesting design knowledge from internet: High-dimensional performance trade-off modeling for large-scale analog circuits, *IEEE Trans. CAD Int. Circuits Syst.*, vol. 35, no. 1, pp. 23–36.

Tikhonov, A.N., Arsenin, V.Y. (1977). *Solutions to Ill-Posed Problems*, New York, Wiley.

Toal, D.J.J., Keane, A.J. (2011). Efficient multipoint aerodynamic design optimization via cokriging, *Journal of Aircraft*, vol. 48, pp. 1685–1695.

Tong, C. (2006). Refinement strategies for stratified sampling algorithms, *Reliab. Eng. Syst. Saf.*, vol. 91, no. 10–11, pp. 1257–1265.

van Dam, E.R., Husslage, B., den Hertog, D., Melissen, H. (2007). Maximin Latin hypercube designs in two dimensions, *Operations Research*, vol. 55, no. 1, pp. 158–169.

Vehtari, A., Gelman, A., Gabry, J. (2017). Practical Bayesian model evaluation using leave-one-out cross-validation and WAIC, *J. Stat. Comput.*, vol. 27, no. 5, pp. 1413–1432.

Viana, F.A.C., Venter, G., Balabanov, V. (2009). An algorithm for fast optimal Latin hypercube design of experiments, *Int. J. Numer. Methods Eng.*, vol. 82, pp. 135–156.

Wang, G.G. (2003). Adaptive response surface algorithm using inherited latin hypercube design points, *J. Mech. Des.*, vol. 125, no. 2, pp. 210–220.

Wang, G., Shan, S.S. (2006). Review of metamodeling techniques in support of engineering design optimization, *ASME. J. Mech. Des.*, vol. 129, no. 4, pp. 370–380.

Wang, F., Cachecho, P., Zhang, W., Sun, S., Li, X., Kanj, R., Gu, C. (2016). Bayesian model fusion: large-scale performance modeling of analog and mixed-signal circuits by reusing early-stage data, *IEEE Trans. CAD Integr. Circuits Systems*, vol. 35, no. 8, pp. 1255–1268.

Wild, S.M., Regis, R.G., Shoemaker, C.A. (2008). ORBIT: Optimization by radial basis function interpolation in trust-regions. *SIAM J. Sci. Comput.*, 30, pp. 3197–3219.

Woods, D.C., Lewis, S.M. (2015). *Design of experiments for Screening*, in Ghanem R., Higdon D., Owhadi H. (eds.), *Handbook of uncertainty quantification*, Cham, Springer.

Wu, X., Peng, X., Chen, W., Zhang W. (2019). A developed surrogate-based optimization framework combining HDMR-based modeling technique and TLBO algorithm for high-dimensional engineering problems, *Struct. Multidisc. Optim.*, vol. 60, no. 2, pp. 663–680.

Xia, L., Xu, R.M., Yan, B. (2007). LTCC interconnect modeling by support vector regression, *Progress In Electromagnetics Research*, vol. 69, pp. 67–75.

Xiong, F., Xiong, Y., Chen W., Yang, S. (2009). Optimizing Latin hypercube design for sequential sampling of computer experiments, *Eng. Optim.*, vol. 41, no. 8, pp. 793–810.

Xiu, D., Karniadakis, G.E. (2002). The Wiener-Askey polynomial chaos for stochastic differential equations, *SIAM J. Scientific Comput.*, vol. 24, no. 2, pp. 619–644.

Xu, J.J., Yagoub, M.C.E., Ding, R., Zhang, Q.J. (2002). Neural-based dynamic modeling of nonlinear microwave circuits, *IEEE Trans. Microwave Theory Tech.*, vol. 50, no. 12, pp. 2769–2780.

Yan, Z., Wang, J. (2015). Nonlinear model predictive control based on collective neurodynamic optimization, *IEEE Trans. Neural Netw. Learning Syst.*, vol. 26, no. 4, pp. 840–850.

Yang, Y., Hu, S.M., Chen, R.S. (2005). A combination of FDTD and least-squares support vector machines for analysis of microwave integrated circuits, *Microwave Opt. Technol. Lett.*, vol. 44, pp. 296–299.

Ye, K.Q. (1998). Orthogonal column latin hypercubes and their application in computer experiments, *J. American Statistical Association*, vol. 93, pp. 1430–1439.

Zhang, K., Han, Z. (2013). Support vector regression-based multidisciplinary design optimization in aircraft conceptual design, AIAA Aerospace Sciences Meeting, AIAA paper 2013–1160.

Zhang, Q.J., Gupta, K.C. (2000). *Neural Networks for RF and Microwave Design*. Norwood, MA, Artech House.

Zhang, Q.J., Gupta, K.C., Devabhaktuni, V.K. (2003). Artificial neural networks for RF and microwave design: From theory to practice, *IEEE Trans. Microw. Theory Tech.*, vol. 51, no. 4, pp. 1339–1350.

Chapter 2

Fundamentals of Physics-Based Surrogate Modeling

Anna Pietrenko-Dabrowska and Slawomir Koziel

Abstract

Chapter 1 was focused on data-driven (or approximation-based) modeling methods. The second major class of surrogates are physics-based models outlined in this chapter. Although they are not as popular, their importance is growing because of the challenges related to construction and handling of approximation surrogates for many real-world problems. The high cost of evaluating computational models, nonlinearity of system responses, dimensionality issues as well as combinations of these factors, may lead to a situation, where setting up a data-driven model is not possible or at least not practical. On the other hand, incorporation of the problem-specific knowledge, typically in the form of a lower-fidelity computational model, often alleviates the aforementioned difficulties. The enhancement of the low-fidelity models using a limited amount of high-fidelity data is the essence of physics-based surrogate modeling. This chapter provides a brief characterization of this class of surrogates, explains the concept and various types of low-fidelity models, as well as outlines several specific modeling approaches, also in the context of surrogate-assisted optimization.

Keywords: Surrogate modeling, physics-based models, variable-resolution modeling, response correction, surrogate-assisted optimization

2.1 Physics-Based Surrogates: Overview

Data-driven surrogates (Chapter 1) exhibit some clear advantages, including versatility, low evaluation cost, numerous well-established techniques, as well as widespread availability (toolboxes implemented in popular programming environments, e.g., Matlab). Notwithstanding, there are several problems that limit their applicability when it comes to real-world modeling tasks. These are related to high cost of training data acquisition, the curse of dimensionality, and wide parameter ranges that the models should be valid for in order to represent reasonable ranges of operating conditions, material parameters, etc. In the case of high-frequency structures such as microwave and antenna components, an additional difficulty is high nonlinearity of the system responses but also the necessity of handling vector-valued outputs (Goudos, 2017; Koziel and Ogurtsov, 2014; Petosa, 2007; Rayas-Sanchez *et al.*, 2017).

Physics-based models can be considered a remedy to some of these difficulties (Bandler *et al.*, 2003; Cheng *et al.*, 2006; Dorica and Giannacopoulos, 2006; Koziel *et al.*, 2016; Zhang *et al.*, 2018). For this class of techniques, an important component of the modeling process is the problem-specific knowledge embedded into the surrogate, typically in the form of a lower-fidelity model that undergoes a suitable correction (Bandler *et al.*, 2004a). The latter is realized using a limited amount of high-fidelity data (Bandler *et al.*, 2008), in some cases, even a single sample point (Fernández-Godino *et al.*, 2019). The correction process aims at improving the alignment between the low-fidelity model and the high-fidelity model, either locally or across the entire design space. For high-frequency structures, the low-fidelity model may be implemented as an equivalent circuit (Bandler *et al.*, 2004a), coarse-discretization EM simulation model (Koziel and Ogurtsov, 2014), or, in rare cases, analytical or semi-empirical formulas (Koziel *et al.*, 2014).

The fundamental features of physics-based surrogates include:

- The models incorporate the problem specific knowledge, typically in the form of an underlying low-fidelity model.
- The surrogate is constructed by aligning the low-fidelity model with the high-fidelity one using a limited amount of information from the latter, usually through appropriately formulated nonlinear regression.
- The low- and high-fidelity models are supposed to be well correlated to ensure good generalization capability of the surrogate.
- As low-fidelity model often involves computer simulation, physics-based surrogates are more expensive than data-driven ones.

- Physics-based models are more immune to the curse of dimensionality and can often be established over wider ranges of parameters as compared to the approximation models.
- Their versatility is limited (i.e., the surrogates are not easily transferrable between application areas) because each problem entails the development of dedicated low-fidelity models.
- Physics-based surrogates are more difficult to handle than data-driven models because the establishment of the low-fidelity model as well as its appropriate correction requires certain experience and engineering insight.
- Availability of surrogates of this class is rather limited (e.g., there are no general-purpose Matlab toolboxes, etc.).

The fundamental advantage of physics-based surrogates is to overcome some of the issues affecting approximation models at the expense of limited versatility and computational efficiency. The primary application area for these models is surrogate-based optimization (SBO) where the alignment between the coarse model and the high-fidelity model is of concern mostly along the optimization path (Koziel et al., 2011a). In this setup, the surrogate is often updated before each iteration of the optimization algorithm and using a very small number of high-fidelity data points (often, just one) (Bandler et al., 2003, 2004a).

Here are important stages of physics-based surrogate model construction:

- **Selection of the low-fidelity model.** This step requires engineering insight and experience as a sufficient accuracy of the low-fidelity model may be difficult to quantify and often entails visual assessment of the system outputs, grid convergence studies, etc. (Leifsson et al., 2012; Zhu et al., 2007).
- **Selection of the appropriate correction technique.** This step is also area- and experience-dependent. The major types of correction techniques are outlined later in this chapter.
- **Design of experiments.** In the case of physics-based surrogates, simple and economical DoE strategies are often utilized such as factorial designs (Kleijnen, 2018) with a notable example of the star distribution (Giunta et al., 2003).
- **Acquisition** of the high-fidelity data and model identification.

- **Model validation,** typically carried out similarly as for the data-driven surrogates (cf. Chapter 1). However, in the cases of surrogates constructed for the purpose of local optimization, validation is often omitted as long as the models ensure (e.g., by formulation) satisfaction of zero- and first-order consistency conditions (Alexandrov and Lewis, 2001; Eldred and Dunlavy, 2006).

This chapter only outlines the main concepts of physics-based surrogate modeling. More details can be found in Koziel *et al.* (2011b), Koziel and Ogurtsov (2014), Bekasiewicz *et al.* (2014), Robinson *et al.* (2008), Sarkar *et al.* (2019), Bakr *et al.* (2000).

2.2 Low-Fidelity Models

The low-fidelity (or coarse) model is the fundamental component of any physics-based surrogate. It is a simplified representation of the original, high-fidelity (or fine) computational model of the system of interest. The simplifications are intentional and introduced to make the model computationally cheaper than the high-fidelity one. Clearly, making the model faster leads to a degradation of its accuracy. Maintaining appropriate balance between the speed and accuracy is one of the considerations that are important for physics-based modeling and optimization (Hazaveh *et al.*, 2017; Koziel and Bekasiewicz, 2017). This section provides a general overview of the low-fidelity modeling, providing illustrations of model development through examples from the areas of high-frequency electronics (including microwave and antenna engineering) as well as aerodynamics.

2.2.1 *Overview*

The major reason for using the low-fidelity model is a reduced evaluation cost. However, the low-fidelity model should adequately represent the important features of the system of interest, which can be further translated into the reliability of the physics-based surrogate. In particular, in the context of design optimization, information from the surrogate is used to navigate the high-fidelity design space and search for the optimum. Therefore, the low-fidelity model needs to be able to capture the changes in the measures of merit with respect to the design parameters in the same way as the high-fidelity model. In other words, the low- and the high-fidelity model have to be sufficiently well correlated (Perdikaris et al., 2017).

One can distinguish the following generic approaches to low-fidelity model development (Alexandrov et al., 1998):

1. **Simplified physics:** The governing equations pertinent to the high-fidelity model are replaced by a set of simplified equations. These are often referred to as *variable-fidelity physics models*.
2. **Coarse-discretization:** In this approach, the same governing equations are used as in the high-fidelity model, but with a coarser computational grid. Often referred to as *variable-resolution models*.
3. **Relaxed convergence criteria:** Reduce/relax the solver convergence criteria. Sometimes referred to as *variable-accuracy models*.
4. A combination of (1)–(3).

The above is a general classification. In practice, implementation of (1)–(4) is domain specific. In the area of high-frequency electronics, reducing discretization density can be accompanied by other simplifications:

1. Reduction of the computational domain and applying simple absorbing boundaries with the finite-volume methods implemented in full-wave EM simulations (Baumann et al., 2004; Liu and Gedney, 2000).
2. Using low-order basis functions with the finite-element and moment method solvers (Kolundzija and Sumic, 2004; Schmidthausler and Clemens, 2012).
3. Using relaxed solution termination criteria such as the scattering parameter error for the frequency domain methods with adaptive meshing, and residue energy for the time-domain solvers (Caratelli and Yarovoy, 2010; Petrides and Demkowicz, 2017).

Simplifications of the physics for modeling of high-frequency structures, aside from using different governing physics, include (Obaidat et al., 2019):

1. Ignoring dielectric and metal losses as well as material dispersion if their impact to the simulated response is not significant.
2. Using zero metallization thickness for traces, strips, and patches.

3. Ignoring moderate anisotropy of substrates.
4. Energizing the antenna with discrete sources rather than waveguide ports.

The provided lists are pertinent to full-wave electromagnetic (EM) models, whereas, in general, the particular simplification approaches depend on the engineering discipline and the specific type of simulation utilized in there.

Sections 2.2.2 and 2.2.3 demonstrate the various low-fidelity modeling approaches using the examples from the areas of high-frequency electronics.

2.2.2 *Variable-resolution and variable-accuracy models*

Discretization density, local as well as overall, has the strongest effect on accuracy and computational time of the models evaluated by the EM solver. At the same discretization density is the most efficient parameter allowing to trade accuracy for speed. Thus, a straightforward way to create a low-fidelity model of an antenna is to relax mesh settings compared to those of the high-fidelity model. For the sake of illustration, consider a dielectric resonator antenna (DRA) shown in Fig. 2.1. The low-fidelity model is faster than the high-fidelity one due to introduced simplifications. Typically, in EM simulations, the low-fidelity model can be made 5 to 50 times faster than the high-fidelity model. This is, however, achieved at the expense of accuracy loss. Thus, the low-fidelity model cannot simply replace the high-fidelity model in the design optimization process. Figure 2.2(a) shows the high- and low-fidelity model responses simulated with different hexahedral meshes at the same design of the antenna of Fig. 2.1. Figure 2.2(b) shows the relationship between overall mesh coarseness and antenna simulation time.

Both local and global mesh coarseness strongly affect model simulation time, model accuracy and, thus, its applicability to construct the physics-based surrogate for generic modeling or optimization purposes. In the context of design optimization, the coarser model is faster and allows a lower numerical cost per iteration of the surrogate-assisted process. The coarser model, however, is less accurate; therefore, it results in a larger number of iterations necessary to find a satisfactory design.

Fundamentals of Physics-Based Surrogate Modeling 45

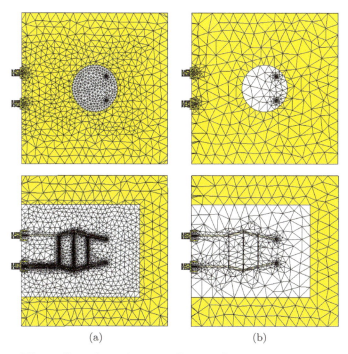

Fig. 2.1. Elliptically polarized 2.0–2.4 GHz DRA with the dielectric resonator core excited through a microstrip branchline coupler: (a) tetrahedral mesh of the high-fidelity model, (b) coarse mesh of the low-fidelity model. The upper views are for the antenna front, and the lower views are for the back (Koziel and Ogurtsov, 2019).

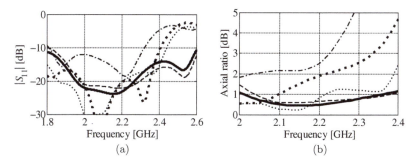

Fig. 2.2. Antenna of Fig. 2.1 at a certain design simulated using the CST MWS transient solver (CST, 2018): (a) reflection coefficient and (b) axial ratio toward zenith. Simulations with hexahedral meshes comprising different numbers of cells: 800,000 ($\cdot - \cdot$); 2,000,000 ($\bullet \bullet \bullet$); 3,700,000 ($\cdots$); 5,600,000 (---); and 15,800,000 (—) (Koziel and Ogurtsov, 2019).

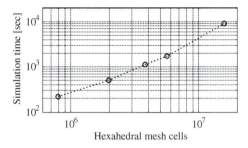

Fig. 2.3. Antenna of Fig. 2.1 at a certain design simulated using the CST MWS transient solver (CST, 2016) on a 2.00 GHz 6-core CPU 64 GB RAM computer: model simulation time versus mesh size. Data points (o) correspond to meshes listed in Fig. 2.2 caption (Koziel and Ogurtsov, 2019).

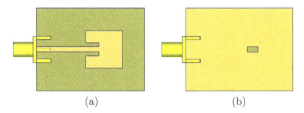

Fig. 2.4. Microstrip patch antenna model: front (a) and back (b).

In addition, there is an increased risk for the optimization algorithm utilizing a coarser model to fail to obtain a satisfactory design (Koziel and Ogurtsov, 2012). On the other hand, finer models are more expensive but they are more likely to produce an improved and reliable design with a smaller number of iteration. One can infer from Fig. 2.3 that the "finest" coarse-discretization model (5,600,000 mesh cells) represents the high-fidelity model responses (shown with thick solid lines) very closely in terms of reflection coefficient and axial ratio. The model with 3,700,000 cells can be considered as a borderline in terms of reflection coefficient and axial ratio. The two remaining models could be considered inaccurate; in particular, the model with 800,000 cells is essentially unreliable.

For another illustration, consider a 5.8 GHz microstrip patch antenna shown in Fig. 2.4 (Koziel and Ogurtsov, 2019), which again illustrates differences of its responses evaluated with models of different fidelities. Both of the models are defined, discretized, and simulated using CST MWS (CST, 2016). The reference, fine-model, is

Fundamentals of Physics-Based Surrogate Modeling

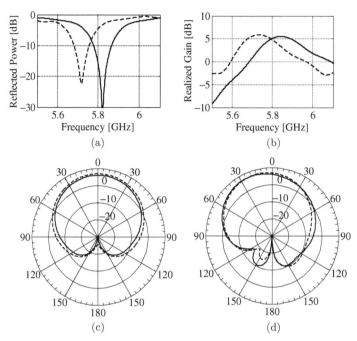

Fig. 2.5. Microstrip patch antenna evaluated with the high-fidelity model (—) and low-fidelity model (- - -): (a) reflection coefficient, (b) realized gain toward zenith, (c) H-plane directivity pattern at 5.8 GHz, (d) E-plane directivity pattern at 5.8 GHz (Koziel and Ogurtsov, 2019).

discretized with about 1,600,000 mesh cells and with the minimal mesh step of 0.10 mm; it is evaluated in 10 min 6 s with the CST MWS transient solver on a 2.00 GHz 6-core CPU 64 GB RAM computer. A quite dense discretization of the fine-model ensures that no noticeable changes of the response occur with increasing the discretization density. At the same time, the low-fidelity model is discretized with about 150,000 mesh cells and with the minimal mesh step of 0.3 mm; it is simulated on the same computer in only 2 min 15 s.

For this antenna — supposed to operate around 5.8 GHz — the reflection coefficient response of the low-fidelity model, shown in Fig 2.5(a), is substantially misaligned with that of the high-fidelity model. Thus, tuning dimensions of the antenna using the low-fidelity model can result in an unreliable design in terms of reflection coefficient. At the same time, the direct utilization of the high-fidelity model for design optimization will be associated with a substantial total design time. On the other hand, the model fidelity is less critical for the realized gain value about 5.8 GHz though the realized gain responses are also misaligned in frequency, as shown in Fig. 2.5(b). Finally, the two models have very similar directivity patterns, as shown with Figs. 2.5(c) and 2.5(b).

2.2.3 Variable-fidelity physics models

Variable-fidelity physics models are constructed by replacing the set of high-fidelity governing equations by a set of simplified or modified equations. Let us consider an illustrative example demonstrating this process for a specific case study in microwave engineering.

Figure 2.6 shows an example of a dual-band microstrip filter (Guan et al., 2008). The high-fidelity model of this device determines

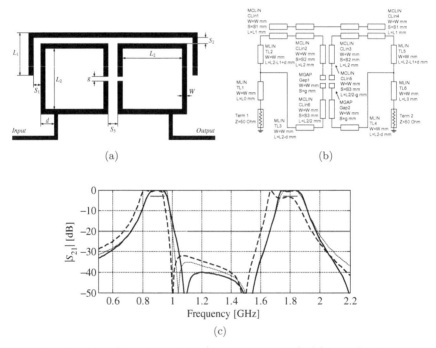

Fig. 2.6. Dual-band bandpass filter (Guan et al., 2008): (a) high-fidelity model, (b) low-fidelity model, (c) responses at two designs: dashed line = high-fidelity model at design 1, thin dashed line = low-fidelity at design 2, solid line = high-fidelity model at design 2, thick horizontal lines = minimax design specifications (Koziel et al., 2016).

the scattering parameters of the filter using a commercial planar-3D solver (Sonnet, 2018). Simulation time is a few minutes per design. The low-fidelity model (Fig. 2.6(b)) is an equivalent circuit implemented in ADS (ADS, 2019). The low-fidelity model is a simplified representation of the filter, where the lumped microstrip line models are connected together using the circuit theory rules. The low-fidelity model is very fast (simulation time is in the range of milliseconds). A comparison of the model responses is given in Fig. 2.6(c).

In should be emphasized that while equivalent circuit models are often acceptable for relatively simple structures such as the filter in Fig. 2.6(a), they are usually not sufficiently accurate for complex circuits featuring considerable EM cross-couplings (e.g., highly miniaturized microstrip passives; Koziel (2015)).

2.2.4 Selecting low-fidelity models

Appropriate selection of the low-fidelity model is one of the important aspects of physics-based modeling. Often, there are many options available, characterized by different trade-offs between the accuracy and evaluation cost. While choosing the faster model may be potentially attractive, one needs to make sure that the model accuracy is sufficient, in particular, that it properly represents the important characteristics of the system output. Also, relaxing accuracy has to be balanced by a larger number of high-fidelity data points for subsequent model correction. In a particular context of surrogate-based optimization, the selection of the model coarseness strongly affects the simulation time, and, therefore, the performance of the design optimization process. Coarser models are faster, which turns into a lower cost per design iteration of the optimization process. However, lower accuracy of the coarser models typically results in a larger number of iterations necessary to find a satisfactory design. Furthermore, there is an increased risk of failure for the optimization algorithm to find a satisfactory design. Finer models, on the other hand, are more expensive but they are more likely to produce a useful design with a smaller number of iterations. Experimental studies have been carried out to investigate the issues of the low-fidelity model selection and its impact on the performance of the optimization process (Leifsson and Koziel, 2015a).

It should be mentioned that up to now, no general-purpose methods for automated selection of the physics-based low-fidelity model are available. The decision about the "right" low-fidelity model setup is normally made based on grid convergence studies such as those presented in Fig. 2.2, engineering experience, and visual inspection of the low- and high-fidelity model responses. Again, in the context of surrogate-assisted optimization, some results reported by Leifsson *et al.* (2014b) indicate that certain parameters of the low-fidelity model (e.g., those controlling grid density) may be automatically adjusted using optimization methods.

2.3 Physics-Based Surrogates — Basic Techniques

This section outlines basic physics-based modeling techniques. More detailed treatment of the subject can be found in Bakr *et al.* (2000); Bekasiewicz *et al.* (2014); Koziel *et al.* (2011b); Koziel and Ogurtsov (2014); Robinson *et al.* (2008) and Sarkar *et al.* (2019) as well as in the remaining sections of this chapter.

The low-fidelity model of the system of interest is denoted as $c(\boldsymbol{x})$. Because physics-based surrogates are often used within surrogate-assisted optimization frameworks, such a context will be considered when explaining some of the modeling methodologies. In particular, we consider an iterative SBO procedure

$$\boldsymbol{x}^{(i+1)} = \arg\min_{\boldsymbol{x}} U(s^{(i)}(\boldsymbol{x})), \qquad (2.1)$$

producing a series $\boldsymbol{x}^{(i)}, i = 0, 1, \ldots$, of approximations to the original design problem $\boldsymbol{x}^* = \arg\min\{\boldsymbol{x} : U(f(\boldsymbol{x}))\}$. The high-fidelity model is denoted as $f(\boldsymbol{x})$, whereas U is the merit function to be minimized; $s^{(i)}$ stands for the surrogate model established at the current iteration point $\boldsymbol{x}^{(i)}$ (i.e., the starting point for the next iteration of (2.1)).

First, let us discuss a simple case of multiplicative response correction. For ensuring convergence of a sequence $\{\boldsymbol{x}^{(i)}\}$ to $\boldsymbol{x}^*$, it is imperative to have the surrogate and the high-fidelity model well (at least locally) aligned. The surrogate $s^{(i)}(\boldsymbol{x})$ at the iteration i can be constructed with response correction

$$s^{(i)}(\boldsymbol{x}) = \beta_k(\boldsymbol{x})c(\boldsymbol{x}), \qquad (2.2)$$

where $\beta_k(\boldsymbol{x}) = \beta_k(\boldsymbol{x}^{(i)}) + \nabla \beta(\boldsymbol{x}^{(i)})^T(\boldsymbol{x}-\boldsymbol{x}^{(i)})$, and $\beta(\boldsymbol{x}) = f(\boldsymbol{x})/c(\boldsymbol{x})$. This ensures zero- and first-order consistency, i.e., $s^{(i)}(\boldsymbol{x}^{(i)}) = f(\boldsymbol{x}^{(i)})$ and $\nabla s^{(i)}(\boldsymbol{x}^{(i)}) = \nabla f(\boldsymbol{x}^{(i)})$ (Alexandrov and Lewis, 2001). At the same time, the problem-specific knowledge embedded in the low-fidelity model ensures that the surrogate is more accurate than a purely data-driven model established using the same high-fidelity model information (e.g., the first-order Taylor model) as conceptually illustrated in Fig. 2.7.

A generalization of this concept is a multi-point response correction. For example (Koziel and Leifsson, 2013b), the surrogate can be defined for vector-valued models as

$$s^{(i)}(\boldsymbol{x}) = \boldsymbol{\Lambda}^{(i)} \circ \boldsymbol{c}(\boldsymbol{x}) + \boldsymbol{\Delta}_r^{(i)} + \boldsymbol{\delta}^{(i)}, \qquad (2.3)$$

with column vectors $\boldsymbol{\Lambda}^{(i)}$, $\boldsymbol{\Delta}_r^{(i)}$, and $\boldsymbol{\delta}^{(i)}$ ($\circ$ denotes a component-wise multiplication). The global response correction parameters $\boldsymbol{\Lambda}^{(i)}$ and $\boldsymbol{\Delta}_r^{(i)}$ are obtained as

$$[\boldsymbol{\Lambda}^{(i)}, \boldsymbol{\Delta}_r^{(i)}] = \arg \min_{[\boldsymbol{\Lambda}, \boldsymbol{\Delta}_r]} \sum_{k=0}^{i} \|\boldsymbol{f}(\boldsymbol{x}^{(k)}) - \boldsymbol{\Lambda} \circ \boldsymbol{c}(\boldsymbol{x}^{(k)}) + \boldsymbol{\Delta}_r\|^2. \qquad (2.4)$$

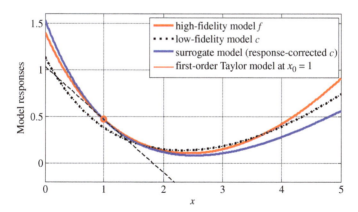

Fig. 2.7. Visualization of the response correction (2.2) for the example analytical functions c (low-fidelity model) and f (high-fidelity model). The correction is established at $\boldsymbol{x}_0 = 1$. The surrogate exhibits good alignment with the high-fidelity model in a broad vicinity of $\boldsymbol{x}_0$ when compared to the first-order Taylor model of f set up at $\boldsymbol{x}_0$.

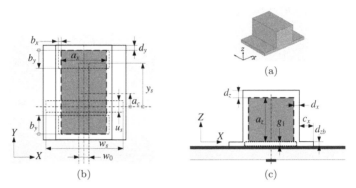

Fig. 2.8. Suspended DRA (a) 3D view of its housing, top (b) and front (c) views.

The scaling (2.4) is supplemented by the (local) term $\boldsymbol{\delta}^{(i)}$ defined as

$$\boldsymbol{\delta}^{(i)} = \boldsymbol{f}(\boldsymbol{x}^{(i)}) - \left[\boldsymbol{\Lambda}^{(i)} \circ \boldsymbol{c}(\boldsymbol{x}^{(i)}) + \boldsymbol{\Delta}_r^{(i)} \right]. \tag{2.5}$$

Note that (2.5) ensures zero-order consistency, i.e., $\boldsymbol{s}^{(i)}(\boldsymbol{x}^{(i)}) = \boldsymbol{f}(\boldsymbol{x}^{(i)})$. The correction terms $\boldsymbol{\Lambda}^{(i)}$, $\boldsymbol{\Delta}_r^{(i)}$ and $\boldsymbol{\delta}^{(i)}$ can be obtained analytically by solving appropriate linear regression problems (Koziel and Leifsson, 2013b).

For illustration, consider a DRA shown in Fig. 2.8. The high-fidelity model $\boldsymbol{f}$ is simulated using the CST (CST, 2018) (~800,000 mesh cells, evaluation time 20 min). The low-fidelity model $\boldsymbol{c}$ is also evaluated in CST (~30,000 mesh cells, evaluation time 40 s). Figure 2.9(a) shows the responses of the low- and high-fidelity DRA model at several designs. Conventional one-point response correction (Fig. 2.9(b)) perfectly aligns $\boldsymbol{c}$ and $\boldsymbol{f}$ at the design where it is established, but the alignment is not as good for other designs. Multi-point response correction improves model alignment at all designs involved in the model construction (note that Fig. 2.9(c) only shows the global part $\boldsymbol{\Lambda}^{(i)} \circ \boldsymbol{c}(\boldsymbol{x}) + \boldsymbol{\Delta}_r^{(i)}$ without $\boldsymbol{\delta}^{(i)}$ which would give a perfect alignment at $\boldsymbol{x}^{(i)}$).

Another approach to low-fidelity model correction is a manipulation of the model domain. Perhaps the most popular example of such a procedure is input space mapping (ISM) (Bandler et al., 2004a), where the surrogate is rendered as

$$\boldsymbol{s}^{(i)}(\boldsymbol{x}) = \boldsymbol{c}\left(\boldsymbol{x} + \boldsymbol{q}^{(i)}\right). \tag{2.6}$$

Fundamentals of Physics-Based Surrogate Modeling

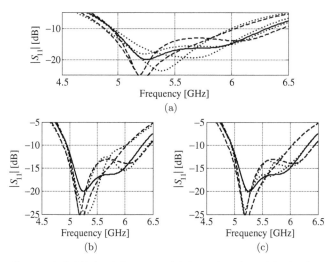

Fig. 2.9. Suspended DRA of Fig. 2.8: (a) low- (····) and high- (- - - and —) fidelity model responses at three designs, (b) OSM-corrected low- (····) and high- (- - - and —) fidelity model responses at the same designs (OSM correction at the design marked —), (c) multi-point-corrected low- (····) and high- (- - - and —) fidelity model responses.

In (2.6), both the low-fidelity model and the surrogate are vector-valued. Model identification is realized by minimizing the misalignment between the surrogate and the high-fidelity model, $||f(x^{(i)}) - c(x^{(i)} + q^{(i)})||$; $x^{(i)}$ is a reference design (e.g., the most recent design encountered during the optimization run) at which the surrogate is established. Figure 2.10 shows the advantages of ISM in the design of a microwave filter (Hong and Lancaster, 2001). The high-fidelity model is evaluated using EM simulation, whereas the low-fidelity model is an equivalent circuit. The response of interest is the reflection coefficient $|S_{11}|$ as a function of frequency. For this particular example, ISM offers both good approximation and generalization capability (cf. Fig. 2.10(d)).

An alternative correction method is to exploit parameters that are normally fixed in the high-fidelity model, e.g., substrate height and/or dielectric permittivity (Bandler et al., 2004b; Koziel et al., 2008a). As the surrogate is an auxiliary tool and it is not supposed to be built, such parameters can be freely adjusted in the low-fidelity model to improve its alignment with the high-fidelity one. Implicit

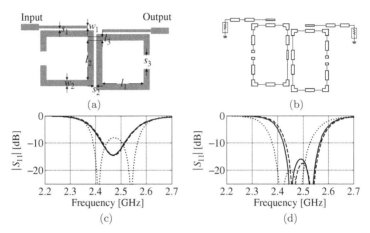

Fig. 2.10. Low-fidelity model correction through parameter shift (ISM): (a) microstrip filter geometry (high-fidelity model f evaluated using EM simulation), (b) low-fidelity model c (equivalent circuit), (c) response of f (—) and c ($\cdots$), as well as response of the surrogate model s (- -) created using ISM, (d) surrogate model verification at a different design (other than that at which the model was created) f (—), c ($\cdots$), and s (- -). Good alignment indicates excellent generalization of the model (Koziel and Bekasiewicz, 2016a).

space mapping (SM) (Bandler et al., 2003, 2004b; Koziel et al., 2010) utilizes this concept by defining the surrogate as

$$s^{(i)}(x) = c_I\left(x, p^{(i)}\right). \tag{2.7}$$

In (2.7), c_I denotes the low-fidelity model with the explicit dependence on the implicit (preassigned) parameter vector p. If the surrogate is established at the design $x^{(i)}$, the vector $p^{(i)}$ is typically obtained by minimizing the norm-wise discrepancy between the models, $\|f(x^{(i)}) - c_I(x^{(i)}, p)\|$. Figure 2.11 provides an illustration of this concept. The mapping parameters are dielectric permittivities of the microstrip-line components (rectangle elements in Fig. 2.11(b)).

Another type of low-fidelity model enhancement is derived from the observation of the vector-valued responses of the system, which are evaluations of the same design but at different values of certain parameters such as the time or frequency. Based on such observations, the low-fidelity model can be enhanced by applying a linear or nonlinear scaling to these parameters. A representative example

Fundamentals of Physics-Based Surrogate Modeling

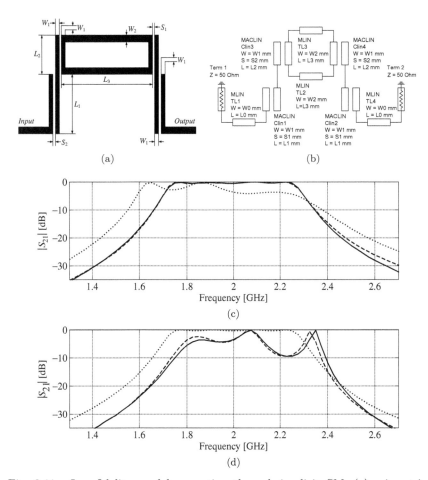

Fig. 2.11. Low-fidelity model correction through implicit SM: (a) microstrip filter geometry (high-fidelity model f evaluated using EM simulation), (b) low-fidelity model of the filter, (c) response of f (—) and c (· · · ·), as well as response of the surrogate model s (- - -) created using implicit SM, and (d) surrogate model verification at a different design (other than that at which the model was created) f (—), c (· · · ·), and s (- - -) (Koziel et al., 2016).

is frequency scaling. It is useful whenever the figures of interest are frequency characteristics, e.g., S-parameters (Koziel et al., 2006a; Koziel and Ogurtsov, 2014).

Let $\boldsymbol{f}(\boldsymbol{x}) = [f(\boldsymbol{x}, \omega_1) \; f(\boldsymbol{x}, \omega_2) \cdots f(\boldsymbol{x}, \omega_m)]^T$ where $f(\boldsymbol{x}, \omega_k)$ is the evaluation of the high-fidelity model at a frequency ω_k; similarly,

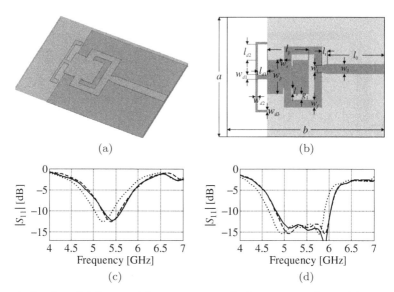

Fig. 2.12. Low-fidelity model correction with frequency scaling: (a) antenna view, (b) antenna geometry (both f and c evaluated using EM simulation, coarse-discretization used for c). Response of f (—) and c (· · · ·), as well as response of the surrogate model s (- - -) created using frequency scaling at: (c) a certain reference design, and (d) another (test) design. Note that the surrogate properly accounts for the frequency shifts between c and f (Koziel and Bekasiewicz, 2016a).

$c(x) = [c(x, \omega_1) c(x, \omega_2) \cdots c(x, \omega_m)]^T$. The frequency-scaled surrogate $s_F(x)$ is defined as

$$s_F(x, [F_0\ F_1]) = [c(x, F_0 + F_1\omega_1) \cdots c(x, F_0 + F_1\omega_m)]^T. \quad (2.8)$$

The scaling parameters F_0 and F_1 are obtained as $[F_0, F_1]$ = argmin$\{[F_0, F_1] : \|f(x^{(j)}) - s_F(x^{(j)}, [F_0 F_1])\|\}$. Figure 2.12 shows an example of frequency scaling applied to the low-fidelity model of a substrate-integrated cavity antenna (Bekasiewicz and Koziel, 2016). Here, both the low- and high-fidelity models are evaluated using coarse- and fine-discretization EM simulations.

More extensive discussion on physics-based surrogates can be found in the literature e.g., Bandler *et al.* (2004a); Koziel *et al.* (2013); Koziel and Ogurtsov (2014) and Leifsson and Koziel (2015b).

2.4 Response Correction Techniques

This section briefly discusses selected physics-based models involving response correction. The high-fidelity model (and, consequently, the surrogate) are generally assumed to be vector-valued, i.e., $f\colon X_f \to R^m$, $X_f \subseteq R^n$. The surrogate s is to match f as well as possible in the region of interest $X_R \subseteq X_f$, which is typically, an n-dimensional interval in R^n with center at the reference point $x^0 = [x_{0.1} \cdots x_{0.n}]^T \in R^n$ and the size $\boldsymbol{\delta} = [\delta_1 \cdots \delta_n]^T$.

2.4.1 Quasi-global modeling using multi-point space mapping

Let $c\colon X_c \to R^m$, $X_c \subseteq R^n$ be the (vector-valued) low-fidelity model. The training set is denoted as $X_B = \{x^1, x^2, \ldots, x^N\} \subset X_R$ with the high-fidelity model responses known at all x_j, $j = 1, 2, \ldots, N$. A standard SM model (SM-Standard) is defined as

$$s_{\mathrm{SM}}(x) = \bar{s}_{\mathrm{SM}}(x, p), \tag{2.9}$$

where the SM parameters p are obtained using the parameter extraction process

$$p = \arg\min_r \sum_{k=1}^N \| f(x^k) - s_{\mathrm{SM}}(x^k, r) \|, \tag{2.10}$$

while $\bar{s}$ is a generic space-mapping model, i.e., the low-fidelity model composed with some suitable mappings, e.g.,

$$\bar{s}_{\mathrm{SM}}(x, p) = \bar{s}_{\mathrm{SM}}(x, A, B, c, d) = A \cdot c(B \cdot x + q) + d, \tag{2.11}$$

where A, B, c, and d are matrices of appropriate sizes (Koziel and Bandler, 2007a,b). The flexibility of the model (2.9)–(2.11) can be enhanced in many ways, e.g., by exploiting so-called "pre-assigned" or "implicit" parameters (Bandler et al., 2004b; Cheng et al., 2008; Koziel et al., 2006a, 2008a).

Note that the linear mappings of the standard SM surrogate may only provide a limited modification of the range of the low-fidelity model, and this modification is basically independent of the design variables. Due to the fixed number of parameters, which are extracted in one shot for the entire region of interest, the surrogate is, in fact, a nonlinear regression model. Consequently, the modeling error might not decrease below certain, problem dependent, non-zero limits even if the number of training points becomes unlimited (cf. Koziel et al., 2006b). This limits the practical applicability of the method.

2.4.2 Space mapping with a function approximation layer

The limitations of the standard SM model s_{SM} may be addressed to some extent by incorporating an approximation layer. The surrogate is defined as

$$s(x) = s_{SM}(x) + \tilde{s}(x), \qquad (2.12)$$

where $\tilde{s}$ is a data-driven model accounting for the residuals between the high-fidelity model and s_{SM} (Koziel and Bandler, 2007a). With this method:

1. A relatively good accuracy can be obtained using a limited amount of high-fidelity model data due to underlying low-fidelity model.
2. The surrogate is as fast as the low-fidelity model because the function approximation layer typically exploits analytical formulas.
3. Any number of available high-fidelity data points can be effectively incorporated into the surrogate.

Several implementations of (2.12) have been proposed, including $\tilde{s}$ realized by radial basis functions (Koziel and Bandler, 2007a), fuzzy systems (Koziel and Bandler, 2007b), and kriging (Koziel and Bandler, 2012). For illustration, consider the microstrip bandpass filter shown in Fig. 2.13(a) (Hsieh and Chang, 2003). The design parameters are $x = [L \ g \ s \ d]^T$. The high-fidelity model is simulated in FEKO (Altair FEKO, 2018). The surrogate model domain is an interval $[x^0 - d, x^0 + d]$ with $x^0 = [7.0 \ 0.1 \ 0.5 \ 1.5]^T$ mm and the region size $d = [0.25 \ 0.05 \ 0.1 \ 0.5]^T$ mm. The low-fidelity model (Fig. 2.13(b)) is implemented in ADS (2019). The base set contains 100 points allocated using LHS. The standard SM surrogate s_{SM} (SM-Standard) is the model (2.9)–(2.11) enhanced by implicit SM (Koziel and Ogurtsov, 2012).

The average and maximum relative RMS errors are gathered in Table 2.1. For comparison, the results obtained for the SM model

Fundamentals of Physics-Based Surrogate Modeling 59

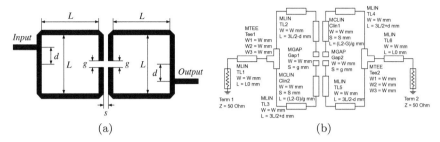

Fig. 2.13. Bandpass filter with two transmission zeros: geometry (Hsieh and Chang, 2003): (a) geometry, (b) low-fidelity model (ADS, 2019).

Table 2.1. Modeling results for the filter of Fig. 2.13.

Model type	Model name	Average error (%)	Maximum error (%)
Space mapping	SM-Standard	3.8	6.8
Enhanced space mapping	SM-RBF	1.8	5.9
	SM-Fuzzy	3.4	6.7
	SM-Kriging	1.5	4.1
Data-driven surrogates	RBF	6.9	24.8
	Fuzzy	9.5	22.0
	Kriging	6.1	20.1

enhanced by RBF (Koziel and Bandler, 2007a) and by Fuzzy System surrogate (Koziel and Bandler, 2007b) are also included. Note that these surrogates are significantly more accurate than the purely data-driven models set up using the same amounts of high-fidelity data.

2.4.3 Multi-point output space mapping

The multi-point OSM applies correction terms directly to the low-fidelity model output components. The technique is explained assuming the scalar low-fidelity model output $c(\boldsymbol{x})$. The OSM surrogate model is defined as (Leifsson and Koziel, 2015a)

$$s(\boldsymbol{x}) = A(\boldsymbol{x})c(\boldsymbol{x}) + D(\boldsymbol{x}), \qquad (2.13)$$

where

$$A(\boldsymbol{x}) = a_0 + [a_1 \ a_2 \ \cdots \ a_n] \cdot (\boldsymbol{x} - \boldsymbol{x}^0), \quad (2.14)$$

$$D(\boldsymbol{x}) = d_0 + [d_1 \ d_2 \ \cdots \ d_n] \cdot (\boldsymbol{x} - \boldsymbol{x}^0), \quad (2.15)$$

and $\boldsymbol{x}^0$ is a reference point (typically, the center of the design space). Response correction parameters A and D are obtained as

$$[A, D] = \arg\min_{[\bar{A}, \bar{D}]} \sum_{k=1}^{N} \|f(\boldsymbol{x}^k) - (\bar{A}(\boldsymbol{x}^k) c(\boldsymbol{x}^k) + \bar{D}(\boldsymbol{x}^k))\|^2, \quad (2.16)$$

i.e., the response scaling is supposed to globally improve the matching for all training points $\boldsymbol{x}^k$, $k = 1, \ldots, N$. The problem (2.16) is equivalent to a linear regression problem $\boldsymbol{C}[a_0 \ a_1 \cdots a_n \ d_0 \ d_1 \cdots d_n]^T = \boldsymbol{F}$, the solution of which can be found as (Leifsson and Koziel, 2015a)

$$[a_0 \ a_1 \ \cdots \ a_n \ d_0 \ d_1 \ \cdots \ d_n]^T = (\boldsymbol{C}^T \boldsymbol{C}) \boldsymbol{C}^T \boldsymbol{F}, \quad (2.17)$$

where

$$\boldsymbol{C} = \begin{bmatrix} c(\boldsymbol{x}^1) & c(\boldsymbol{x}^1) \cdot (x_1^1 - x_1^0) & \cdots & c(\boldsymbol{x}^1) \cdot (x_n^1 - x_n^0) & 1 & (x_1^1 - x_1^0) & \cdots & (x_n^1 - x_n^0) \\ c(\boldsymbol{x}^2) & c(\boldsymbol{x}^2) \cdot (x_1^2 - x_1^0) & \cdots & c(\boldsymbol{x}^2) \cdot (x_n^2 - x_n^0) & 1 & (x_1^1 - x_1^0) & \cdots & (x_n^1 - x_n^0) \\ \vdots & \vdots & \ddots & \vdots & \vdots & \vdots & \ddots & \vdots \\ c(\boldsymbol{x}^N) & c(\boldsymbol{x}^N) \cdot (x_1^N - x_1^0) & \cdots & c(\boldsymbol{x}^N) \cdot (x_n^N - x_n^0) & 1 & (x_1^1 - x_1^0) & \cdots & (x_n^1 - x_n^0) \end{bmatrix},$$

$$(2.18)$$

$$\boldsymbol{F} = \begin{bmatrix} f(\boldsymbol{x}^1) & f(\boldsymbol{x}^2) & \cdots & f(\boldsymbol{x}^N) \end{bmatrix}^T. \quad (2.19)$$

Note that the matrices $\boldsymbol{C}^T \boldsymbol{C}$ are non-singular for $N > n + 1$ assuming that all training points are distinct. Choosing a star distribution training set (Giunta et al., 2003) satisfies this condition and is sufficient in many modeling cases. The star distribution training set consists of $N = 2n + 1$ points allocated at the center of the design space $\boldsymbol{x}^0 = (\boldsymbol{l} + \boldsymbol{u})/2$ ($\boldsymbol{l}$ and $\boldsymbol{u}$ being the lower and upper bound for the design variables, respectively), and the centers of its faces, i.e., points with all coordinates but one equal to those of $\boldsymbol{x}^0$, and the remaining one equal to the corresponding component of $\boldsymbol{l}$ or $\boldsymbol{u}$. Generalization of the model (2.13)–(2.19) for vector-valued outputs is straightforward.

2.4.4 Generalized shape-preserving response prediction

As an example of a slightly more involved physics-based surrogate modeling approach, let us consider generalized shape-preserving response prediction (GSPRP) methods. The shape-preserving response prediction (SPRP) technology has been initially developed for numerical optimization purposes (Koziel, 2010a). SPRP is based on so-called characteristic points that describe the most critical parts of the system output and are employed to set up a mapping between the low- and high-fidelity models. GSPRP is a version of SPRP developed for quasi-global modeling purposes (Koziel and Ogurtsov, 2012). Let $X_B = \{x^1, x^2, \ldots, x^N\}$ be the training data set. The high-fidelity model responses at the base designs, $f(x^k)$, are assumed to be known. The GSPRP concept is explained below for high-frequency structures with the outputs being frequency characteristics. Consequently, $f(x) = [f(x, \omega_1) \cdots f(x, \omega_m)]^T$, where ω_i, $i = 1, \ldots, m$, are discrete frequency values.

Figure 2.14 shows the response of the high-fidelity model of an exemplary microstrip filter at a certain number of training points (only a few points are shown for clarity). A set of characteristic points is distinguished on each of the plots, in this case corresponding to $|S_{21}| = -6\,\text{dB}$, $-15\,\text{dB}$, as well as local $|S_{21}|$ minima within the pass band. A discussion on selecting a set of characteristic points for a given design case can be found in Koziel (2010b). The notation $p_k^j = [\omega_k^j \ \lambda_k^j]^T, j = 1, \ldots, K$, is used to denote the characteristic

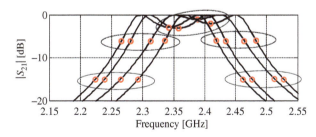

Fig. 2.14. Example response of the high-fidelity model f at several training points (solid lines). Circles indicate the characteristic points of the responses. Ellipses indicate the groups of corresponding characteristic points (Koziel and Ogurtsov, 2012).

points of $\boldsymbol{f}(\boldsymbol{x}^k)$. Let ω_k^j and λ_k^j denote the frequency and magnitude components of $\boldsymbol{p}_k^j$, respectively.

GSPRP predicts the response of the high-fidelity model at any design variable vector $\boldsymbol{x}$ using the information contained in the training points. The model is initialized by constructing auxiliary models $s_{\omega.j}(\boldsymbol{x})$ and $s_{\lambda.j}(\boldsymbol{x})$, $j = 1,\ldots,K$, of the sets of corresponding characteristic points for all training points, $\{\boldsymbol{p}_1^j, \boldsymbol{p}_2^j, \ldots, \boldsymbol{p}_N^j\}$, $j = 1,\ldots,K$. For this purpose, kriging interpolation is used (Lophaven et al., 2002).

Evaluation of the GSPRP model is a three-step process. In the first step, the characteristic points corresponding to the vector $\boldsymbol{x}$ are obtained as

$$\boldsymbol{p}^j(\boldsymbol{x}) = [s_{\omega.j}(\boldsymbol{x})\ s_{\lambda.j}(\boldsymbol{x})]^T, \qquad (2.20)$$

where $j = 1,\ldots,K$. In the second step, the index $k_{\min}(\boldsymbol{x})$ of the training point is identified as being the closest to $\boldsymbol{x}$, i.e.,

$$k_{\min}(\boldsymbol{x}) = \arg\min\{k \in \{1,\ldots,N\} : \|\boldsymbol{x}^k - \boldsymbol{x}\|\}, \qquad (2.21)$$

and then the translation vectors defined as

$$\boldsymbol{t}^j = [\omega_t^j\ \lambda_t^j]^T = [s_{\omega.j}(\boldsymbol{x}) - \omega_{k_{\min}(\boldsymbol{x})}^j\ \ s_{\lambda.j}(\boldsymbol{x}) - \lambda_{k_{\min}(\boldsymbol{x})}^j]^T, \qquad (2.22)$$

$j = 1,\ldots,K$, are calculated. These vectors indicate the change of the characteristic points of the $\boldsymbol{f}$ response while moving from $\boldsymbol{x}^{k_{\min}(\boldsymbol{x})}$ to $\boldsymbol{x}$. Figure 2.15 shows the high-fidelity model responses at $\boldsymbol{x}^{k_{\min}(\boldsymbol{x})}$ and $\boldsymbol{x}$,

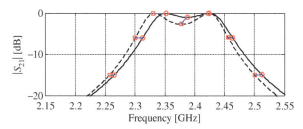

Fig. 2.15. The high-fidelity model $\boldsymbol{f}$ at $\boldsymbol{x}^{k_{\min}(\boldsymbol{x})}$ (the training point closest to the evaluation point $\boldsymbol{x}$), $\boldsymbol{f}(\boldsymbol{x}^{k_{\min}(\boldsymbol{x})})$, (—), the characteristic points of $\boldsymbol{f}(\boldsymbol{x}^{k_{\min}(\boldsymbol{x})})$ (o), the characteristic points $\boldsymbol{p}^j(\boldsymbol{x})$ corresponding to $\boldsymbol{x}$ obtained using (2.20) (□), as well as the translation vectors $\boldsymbol{t}^j$ (2.22) (short line segments). The GSPRP-predicted high-fidelity model response at $\boldsymbol{x}$, $\boldsymbol{s}(\boldsymbol{x})$ (- - -) is obtained using $\boldsymbol{p}^j(\boldsymbol{x})$, $\boldsymbol{t}^j$, and $\boldsymbol{f}(\boldsymbol{x}^{k_{\min}(\boldsymbol{x})})$ from (2.20)–(2.25).

the translation vectors $\boldsymbol{t}^j$, as well as the GSPRP model response at $\boldsymbol{x}$.

Using the translation vectors $\boldsymbol{t}^j$, defined in (2.22), the GSPRP surrogate model $\boldsymbol{s}$ of $\boldsymbol{f}$ can be defined as

$$\boldsymbol{t}^j = [\omega_t^j \ \lambda_t^j]^T = \left[\boldsymbol{s}_{\omega.j}(\boldsymbol{x}) - \omega_{k_{\min}(\boldsymbol{x})}^j \ \ \boldsymbol{s}_{\lambda.j}(\boldsymbol{x}) - \lambda_{k_{\min}(\boldsymbol{x})}^j \right]^T, \quad (2.23)$$

where $\boldsymbol{s}$ is determined at frequencies $\omega_{k_{\min}(\boldsymbol{x})}{}^j + \omega_t^j$, $j = 0, 1, \ldots, K, K+1$, as (with $\omega_{k_{\min}(\boldsymbol{x})}{}^0 = \omega_1$, $\omega_{k_{\min}(\boldsymbol{x})}{}^{K+1} = \omega_m$, and $\omega_t^0 = \omega_t^{K+1} = 0$)

$$\boldsymbol{s}(\boldsymbol{x}, \omega_{k_{\min}(\boldsymbol{x})}^j + \omega_t^j) = \bar{\boldsymbol{f}}(\boldsymbol{x}^{k_{\min}(\boldsymbol{x})}, \omega_{k_{\min}(\boldsymbol{x})}^j) + \lambda_t^j, \quad (2.24)$$

for $j = 1, \ldots, m$. For other frequencies, the model $\boldsymbol{s}$ is obtained through linear interpolation

$$\boldsymbol{s}(\boldsymbol{x}, \omega) = \bar{\boldsymbol{f}}(\boldsymbol{x}^0, (1-\alpha)\omega_{k_{\min}(\boldsymbol{x})}^j + \alpha\omega_{k_{\min}(\boldsymbol{x})}^{j+1}) + [(1-\alpha)\lambda_t^j + \alpha\lambda_t^{j+1}], \quad (2.25)$$

where $\omega_{k_{\min}(\boldsymbol{x})}{}^j + \omega_t^j \leq \omega \leq \omega_{k_{\min}(\boldsymbol{x})}{}^{j+1} + \omega_t^{j+1}$ and $\alpha = [\omega - (\omega_{k_{\min}(\boldsymbol{x})}{}^j + \omega_t^j)]/[(\omega_{k_{\min}(\boldsymbol{x})}{}^{j+1} + \omega_t^{j+1}) - (\omega_{k_{\min}(\boldsymbol{x})}{}^j + \omega_t^j)]$. $\bar{\boldsymbol{f}}(\boldsymbol{x}^{k_{\min}(\boldsymbol{x})}, \omega)$ is an interpolation of $\{\boldsymbol{f}(\boldsymbol{x}^{k_{\min}(\boldsymbol{x})}, \omega_1), \ldots, \boldsymbol{f}(\boldsymbol{x}^{k_{\min}(\boldsymbol{x})}, \omega_m)\}$ onto the frequency interval $[\omega_1, \omega_m]$.

The operation of GSPRP is demonstrated using two examples of microstrip filters: the stacked slotted resonators bandpass filter (Huang et al., 2008) shown in Fig. 2.16(a) (Filter 1), and the microstrip bandpass filter with open stub inverter

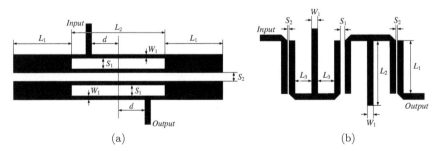

Fig. 2.16. Filter structures for GSPRP modeling: (a) stacked slotted resonators filter (Huang et al., 2008), (b) bandpass filter with open stub inverter (Lee et al., 2000).

Table 2.2. GSPRP modeling results for filters of Fig. 2.16.

Number of training points[a]	Average RMS error (Filter 1)		Average RMS error (Filter 2)	
	GSPRP (%)	Kriging interpolation (%)	GSPRP (%)	Kriging interpolation (%)
20	3.9	15.6	7.7	11.6
50	1.8	11.9	2.7	8.8
100	0.9	11.0	1.8	6.1
200	0.6	9.3	1.4	4.8
400	0.4	6.7	1.2	3.1

Note: [a] Training points allocated using Latin Hypercube Sampling (Beachkofski and Grandhi, 2002).

(Lee et al., 2000), see Fig. 2.16(b) (Filter 2). Design variables are $x = [L_1\ L_2\ W_1\ S_1\ S_2\ d]^T$ (Filter 1), and $x = [L_1\ L_2\ L_3\ S_1\ S_2\ W_1]^T$ (Filter 2). The surrogate model domain for Filter 1 is defined by the reference point $x^0 = [6\ 9.6\ 1\ 1\ 2\ 2]^T$ mm, and the region size $d = [0.8\ 0.8\ 0.2\ 0.2\ 0.4\ 0.4]^T$ mm. For Filter 2, $x^0 = [24\ 10\ 2\ 0.6\ 0.2\ 0.5]^T$ mm, and $d = [2\ 2\ 1\ 0.4\ 0.1\ 0.4]^T$ mm. Table 2.2 shows the average (relative) RMS errors for the GSPRP and kriging surrogates (Forrester and Keane, 2009), with different number of training points from 20 to 400. For both filters, the accuracy of the GSPRP model is better than the accuracy of the kriging surrogate for the corresponding number of training points. A comparison of the high-fidelity model data with kriging interpolation reveals that the comparable error level can be obtained for the training set that is several times smaller than for kriging interpolation.

2.4.5 *Feature-based modeling*

Modeling of high-frequency structures entails handling vector-valued outputs, typically frequency characteristics (Koziel and Ogurtsov, 2019). The responses are often highly nonlinear, both as a function of the frequency and geometry and/or material parameters of the structure at hand. This poses significant challenges when it comes to constructing the surrogate models. Feature-based modeling (Koziel and Bekasiewicz, 2015, 2018) offers a workaround these issues by shifting the interest from the entire characteristics into a set of carefully defined points, so-called response features, that are sufficient for a particular design task considered, yet, their functional dependence on the structure parameters is much less nonlinear than that of the original characteristics (Koziel and Bekasiewicz, 2017). Response feature modeling belongs to the category of physics-based models because the allocation of the feature points is strictly related to the particular characteristics of the structure (e.g., resonances of the antenna, or the poles of the microwave filter), which are, in turn, determined by the physical operation of the device.

In this section, the concept of feature-based modeling is briefly outlined and illustrated through modeling of antenna input characteristics. The high-fidelity model is assumed to be obtained from full-wave EM simulation and denoted as $\boldsymbol{f}(\boldsymbol{x})$. Here, it represents the reflection coefficient $|S_{11}|$ of the antenna evaluated at m frequencies, ω_1 to ω_m, thus $\boldsymbol{f}(\boldsymbol{x}) = [f(\boldsymbol{x}, \omega_1) \cdots f(\boldsymbol{x}, \omega_m)]^T$. The objective is to build a replacement model (surrogate) s. The surrogate should represent the EM model over a given region X of the design space.

The training set is denoted as $X_B = \{\boldsymbol{x}^1, \boldsymbol{x}^2, \ldots, \boldsymbol{x}^N\} \subset X$. The corresponding EM model responses $\boldsymbol{f}(\boldsymbol{x}^k)$ are acquired beforehand. According to the conventional approach to data-driven surrogate construction, the responses $f(\boldsymbol{x}, \omega_j), j = 1, \ldots, m$, are approximated directly (either separately for each frequency or by treating the frequency as an additional input parameter of the model). The problem is nonlinearity of the responses, particularly for narrow-band antennas (Koziel and Bekasiewicz, 2017). Typical responses of such structures are shown in Fig. 2.17 (reflection responses of a DRA of Fig. 2.20). For clarity, the response evaluated in the region corresponding to $7.0 \leq a_x \leq 9.0$ and $13.0 \leq a_y \leq 15.0$ (cf. Fig. 2.20) at frequency of 5.5 GHz is shown. Accurate modeling of such landscapes is only possible within limited parameter ranges and requires a large number of training samples.

The objective is to reduce the number of training data samples necessary to construct an accurate surrogate model. It is achieved by reformulating the modeling process and conducting it at the level of appropriately defined response features. Figure 2.18 clarifies the definition of the feature points in the case of narrow-band antennas. The characteristic point set is constructed sequentially as follows: (i) identification of the primary point which corresponds to the center frequency (antenna resonance) and the response level at that

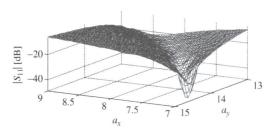

Fig. 2.17. Exemplary responses of the DRA of Fig. 2.20 (reflection coefficient $|S_{11}|$). The response is evaluated in the region $7.0 \leq a_x \leq 9.0$ and $13.0 \leq a_y \leq 15.0$ at the frequency of 5.5 GHz. Other variables are fixed to the following values: $a_z = 9, a_c = 0, u_s = 2, w_s = 10, y_s = 8$ (all in mm) (Koziel and Bekasiewicz, 2017).

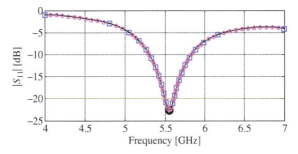

Fig. 2.18. Definition of response features in the case of a narrow-band antenna reflection characteristic: the primary point (corresponding to the antenna resonance) is represented as O; □ represent supplemental points distributed equally with respect to response level, (○) denote infill points distributed equally in frequency between the main and supplemental points (note that the number of points may be different for various intervals) (Koziel and Bekasiewicz, 2017).

frequency, (ii) allocation of the supplemental points (in this case, uniformly with respect to the level and separately on the left- and right-hand-side of the primary point), and (iii) allocation of the infill points uniformly in frequency in between the supplemental points. Clearly, one needs to ensure that the number of characteristic points is sufficient so as to allow a reliable synthesis of the antenna response (through interpolation).

On the other hand, although it is important that the major features of the response (e.g., antenna resonance) are accounted for, particular point allocation is not critical. The response features, once defined, can be easily extracted from the EM simulation

Fundamentals of Physics-Based Surrogate Modeling 67

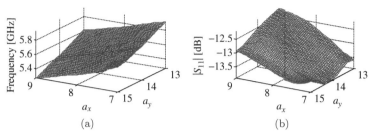

Fig. 2.19. Frequency (a) and level (b) component as functions of geometry parameters a_x and a_y (with other variables fixed) for the selected feature point. The responses are evaluated over the same design space region as that considered in Fig. 2.17. The functional landscapes of the feature point coordinates are considerably less nonlinear than those for original responses (cf. Fig. 2.17) (Koziel and Bekasiewicz, 2017).

data. For subsequent considerations, the jth feature point of $f(x^k)$ ($j = 1, \ldots, K, k = 1, \ldots, N$) will be denoted as $f_k^j = [\omega_k^j\ \lambda_k^j]$, where ω_k^j and λ_k^j represent the frequency and the magnitude (level) components of f_k^j, respectively.

For the sake of illustration, the frequency and level components of the selected feature point have been shown in Fig. 2.19. The considered design space region is the same as the one shown in Fig. 2.18. The functional landscapes of the feature points are not as nonlinear as those shown in Fig. 2.17. It is expected that construction of a reliable surrogate model at the feature point level will require a smaller number of training samples than modeling the reflection response in a traditional manner.

Having the response features defined, the surrogate modeling process works as follows. First, the data-driven models $s_{\omega.j}(x)$ and $s_{\lambda.j}(x)$, $j = 1, \ldots, K$, of the sets of corresponding feature points are constructed using available training designs, $\{f_1^j, f_2^j, \ldots, f_N^j\}$, $j = 1, \ldots, K$ (Koziel and Bekasiewicz, 2017). At this stage, kriging interpolation is utilized (Kleijnen, 2009). The surrogate model itself is defined as

$$s(x) = [s(x, \omega_1) \cdots s(x, \omega_m)]^T, \qquad (2.26)$$

where its jth component is given as (Koziel and Bekasiewicz, 2017)

$$s(x, \omega_j) = I(\Omega(x), \Lambda(x), \omega_j). \qquad (2.27)$$

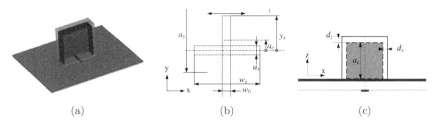

Fig. 2.20. DRA (Sim et al., 2014): (a) 3D view of the structure with housing, (b) top, and (c) front views with marked dimensions.

In (2.27), $\Lambda(\boldsymbol{x}) = [s_{\lambda.1}(\boldsymbol{x})\, s_{\lambda.2}(\boldsymbol{x}) \cdots s_{\lambda.K}(\boldsymbol{x})]$, and $\Omega(\boldsymbol{x}) = [s_{\omega.1}(\boldsymbol{x})\, s_{\omega.2}(\boldsymbol{x}) \cdots s_{\omega.K}(\boldsymbol{x})]$ are the predicted feature point locations corresponding to the evaluation design $\boldsymbol{x}$. The aim is to evaluate the antenna response at a discrete set of frequencies ω_1 through ω_m, it is necessary to interpolate both the level vector Λ and frequency vector Ω into the response at the above set of frequencies. This interpolation is represented as $I(\Omega, \Lambda, \omega)$.

For demonstration, consider a rectangular DRA (Sim et al., 2014) shown in Fig. 2.20. The dielectric resonator is implemented using a material with permittivity $\varepsilon_r = 10$. The substrate is Rogers RO4003 ($h = 0.5$ mm, $\varepsilon_r = 3.3$). The computational model of the antenna also includes a polycarbonate housing with dielectric permittivity of 2.8.

There are seven geometry parameters describing the DRA structure $\boldsymbol{x} = [a_x\ a_y\ a_z\ a_c\ u_s\ w_s\ y_s]^T$. The computational model of the DRA is implemented in CST Microwave Studio (CST, 2018). The physics-based model of the structure denoted as $\boldsymbol{f}(\boldsymbol{x})$ has been implemented in the package and then discretized using tetrahedral mesh ($\sim$420,000 mesh cells). The average simulation time of the model on a dual Intel Xeon E5540 machine with 6 GB RAM is 19 min. The frequency range is 4–7 GHz.

The objective is to construct the surrogate model of the reflection coefficient $|S_{11}|$. The model domain is defined as $X = [\boldsymbol{x}^{(0)} - \boldsymbol{d}, \boldsymbol{x}^{(0)} + \boldsymbol{d}]$, with the center $\boldsymbol{x}^{(0)} = [8.0\ 14.0\ 9.0\ 0.0\ 2.0\ 10.0\ 8.0]^T$ mm, and size $\boldsymbol{d} = [2.0\ 2.0\ 2.0\ 2.0\ 1.0\ 2.0\ 2.0]^T$ mm. Table 2.3 shows the average RMS errors of the feature-based surrogates obtained for various numbers of training data samples. For comparison, conventional (kriging) models of S_{11} are also included.

Table 2.3. Modeling results of the DRA (Koziel and Bekasiewicz, 2017).

Model	Average error (%)					
Number of training points	20	50	100	200	400	800
Feature-based surrogate	23.4	10.8	7.8	6.7	3.9	3.0
Kriging interpolation[a]	43.2	36.9	29.1	24.0	11.1	8.0

Note: [a]Direct kriging interpolation of high-fidelity model $|S_{11}|$ responses.

It can be observed that the feature-based technique allows for a considerable reduction of the training data set while ensuring comparable predictive power of the surrogate. Furthermore, even for the largest considered training set (800 samples), accuracy of the conventional model is still insufficient for conducting practical design tasks such as parametric optimization. At the same time, accuracy of the feature-based model is already acceptable when constructed using 200 samples.

2.5 Physics-Based Surrogates for Design Optimization

One of the most important applications of physics-based models is (local) surrogate-assisted optimization (Koziel et al., 2008b, 2016b; Feng et al., 2019; Koziel and Leifsson, 2013; Robinson et al., 2008; Eldred and Dunlavy, 2006; Leifsson et al., 2014a). The considered problem is typically formulated as a minimization task

$$\boldsymbol{x}^* = \arg\min_{\boldsymbol{x} \in X} U(\boldsymbol{f}(\boldsymbol{x})), \qquad (2.28)$$

where $\boldsymbol{f}(\boldsymbol{x})$ is the (generally a vector-valued) output of the high-fidelity model, U is a scalar objective function, whereas $\boldsymbol{x} \in X \subseteq R^n$. The surrogate-assisted algorithm is an iterative procedure producing a series $\boldsymbol{x}^{(i)}$, $i = 0, 1, 2, \ldots$, of approximations to $\boldsymbol{x}^*$ as (a vector-valued version of (2.1))

$$\boldsymbol{x}^{(i+1)} = \arg\min_{\boldsymbol{x}} U(\boldsymbol{s}^{(i)}(\boldsymbol{x})), \qquad (2.29)$$

where $\boldsymbol{s}^{(i)}$ is the surrogate model at the iteration i. The starting point for solving (2.29) is $\boldsymbol{x}^{(i)}$, and the surrogate is supposed to be at least zero-order consistent with the high-fidelity model, i.e., $\boldsymbol{s}^{(i)}(\boldsymbol{x}^{(i)}) = \boldsymbol{f}(\boldsymbol{x}^{(i)})$ (Alexandrov et al., 1998). First-order consistency is required to ensure the algorithm convergence in classical terms, however, the problem-specific knowledge embedded in the underlying low-fidelity model (the surrogate is constructed from) is often sufficient to ensure satisfactory performance (Koziel, 2010a; Koziel and Leifsson, 2013a; Robinson et al., 2008).

A number of modeling techniques have been developed in this particular context. Their common feature is that the surrogate is usually constructed using a single high-fidelity model sample, typically, evaluated at the most recent design along the optimization path. These methods include, among others, SM (Bandler et al., 2004a), approximation model management optimization (AMMO) (Alexandrov and Lewis, 2001), manifold mapping (MM) (Echeverria and Hemker, 2005), SPRP (Koziel, 2010a), adaptively adjusted design specifications (AADS) (Koziel, 2010c), feature-based optimization (FBO) (Koziel, 2015), cognition-driven design (Zhang et al., 2015), neuro-transfer function optimization (Feng et al., 2020), or adaptive response scaling (ARS) (Koziel and Bekasiewicz, 2016b). The details concerning these techniques are omitted here for the sake of brevity. Interested reader can find them in the cited literature. For illustration, one of these methods, manifold mapping, is discussed below.

The MM surrogate model is defined as (Echeverria and Hemker, 2005)

$$s^{(i)}(\boldsymbol{x}) = \boldsymbol{f}\left(\boldsymbol{x}^{(i)}\right) + \boldsymbol{S}^{(i)}\left(\boldsymbol{c}(\boldsymbol{x}) - \boldsymbol{c}(\boldsymbol{x}^{(i)})\right), \qquad (2.30)$$

with $\boldsymbol{S}^{(i)}$, being the $m \times m$ correction matrix, defined as

$$\boldsymbol{S}^{(i)} = \boldsymbol{\Delta F} \cdot \boldsymbol{\Delta C}^{\dagger}, \qquad (2.31)$$

where

$$\boldsymbol{\Delta F} = \left[\boldsymbol{f}\left(\boldsymbol{x}^{(i)}\right) - \boldsymbol{f}\left(\boldsymbol{x}^{(i-1)}\right) \quad \cdots \quad \boldsymbol{f}\left(\boldsymbol{x}^{(i)}\right) - \boldsymbol{f}\left(\boldsymbol{x}^{(\max\{i-n,0\})}\right)\right], \quad (2.32)$$

$$\boldsymbol{\Delta C} = \left[\boldsymbol{c}\left(\boldsymbol{x}^{(i)}\right) - \boldsymbol{c}\left(\boldsymbol{x}^{(i-1)}\right) \quad \cdots \quad \boldsymbol{c}\left(\boldsymbol{x}^{(i)}\right) - \boldsymbol{c}\left(\boldsymbol{x}^{(\max\{i-n,0\})}\right)\right]. \quad (2.33)$$

The pseudoinverse, denoted by †, is defined as

$$\boldsymbol{\Delta C}^{\dagger} = \boldsymbol{V}_{\boldsymbol{\Delta C}} \boldsymbol{\Sigma}^{\dagger}_{\boldsymbol{\Delta C}} \boldsymbol{U}^T_{\boldsymbol{\Delta C}}, \qquad (2.34)$$

where $\boldsymbol{U}_{\boldsymbol{\Delta C}}$, $\sum_{\boldsymbol{\Delta C}}$, and $\boldsymbol{V}_{\boldsymbol{\Delta C}}$ are the factors in the singular value decomposition of the matrix $\boldsymbol{\Delta C}$. The matrix $\sum^{\dagger}_{\boldsymbol{\Delta C}}$ is the result of

Fundamentals of Physics-Based Surrogate Modeling

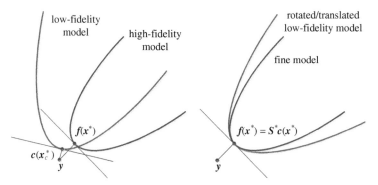

Fig. 2.21. The concept of the MM model alignment for a least-squares optimization problem: x_c^* is the low-fidelity model minimizer, and y is the vector of design specifications. The straight lines denote the tangent planes for f and c at their optimal designs, respectively. By the linear correction S^*, the point $c(x^*)$ is mapped to $f(x^*)$, and the tangent plane for $c(x)$ at $c(x^*)$ to the tangent plane for $f(x)$ at $f(x^*)$ (Koziel and Ogurtsov, 2011).

inverting the nonzero entries in $\Sigma_{\Delta C}$, leaving the zeroes invariant. Upon convergence, the linear correction S^* (being the limit of $S^{(i)}$ with $i \to \infty$) maps the point $c(x^*)$ to $f(x^*)$, and the tangent plane for $c(x)$ at $c(x^*)$ to the tangent plane for $f(x)$ at $f(x^*)$, cf. Fig. 2.21 (Echeverria and Hemker, 2005).

Although MM does not explicitly use sensitivity information, the surrogate and the high-fidelity model Jacobians start to coincide upon convergence of the MM process (i.e., when $\|x^{(i)} - x^{(i-1)}\| \to 0$), so that the surrogate (approximately) satisfies both zero- and first-order consistency conditions (Alexandrov and Lewis, 2001) with f. This allows for a more precise identification of the high-fidelity model optimum. On the other hand, the correction matrix $S^{(i)}$ can be defined using exact Jacobians of the low- and high-fidelity models if available.

The MM algorithm is illustrated using an ultra-wideband (UWB) monopole shown in Fig. 2.22. Design variables are $x = [h_0\ w_0\ a_0\ s_0\ h_1\ w_1\ l_{gnd}\ w_s]^T$. Both the high- and low-fidelity models are implemented in CST Microwave Studio (CST, 2018). The simulation time ratio for the models is around twenty. The design specifications for antenna reflection are $|S_{11}| \leq -10$ dB for 3.1–10.6 GHz. The initial design is $x^{(0)} = [18\ 12\ 2\ 0.5\ 1\ 15\ 40]^T$ mm. Because the low-fidelity model is relatively expensive, the MM algorithm is using the

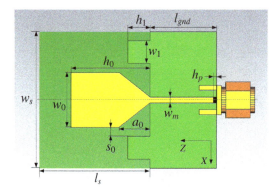

Fig. 2.22. UWB monopole: top view, substrate shown transparent. Magnetic-symmetry wall is shown with the dash-dot line (Koziel *et al.*, 2016).

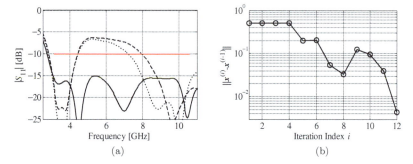

Fig. 2.23. UWB monopole: (a) high- (- - -) and low-fidelity ($\cdots$) model responses at the initial design, as well as high-fidelity model (—) at the final design x^*, and (b) convergence plot of the MM algorithm (Koziel *et al.*, 2016).

underlying kriging interpolation model c_{kr} created in the vicinity of the approximate optimum of c (obtained at the cost of 100 c evaluations). Optimization performed using the MM algorithm yields the final design $x^* = [19.13\ 20.13\ 1.95\ 1.33\ 1.79\ 6.32\ 15.03\ 36.36]^T$ mm ($|S_{11}| < -15$ dB in the frequency band of interest). The total design cost is about 21 high-fidelity model evaluations (cf. Fig. 2.23).

Acknowledgments

This chapter was supported in part by the Icelandic Centre for Research (RANNIS) Grant 206606 and by National Science Centre of Poland Grant 2018/31/B/ST7/02369.

References

Advanced Design System (ADS) (2019). Keysight Technologies, Fountaingrove Parkway 1400, Santa Rosa, CA 95403-1799.

Alexandrov, N.M., Lewis, R.M. (2001). An overview of first-order model management for engineering optimization, *Opt. Eng.*, vol. 2, no. 4, pp. 413–430.

Alexandrov, N.M., Dennis, J.E., Lewis, R.M., Torczon, V. (1998). A trust-region framework for managing the use of approximation models in optimization, *Struct. Optimiz.*, vol. 15, no. 1, pp. 16–23.

Altair FEKO (2018). *Altair HyperWorks*, 1820 E Big Beaver Rd, Troy, MI 48083, USA.

Bakr, M.H., Bandler, J.W., Madsen, K., Rayas-Sanchez, J.E., Sondergaard, J. (2000). Space-mapping optimization of microwave circuits exploiting surrogate models, *IEEE Trans. Microwave Theory Tech.*, vol. 48, no. 12, pp. 2297–2306.

Bandler, J.W., Cheng, Q.S., Gebre-Mariam, D.H., Madsen, K., Pedersen, F., Søndergaard, J. (2003). EM-based surrogate modeling and design exploiting implicit, frequency and output space mappings, in *IEEE Int. Microwave Symp. Digest*, Philadelphia, PA, USA, pp. 1003–1006.

Bandler, J.W., Cheng, Q.S., Dakroury, S.A., Mohamed, A.S., Bakr, M.H., Madsen, K., Søndergaard, J. (2004a). Space mapping: The state of the art, *IEEE Trans. Microwave Theory Tech.*, vol. 52, no. 1, pp. 337–361.

Bandler, J.W., Cheng, Q.S., Nikolova, N.K., Ismail, M.A. (2004b). Implicit space mapping optimization exploiting preassigned parameters, *IEEE Trans. Microwave Theory Tech.*, vol. 52, no. 11, pp. 378–385.

Bandler, J.W., Koziel, S., Madsen, K. (2008). Editorial — surrogate modeling and space mapping for engineering optimization, *Optim. Eng.*, vol. 9, no. 4, pp. 307–310.

Baumann, D., Fumeaux, C., Leuchtmann, P., Vahldieck, R. (2004). Finite-volume time-domain (FVTD) modelling of a broadband double-ridged horn antenna, *Int. J. Num. Model.: Electr. Devices Fields*, vol. 17, no. 3, pp. 285–298.

Bekasiewicz, A., Koziel, S. (2016). Cost-efficient design optimization of compact patch antennas with improved bandwidth, *IEEE Ant. Wireless Prop. Lett.*, vol. 15, pp. 270–273.

Bekasiewicz, A., Koziel, S., Zieniutycz, W. (2014). Design space reduction for expedited multi-objective design optimization of antennas in highly-dimensional spaces, in Koziel, S., Leifsson, L., Yang, X.-S. (eds.), *Solving Computationally Expensive Engineering Problems: Methods and Applications*, New York, Springer, pp. 113–147.

Caratelli, D., Yarovoy, A. (2010). Unified time- and frequency-domain approach for accurate modeling of electromagnetic radiation processes in ultrawideband antennas, *IEEE Tran. Ant. Propag.*, vol. 58, no. 10, pp. 3239–3255.

Cheng, Q.S., Koziel, S., Bandler, J.W. (2006). Simplified space mapping approach to enhancement of microwave device models, *Int. J. RF Microwave Computer-Aided Eng.*, vol. 16, no. 5, pp. 518–535.

Cheng, Q.S., Bandler, J.W., Koziel, S. (2008). Combining coarse and fine models for optimal design. *Microwave Magaz.*, vol. 9, pp. 79–88.

CST Microwave Studio (2018). CST AG, Bad Nauheimer Str. 19, D-64289 Darmstadt, Germany.

Dorica, M., Giannacopoulos, D.D. (2006). Response surface space mapping for electromagnetic optimization, *IEEE Trans. Magn.*, vol. 42, no. 4, pp. 1123–1126.

Echeverria, D., Hemker, P.W. (2005). Space mapping and defect correction, *Comput. Meth. Appl. Math.*, vol. 5, no. 2, pp. 107–136.

Eldred, M.S., Dunlavy, D.M. (2006). Formulations for surrogate-based optimization with data fit, multifidelity, and reduced-order models, in *11th AIAA/ISSMO Multidisc. Analysis Optim. Conf.*, Portsmouth, Virginia, AIAA–2006-7117.

Feng, F., Na, W., Liu, W., Yan, S., Zhu, L., Ma, J., Zhang, Q.J. (2020). Multifeature-assisted neuro-transfer function surrogate-based EM optimization exploiting trust-region algorithms for microwave filter design, *IEEE Trans. Microwave Theory Techn.*, vol. 68, no. 2, pp. 531–542.

Feng, F., Zhang, C., Na, W., Zhang, J., Zhang, W., Zhang, Q. (2019). Adaptive feature zero assisted surrogate-based EM optimization for microwave filter design, *IEEE Microw. Wireless Comp. Lett.*, vol. 29, no. 1, pp. 2–4.

Fernández-Godino, M.G., Park, C., Kim, N.H., Haftka, R.T. (2019). Issues in deciding whether to use multifidelity surrogates, *AIAA J.*, vol. 57, no. 5, pp. 2039–2054.

Forrester, A.I.J., Keane, A.J. (2009). Recent advances in surrogate-based optimization, *Prog. Aerospace Sci.*, vol. 45, no. 1, pp. 50–79.

Giunta, A.A., Wojtkiewicz, S.F., Eldred, M.S. (2003). Overview of modern design of experiments methods for computational simulations, *Paper AIAA*, pp. 2003–0649.

Goudos, S. (ed.) (2017). *Microwave Systems and Applications*, London, UK, IntechOpen.

Guan, X., Ma, Z., Cai, P., Anada, T., Hagiwara, G. (2008). A microstrip dual-band bandpass filter with reduced size and improved stopband characteristics, *Microwave Opt. Tech. Lett.*, vol. 50, pp. 618–620.

Hazaveh, P.K., Bergstrom, P.L., Jaszczak, J.A. (2017). Efficient physics-based modeling of a representative semiconducting quantum dot single electron device, in *IEEE 17th Int. Conf. Nanotechnology (IEEE-NANO)*, Pittsburgh, PA, pp. 739–744.

Hong, J.-S., Lancaster, M. (2001). *Microstrip Filters for RF/Microwave Applications*, Hoboken, Wiley.

Hsieh, L.H., Chang, K. (2003). Tunable microstrip bandpass filters with two transmission zeros, *IEEE Trans. Microwave Theory Tech.*, vol. 51, no. 2, pp. 520–525.

Huang, C.L., Chen, Y.B., Tasi, C.F. (2008). New compact microstrip stacked slotted resonators bandpass filter with transmission zeros using high-permittivity ceramics substrate, *Microwave Opt. Tech. Lett.*, vol. 50, no. 5, pp. 1377–1379.

Kleijnen, J.P.C. (2009). Kriging metamodeling in simulation: A review, *European J. Operational Research*, vol. 192, no. 3, pp. 707–716.

Kleijnen, J.P.C. (2018). Design and analysis of simulation experiments, in Pilz, J., Rasch, D., Melas, V., Moder, K. (eds.), *Statistics and Simulation. IWS 2015*, Springer Proceedings in Mathematics & Statistics, vol. 231, Cham, Springer.

Kolundzija, B., Sumic, D. (2004). Hierarchical conjugate gradient method applied to MoM analysis of electrically large structures, in *IEEE Ant. Propag. Society Symp. (APS)*, 2004, Monterey, CA, USA, vol. 4, pp. 4455–4458.

Koziel, S. (2010a). Shape-preserving response prediction for microwave design optimization, *IEEE Trans. Microwave Theory Tech.*, vol. 58, no. 11, pp. 2829–2837.

Koziel, S. (2010b). Shape-preserving response prediction for microwave circuit modeling, in *IEEE MTT-S Int. Microwave Symp. Digest*, Anaheim, CA, USA, pp. 1660–1663.

Koziel, S. (2010c). Adaptively adjusted design specifications for efficient optimization of microwave structures, *Prog. EM Res. B*, vol. 21, pp. 219–234.

Koziel, S. (2015). Fast simulation-driven antenna design using response-feature surrogates, *Int. J. RF & Microwave CAE*, vol. 25, no. 5, pp. 394–402.

Koziel, S. (2017). Space mapping: Performance, reliability, open problems and perspectives, *IEEE MTT-S Int. Microwave Symp. (IMS)*, Honololu, HI, pp. 1512–1514.

Koziel, S., Bandler, J.W. (2007a). Microwave device modeling using space-mapping and radial basis functions, *IEEE MTT-S Int. Microwave Symp. Digest*, Honolulu, HI, USA, pp. 799–802.

Koziel, S., Bandler, J.W. (2007b). A space-mapping approach to microwave device modeling exploiting fuzzy systems, *IEEE Trans. Microwave Theory Tech.*, vol. 55, no. 12, pp. 2539–2547.

Koziel, S., Bandler, J.W. (2012). Accurate modeling of microwave devices using kriging-corrected space mapping, *Int. J. Num. Model.: Electr. Devi. Fields*, vol. 25, no. 1, pp. 1–4.

Koziel, S., Bekasiewicz, A. (2015). Fast simulation-driven feature-based design optimization of compact dual-band microstrip branch-line coupler, *Int. J. RF & Microwave CAE*, vol. 26, no. 1, pp. 13–20.

Koziel, S., Bekasiewicz, A. (2016a). *Multi-Objective Design of Antennas Using Surrogate Models*, World Scientific, Singapore.

Koziel, S., Bekasiewicz, A. (2016b). Rapid microwave design optimization in frequency domain using adaptive response scaling, *IEEE Trans. Microwave Theory Techn.*, vol. 64, no. 9, pp. 2749–2757.

Koziel, S., Bekasiewicz, A. (2017). Computationally feasible narrow-band antenna modeling using response features, *Int. J. RF & Microwave CAE*, vol. 27, no. 4, pp. e21077.

Koziel, S., Bekasiewicz, A. (2018). Simulation-driven size-reduction-oriented design of multi-band antennas by means of response features, *IET Microwave Ant. Propag.*, vol. 12, no. 7, pp. 1093–1098.

Koziel, S., Kurgan, P. (2015). Rapid design of miniaturized branch-line couplers through concurrent cell optimization and surrogate-assisted fine-tuning, *IET Microwaves Ant. Prop.*, vol. 9, no. 9, pp. 957–963.

Koziel, S., Leifsson, L. (2012). Generalized shape-preserving response prediction for accurate modeling of microwave structures, *IET Microwaves, Ant. Prop.*, vol. 6, pp. 1332–1339.

Koziel, S., Leifsson, L. (eds.) (2013a). *Surrogate-Based Modeling and Optimization. Applications in Engineering*, New York, Springer.

Koziel, S., Leifsson, L. (2013b). Multi-level airfoil shape optimization with automated low-fidelity model selection, in *Int. Conf. Comp. Science*, Barcelona, Spain.

Koziel, S., Leifsson, L. (2016). *Simulation-Driven Design by Knowledge-Based Response Correction Techniques*, Cham, Springer.

Koziel, S., Ogurtsov, S. (2011). Simulation-driven design in microwave engineering: Application case studies, in Yang, X.S., Koziel, S. (eds.), *Computational Optimization and Applications in Engineering and Industry*, Series: Studies in Computational Intelligence, Springer-Verlag.

Koziel, S., Ogurtsov, S. (2012). Model management for cost-efficient surrogate-based optimization of antennas using variable-fidelity electromagnetic simulations, *IET Microwaves Ant. Prop.*, vol. 6, pp. 1643–1650.

Koziel, S., Ogurtsov, S. (2014). *Antenna Design by Simulation-Driven Optimization*, Berlin, Springer.

Koziel, S., Ogurtsov, S. (2019). *Simulation-Based Optimization of Antenna Arrays*, Singapore, World Scientific.

Koziel, S., Bandler, J.W., Madsen, K. (2006a). A space mapping framework for engineering optimization: Theory and implementation, *IEEE Trans. Microwave Theory Tech.*, vol. 54, no. 10, pp. 3721–3730.

Koziel, S., Bandler, J.W., Madsen, K. (2006b). Theoretical justification of space-mapping-based modeling utilizing a data base and on-demand parameter extraction, *IEEE Trans. Microwave Theory Tech.*, vol. 54, no. 12, pp. 4316–4322.

Koziel, S., Cheng, Q.S., Bandler, J.W. (2008a). Space mapping, *IEEE Microwave Magaz.*, vol. 9, no. 6, pp. 105–122.

Koziel, S., Bandler, J.W., Madsen, K. (2008b). Quality assessment of coarse models and surrogates for space mapping optimization, *Opt. Eng.*, vol. 9, pp. 375–391.

Koziel, S., Cheng, Q.S., Bandler, J.W. (2010). Implicit space mapping with adaptive selection of preassigned parameters, *IET Microwaves, Ant. Prop.*, vol. 4, pp. 361–373.

Koziel, S., Bandler, J.W., Cheng, Q.S. (2011a). Constrained parameter extraction for microwave design optimisation using implicit space mapping, *IET Microwave Antennas Propag.*, vol. 5, pp. 1156–1163.

Koziel, S., Ciaurri D.E., Leifsson L. (2011b). Surrogate-based methods, in Koziel, S., Yang, X.S. (eds.), *Computational Optimization, Methods and Algorithms. Studies in Computational Intelligence*, vol. 356, Berlin, Heidelberg, Springer, pp. 33–59.

Koziel, S., Yang, X.S., Zhang, Q.J. (eds.) (2013). *Simulation-Driven Design Optimization and Modeling for Microwave Engineering*, London, Imperial College Press.

Koziel, S., Ogurtsov, S., Zieniutycz, W., Sorokosz, L. (2014). Simulation-driven design of microstrip antenna subarrays, *IEEE Trans. Ant. Prop.*, vol. 62, no. 7, pp. 3584–3591.

Koziel, S., Bekasiewicz, A., Leifsson, L. (2016). Cost-efficient modeling of input characteristics of narrow-band antennas using response features, in *10th European Conf. Ant. Propag. (EuCAP)*, Davos, 2016, pp. 1–4.

Lee, J.R., Cho, J.H., Yun, S.W. (2000). New compact bandpass filter using microstrip $\lambda/4$ resonators with open stub inverter, *IEEE Microwave Guided Wave Lett.*, vol. 10, no. 12, pp. 526–527.

Leifsson, L., Koziel, S. (2015a). Variable-resolution shape optimization: Low-fidelity model selection and scalability, *Int. J. Math. Model. Numer. Optim.*, vol. 6, pp. 1–21.

Leifsson, L., Koziel, S. (2015b). *Simulation-Driven Aerodynamic Design Using Variable-Fidelity Models*, London, Imperial College Press.

Leifsson, L., Koziel, S., Ogurtsov, S. (2012). Low-fidelity model mesh density and the performance of variable-resolution shape optimization algorithms, *Procedia Computer Science*, vol. 9, pp. 842–851.

Leifsson, L., Koziel, S., Hosder, S. (2014a). Aerodynamic design optimization: Physics-based surrogate approaches for airfoil and wing design, in *52nd Aerospace Sciences Meeting AIAA SciTech Forum*, National Harbor, Maryland, USA, AIAA 2014-0572.

Leifsson, L., Koziel, S., Kurgan, P. (2014b). Automated low-fidelity model setup for surrogate-based aerodynamic optimization, in Koziel, S., Leifsson, L., Yang, X.S. (eds.), *Solving Computationally Extensive Engineering Problems: Methods and Applications*, Springer, pp. 87–112.

Liu, G., Gedney, S.D. (2000). Perfectly matched layer media for an unconditionally stable three-dimensional ADI-FDTD method, *IEEE Microwave Guided Wave Lett.*, vol. 10, no. 7, pp. 261–263.

Lophaven, S.N., Nielsen, H.B., Søndergaard, J. (2002). *DACE: A Matlab kriging toolbox*, Technical University of Denmark.

Obaidat, M.S., Ören, T., De Floriano R. (eds.) (2019). *Simulation and modeling methodologies, technologies and applications*, in 7th *Int. Conf., SIMULTECH 2017*, Madrid, Spain, Advances in Intelligent Systems and Computing, Springer.

Petosa, A. (2007). *Dielectric Resonator Antenna Handbook*, Artech House, Norwood.

Perdikaris, P., Raissi, M., Damianou, A., Lawrence, N.D., Karniadakis, G.E. (2017). Nonlinear information fusion algorithms for data-efficient multi-fidelity modelling, *Proc. R. Soc. A*, vol. 473, no. 20160751, pp. 1–16.

Petrides, S., Demkowicz, L.F. (2017). An adaptive DPG method for high frequency time-harmonic wave propagation problems, *Comput. Math. Appli.*, vol. 74, no. 8, pp. 1999–2017.

Rayas-Sanchez, J.E., Chávez-Hurtado, J.L., Brito-Brito, Z. (2017). Optimization of full-wave EM models by low-order low-dimension polynomial surrogate functionals, *Int. J. Num. Model.: Electr. Devi. Fields*, vol. 30, no. 3–4, pp. e2094.

Robinson, T.D., Eldred, M.S., Willcox, K.E., Haimes, R. (2008). Surrogate-based optimization using multifidelity models with variable parameterization and corrected space mapping, *AIAA J.*, vol. 46, no. 11, pp. 2814–2822.

Sarkar, T.K., Chen, H., Palma, M.S., Zhu, M. (2019). Lessons learned using a physics based macro model for analysis of radio wave propagation in wireless transmission, *IEEE Trans. Ant. Propag.*, vol. 67, no. 4, pp. 2150–2157.

Schmidthausler, D., Clemens, M. (2012). Low-order electroquasistatic field simulations based on proper orthogonal decomposition, *IEEE Trans. Magn.*, vol. 48, no. 2, pp. 567–570.

Sim, C.Y.D., Chang, M.H., Chen, B.Y. (2014). Microstrip-fed ring slot antenna design with wideband harmonic suppression, *IEEE Trans. Ant. Prop.*, vol. 62, no. 9, pp. 4828–4832.

Sonnet (2018). *em*^{TM} Version 16.56 Sonnet Software, Inc., Sonnet Software, Inc., 126 N. Salina Street, Syracuse, NY 13202, USA.

Zhang, C., Feng, F., Zhang, Q., Bandler, J.W. (2018). Enhanced cognition-driven formulation of space mapping for equal-ripple optimisation of microwave filters, *IET Microwaves Ant. Prop.*, vol. 12, no. 1, pp. 82–91.

Zhang, C., Feng, F., Gongal-Reddy, V.M.R., Zhang, Q.J., Bandler, J.W. (2015). Cognition-driven formulation of space mapping for equal-ripple optimization of microwave filters, *IEEE Trans. Microwave Theory Techn.*, vol. 63, no. 7, pp. 2154–2165.

Zhu, J., Bandler, J.W., Nikolova, N.K., Koziel, S. (2007). Antenna optimization through space mapping, *IEEE Trans. Ant. Propag.*, vol. 55, no. 3, pp. 651–658.

© 2022 World Scientific Publishing Europe Ltd.
https://doi.org/10.1142/9781800610750_0003

Chapter 3

Parametric Modeling of Microwave Components Using Combined Neural Network and Transfer Function

Feng Feng, Jianan Zhang, Weicong Na, Jing Jin, and Qi-Jun Zhang

Abstract

This chapter provides an overview of parametric modeling of microwave components using combined neural network and transfer function (neuro-transfer function or neuro-TF). Transfer functions are used to represent the electromagnetic (EM) responses of passive components versus frequency. With the help of the transfer function, the nonlinearity of the neural network structure can be significantly decreased. We first introduce the neuro-TF modeling approach in rational format. Following that, we review the recent neuro-TF modeling approach in pole/zero format, where a pole/zero-matching algorithm is needed for addressing the issue of mismatch of poles and zeros before training the overall neuro-TF model. The neuro-TF modeling technique in pole/residue format is also reviewed. The orders of the pole–residue transfer functions may vary over different regions of geometrical parameters. A pole–residue tracking technique can be used to solve this order-changing problem. As a further advancement, we discuss the sensitivity analysis-based neuro-TF modeling technique. The purpose is to increase the model accuracy by utilizing EM sensitivity information and to speed up the model development process by reducing the number of training data required for developing the model. After the modeling process, the trained model can be used to provide accurate and fast prediction of the EM responses with respect to the geometrical variables and can be subsequently used in the high-level circuit and system design.

Keywords: Electromagnetic, parametric modeling, neural network, transfer function, sensitivity analysis

3.1 Introduction

Parametric modeling techniques have become more and more important for electromagnetic (EM) design optimizations of microwave components. Parametric models can be constructed based on the relationships between EM responses and geometrical parameters. The constructed parametric models allow fast simulations and optimizations with varying values of geometrical parameters and can be subsequently used in high-level designs.

Artificial neural networks (ANNs) are popular techniques for parametric modeling and design optimization (Zhang and Gupta, 2000; Zhang et al., 2003; Rayas-Sánchez, 2004; Steer et al., 2002; Mkadem and Boumaiza, 2011; Root, 2012; Yu et al., 2019; Liao et al., 2011; Sadrossadat et al., 2013). Through a suitable training process, ANNs can represent the relationship between the EM responses and geometrical parameters. Once trained, the ANN models can provide rapid and accurate predictions of the EM behavior of passive components, and can be further utilized in subsequent high-level microwave circuit designs.

The past decades have witnessed the use of ANNs in parametric modeling of a large variety of microwave applications, including microstrip filters (Na and Zhang, 2013; 2014), antennas (Gosal, 2015), vias (Watson and Gupta, 1996), power amplifiers (Mkadem and Boumaiza, 2011), high-speed interconnects (Zhang and Nakhla, 1994; Veluswami et al., 1997), high electron mobility transistor (HEMT) devices (Huang, 2016), transmission line components (Wang and Zhang, 1997; Devabhaktuni et al., 2002), coplanar waveguide (CPW) components (Watson and Gupta, 1997), waveguide filters (Liao et al., 2011; Zhang et al., 2015; Fedi et al., 2000), embedded passives (Ding et al., 2004; Devabhaktuni et al., 2001), voltage-controlled oscillators (Yu et al., 2019), mixers (Fang et al., 2000), bends (Bandler et al., 1999), and spiral conductors (Creech et al., 1997), etc.

Recent research efforts of ANN-based EM parametric modeling techniques have focused on automated model generation (AMG) methods (Na and Zhang, 2013, 2014; Na et al., 2019), hybrid training methods incorporating parallel processing (Zhang et al., 2015), and multiphysics parametric modeling (Yan et al., 2018; Zhang et al., 2018). In Na and Zhang (2013), an advanced algorithm for AMG using neural networks has been presented, where interpolation techniques are incorporated to avoid redundant training in AMG, accelerating the overall model generation process. As an extension of the work in Na and Zhang (2013), an enhanced AMG algorithm has been presented in Na and Zhang (2014) to automate the development process of knowledge-based neural network models for microwave applications. As a further advance, in Na et al. (2019), the parallel computation method has been incorporated into the AMG algorithm to achieve an additional speedup for neural modeling of microwave devices. In Zhang et al. (2015), a global neural network training method that combines hybrid training algorithm with parallel processing has been provided, where multiple neural network trainings are distributed to different processors and local search is performed in parallel to increase the probability and speed of finding a global optimum. More recently, ANNs have been introduced to multiphysics parametric modeling of microwave components. In Yan et al. (2018), neural networks have been applied to multiphysics parametric modeling of microwave filters. In Zhang et al. (2018), the space mapping (SM) concept has been elevated from pure EM parametric modeling to multiphysics parametric modeling. The coarse model in SM is represented by ANNs trained from the data generated by EM single-physics simulation, while the fine model corresponds to EM-centric multiphysics simulation. EM sensitivity information is further used to reduce the number of training data required for developing the parametric model. The sensitivity analysis has been used on top of ANN to further speed up the modeling process (Xu et al., 2003; Nikolova et al., 2004).

On the other hand, different microwave applications have motivated the research community to investigate and develop ANN-based parametric modeling techniques based on a great variety of ANN structures, such as multilayer perception (MLP) neural networks (Rizzoli et al., 2002), dynamic neural networks (DNNs) (Cao et al., 2003; Xu et al., 2002), radial basis function (RBF) neural networks (Stievano et al., 2002; Isaksson et al., 2005), recurrent neural networks (RNNs) (O'Brien et al., 2006; Fang et al., 2000; Yan et al., 2014), time-delay neural networks (TDNNs) (Liu et al., 2004), state-space dynamic neural networks (SSDNNs) (Cao et al., 2004, 2006), and the recently introduced deep neural networks (Jin et al., 2019).

To enhance the accuracy and reliability of ANN modeling for microwave components, the knowledge-based neural network (KBNN) models have been developed (Na and Zhang, 2014; Cao and Wang, 2007; Rayas-Sánchez and Gutierrez-Ayala, 2006; Devabhaktuni et al., 2003; Kabir et al., 2010). The idea of the knowledge-based model is to exploit existing knowledge in the form of empirical or equivalent circuit models together with neural networks to develop a faster and more accurate model. The knowledge-based approaches combine neural networks with prior knowledge such as equivalent circuits (Na and Zhang, 2014; Cao and Wang, 2007; Rayas-Sánchez and Gutierrez-Ayala, 2006) and analytical expressions (Bandler et al., 1999; Devabhaktuni et al., 2003). This knowledge is used to help improve the capability of learning and generalization of the overall models as well as to accelerate the model development process (Kabir et al., 2010).

SM techniques are recognized in engineering optimization and modeling in microwave area. Different SM techniques have been introduced to map the prior knowledge such as coarse models onto EM behaviors of microwave components (Bandler et al., 2002, 2004a, 2004b, 2004c; Koziel et al., 2006, 2008, 2009, 2011, 2013, 2014; Ayed et al., 2012; Zhang et al., 2005, 2012, 2013; Bakr et al., 2000; Gutierrez-Ayala and Rayas-Sanchez, 2010; Feng and Zhang, 2015a; Gorissen et al., 2011; Meng et al., 2008; Feng et al., 2014a, 2014b, 2015, 2016a; Garcia-Lamperez et al., 2004; Garcia-Lamperez and Salazar-Palma, 2016; Sans et al., 2015). By taking advantage of the vast set of empirical models already available, SM-based neural network models decrease the number of EM simulations required for training, improve model accuracy and generalization ability, and reduce the complexity of the ANN topology with respect to the classical pure ANN modeling approach.

Recently, another advanced parametric modeling approach for microwave components, i.e., combined neural network and transfer function (the neuro-transfer function or neuro-TF) modeling approach, has been reported (Gongal-Reddy et al., 2015; Cao et al., 2009, 2013; Guo et al., 2012; Feng et al., 2016b, 2017; Feng and Zhang, 2015b; Zhang et al., 2020). This approach combines neural networks and transfer functions and can be used even if accurate equivalent circuit models are not available. With this method, the EM behavior of passive components versus frequency can be expressed by transfer functions in a certain form.

This chapter provides an overview of parametric modeling of microwave components using neuro-TF. Transfer functions are used to represent the EM responses of passive components versus frequency. With the help of the transfer function, the nonlinearity of the neural network structure can be significantly decreased. In (Cao et al., 2009), neuro-TF modeling has been presented under the rational transfer function formulation. A constrained neural network training process using eigenvalues of the Hamiltonian matrix to enforce the passivity has been introduced (Guo et al., 2012). In Cao et al. (2013), an alternative constrained optimization strategy has been discussed to enforce passivity of transfer functions and simultaneously guarantee the continuity of coefficients through neuro-TF training process. In Feng et al. (2016b), an alternative pole–residue approach to the transfer function has been presented, which addressed the discontinuity issue of the pole–residue-based transfer function. An order-changing technique is developed for pole–residue-based transfer functions to handle the problem of order variations as geometrical parameters change in a large region. As a further advancement, sensitivity analysis has been introduced into the neuro-TF parametric modeling technique (Feng and Zhang, 2015b; Feng et al., 2017). The model developed by the sensitivity-analysis-based neuro-TF can obtain good accuracy with fewer training data. Different from previous neuro-TF approaches, an alternative way to address the order-changing problem has been provided in Zhang et al. (2020), by using the model order reduction (MOR) technique instead of vector fitting to obtain the coefficients in the transfer function. After the modeling process, the trained model can be used to provide accurate and rapid predictions of the EM responses with respect to the geometrical variables and can be subsequently used in the high-level circuit and system design.

3.2 Parametric Modeling Using Neuro-Transfer Function in Rational Format

3.2.1 *Formulation of the neuro-TF model in rational format*

The neuro-TF model in rational format consists of rational transfer functions and neural networks. The structure of the neuro-TF model in rational format (Cao *et al.*, 2009) is illustrated in Fig. 3.1. Let x represent the inputs of the neural network. Let the coefficients in the transfer function be the outputs of the neural network. The popular vector-fitting approach (Gustavsen and Semlyen, 1999) is used to extract the coefficients of the transfer function from the EM responses. The neural network is trained to learn the nonlinear relationship between x and the coefficients. Let H represent the transfer function output. The transfer function in rational format is formulated as

$$H(\boldsymbol{x}, \boldsymbol{w}, s) = \frac{\sum_{i=0}^{N} a_i(\boldsymbol{x}, \boldsymbol{w}) s^i}{1 + \sum_{i=1}^{N} b_i(\boldsymbol{x}, \boldsymbol{w}) s^i}, \quad (3.1)$$

where s is the Laplace frequency variable; N is the order of the transfer function; a_i and b_i are the coefficients of the rational transfer function; $\boldsymbol{w}$ represents the internal neural network weights. The nonlinear relationship between the coefficients of the rational transfer function and the geometrical variables is learned by the neural networks.

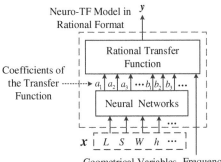

Fig. 3.1. Structure of the neuro-TF model in rational format (based on Cao *et al.* (2009)).

Let y represent a vector of the neuro-TF model outputs, which contains the real and imaginary parts of the output H of the transfer function. Let d represent a vector of training data from the EM simulators. The neural network weights w are adjusted to minimize the error between y and the training data d (Cao et al., 2009).

3.2.2 Two-stage training process of the neuro-TF model

The neuro-TF model is trained using a two-stage training process (Cao et al., 2009). In the first stage, a preliminary training process is performed. During this process, neural networks are trained to learn the relationships between all the coefficients of the transfer function and the geometrical parameters. Let k represent the index of training samples, $k \in T_r$, where T_r represents the index set of training samples at various geometrical parameters. The training data for this phase is (x_k, C_k), i.e., samples of geometrical parameters as model inputs and coefficients of the transfer function as model outputs, where C_k represents the set containing all the coefficients for the kth training sample. After preliminary training of the neural networks, the second stage of training is performed to further refine the final model.

The second stage of the training process performs a model refinement to refine the overall neuro-TF model. The training data for this phase is (x_k, d_k), $k \in T_r$, i.e., samples of geometrical parameters as model inputs and EM responses (e.g., S-parameters) as model outputs. The mechanism for the refinement training process of the overall neuro-TF model is illustrated in Fig. 3.2. The neuro-TF in the refinement training process consists of the transfer function of (3.1) in rational format and the neural networks whose initial values are the optimal solutions from the preliminary training process. This model refinement process consists of both training and testing of the neuro-TF model. Training is performed by optimizing the neural network weights to minimize the error function, formulated as (Cao et al., 2009)

$$E_{Tr}(w) = \frac{1}{2n_s} \sum_{k \in T_r} \sum_{i \in \Omega} \| y(x_k, w, s_i) - d_{k,i} \|^2, \quad (3.2)$$

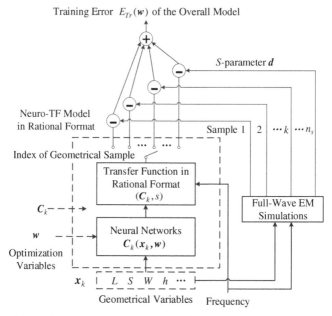

Fig. 3.2. The refinement training process for the neuro-TF model in rational format (based on Cao et al. (2009)).

where T_r is the index set of training samples at various geometrical parameters, and n_s is the total number of training samples; Ω is the index set of frequency samples; y represents the outputs of the overall model, which ultimately is a function of geometrical variables x_k, frequency s_i, and neural network weights w.

When the overall model training error becomes lower than a user defined threshold E_t, the second stage training (i.e., model refinement process) terminates. After the training process, an independent set of testing data which are never used in training is used for testing the quality of the trained neuro-TF model. The testing error E_{Te} is defined as the error between the model response and the testing data. If both the training error and the testing error are lower than the threshold error E_t, the model refinement process terminates, and the trained neuro-TF model is ready to be used for high-level design. Otherwise, the numbers of hidden neurons are adjusted and the overall model training process will be repeated iteratively until the termination condition is satisfied.

3.3 Parametric Modeling Using Neuro-Transfer Function in Pole/Zero Format

3.3.1 *Formulation of the neuro-TF model in pole/zero format*

One of the recent advances on neuro-TF modeling is to apply the MOR technique, more specifically, the matrix Padé via Lanczos (MPVL) algorithm, to compute the poles and zeros in the transfer function to address the order-changing problem. The structure of the neuro-TF model in pole/zero format is illustrated in Fig. 3.3. Two main parts are identified in the model: pole-zero-based transfer functions in the complex propagation space and neural networks used to learn the relationships between poles/zeros/gain and geometrical parameters. The transfer function in pole/zero format can be expressed as follows (Zhang et al., 2020):

$$H(\omega) = \gamma(\omega) K \frac{\prod_{i=1}^{q-1}(\gamma(\omega) - \gamma_0 - z_i)}{\prod_{i=1}^{q}(\gamma(\omega) - \gamma_0 - p_i)} + c, \quad (3.3)$$

where $H(\omega)$ represents the frequency response of the microwave passive component at frequency ω; $\gamma(\omega)$ is a frequency variable (i.e., the propagation constant); p_i, z_i, and K represent the poles, the zeros, and the gain coefficient of the transfer function, respectively; q is the order of the transfer function which is determined by the PVL algorithm; γ_0 is dependent on the expansion point in frequency; and c is a

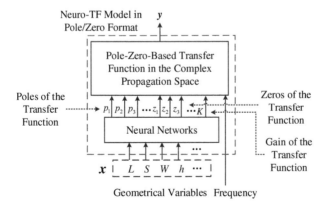

Fig. 3.3. The structure of the neuro-TF model in pole/zero format (based on Zhang et al. (2020)).

constant. The poles and zeros in $H(\omega)$ are referred to as the poles and zeros in the "γ-space" (or the complex propagation space) (Zhang et al., 2020).

Each time the geometrical parameters are changed, the poles and zeros, as well as the gain coefficient in the transfer function need to be recomputed by MPVL and eigendecomposition. However, there may not exist clear correspondences between the indices of the poles/zeros before the recomputation and those after the recomputation. As a result, if these raw poles/zeros are used to predict the poles/zeros for a new change of geometrical parameters, the solution may be wrong. This is referred to as the issue of mismatch of poles and zeros (Zhang et al., 2020).

3.3.2 Pole/zero-matching algorithm for addressing the issue of mismatch of poles and zeros

In this section, a sensitivity-analysis-based pole/zero-matching algorithm that can solve the issue of mismatch of poles and zeros is described. The main purpose is to match the poles and zeros between different geometrical samples correctly so that the indices of the poles/zeros correspond to those at the other geometrical sample. By doing this, the predicted positions of the poles/zeros for a new geometrical sample will be correct, which ultimately leads to an accurate neuro-TF model.

Let x be a m-dimensional vector containing all the geometrical parameters. Let N represent the total number of changes in geometrical samples to generate the training data. Let $X = \{x^{(0)}, x^{(1)}, \ldots, x^{(N)}\}$ be the set containing all the geometrical samples for training the model. Let T_r be the index set of the geometrical samples in the training data, i.e., $T_r = \{0, 1, \ldots, N\}$. Assume that two neighboring geometrical samples, say $x^{(k)}$ and $x^{(l)}$ ($k \in T_r, l \in T_r$), are selected from X to perform pole-matching. Firstly, the new positions of the poles at $x^{(l)}$ are predicted by utilizing the EM sensitivity information of the poles at $x^{(k)}$, formulated as

$$\bar{p}^{(l)} = p^{(k)} + J_p^{(k)} \Delta x, \qquad (3.4)$$

where $\bar{p}^{(l)}$ represents the predicted position of the poles at $x^{(l)}$. $\Delta x = x^{(l)} - x^{(k)}$ and $J_p^{(k)}$ denotes the EM sensitivity information of the poles at the kth geometrical sample $\boldsymbol{x}^{(k)}$. The detailed computation process of $\boldsymbol{J}_p^{(k)}$ is provided by Zhang et al. (2020).

Then, a low-complexity heuristic algorithm is applied to obtain sub-optimal pole-matching solutions. Specifically, we define a matrix $\boldsymbol{D} \in \mathbb{R}^{q \times q}$ which contains the distances between the actual positions ($\boldsymbol{p}^{(l)}$) and the predicted new positions ($\bar{\boldsymbol{p}}^{(l)}$) of the poles at $\boldsymbol{x}^{(l)}$ in the complex propagation space, i.e.,

$$\boldsymbol{D}_{ij} = \left\| \bar{p}_i^{(l)} - p_j^{(l)} \right\|, \tag{3.5}$$

where $i = 1, 2, \ldots, q$ and $j = 1, 2, \ldots, q$. After calculating the distance matrix $\boldsymbol{D}$, the pole-matching process between $\boldsymbol{x}^{(k)}$ and $\boldsymbol{x}^{(l)}$ is executed in q rounds heuristically. In each round (say the rth round, $r = 1, \ldots, q$) of the matching process, a 2D index is found for the smallest element in $\boldsymbol{D}$, formulated as

$$(i_r, j_r) = \arg \min_{\substack{i \in \{1, \ldots, q\}, i \notin \Omega_1^r \\ j \in \{1, \ldots, q\}, j \notin \Omega_2^r}} \{\boldsymbol{D}_{ij}\}, \tag{3.6}$$

where i_r and j_r are used to represent the row index and column index for the smallest element in matrix $\boldsymbol{D}$ in the rth round, respectively. Ω_1^r and Ω_2^r are two sets which are both initialized to be empty at the beginning and updated in subsequent rounds according to

$$\Omega_1^{r+1} = \Omega_1^r \cup \{i_r\}, \tag{3.7}$$
$$\Omega_2^{r+1} = \Omega_2^r \cup \{j_r\}, \tag{3.8}$$

where $r = 1, \ldots, q-1$. After q rounds, the matrix $\boldsymbol{D}$ becomes empty and all the poles at $\boldsymbol{x}^{(l)}$ will be matched with those at $\boldsymbol{x}^{(k)}$. If we use Φ to denote the set which contains the final pole-matching solution, then we have

$$\Phi = \{(i_r, j_r) | r = 1, 2, \ldots, q\}, \tag{3.9}$$

where (i_r, j_r) indicates that, after the pole-matching process, the i_rth pole at $\boldsymbol{x}^{(k)}$ corresponds to the j_rth pole at $\boldsymbol{x}^{(l)}$.

Lastly, all the poles at $\boldsymbol{x}^{(l)}$ are sorted according to Φ. After performing this sorting, we have each pole at $\boldsymbol{x}^{(l)}$ be correctly matched

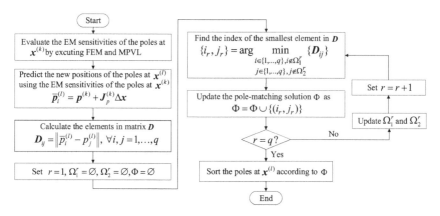

Fig. 3.4. Flow chart of the sensitivity-analysis-based pole-matching algorithm to match the poles between two neighboring geometrical samples $x^{(k)}$ and $x^{(l)}$ (based on Zhang et al. (2020)).

with that at $x^{(k)}$. The flow chart of matching the poles between $x^{(k)}$ and $x^{(l)}$ is shown in Fig. 3.4.

In practice, the geometrical samples in X may not be grid samples. Therefore, an algorithm that can effectively enumerate all the samples in X for reliable pole-matching is applied (Zhang et al., 2020). The algorithm enumerates all the scattered geometrical samples into a sequence of pairs, ensuring that each pair of samples are close to each other in the design parameter space. Moreover, the algorithm guarantees that a common set of reference indices is followed by the poles at all the geometrical samples.

Note that to match the zeros between $x^{(k)}$ and $x^{(l)}$, the same formulas are used as that for the poles, except that the symbol p is replaced by z, and q by $q-1$. The enumeration of all the geometrical samples for zero-matching follows the same procedure as that for pole-matching. Since the pole-matching solution and zero-matching solution are totally independent, the pole-matching algorithm and the zero-matching algorithm should be performed separately. After the pole- and zero-matching algorithms are done, the matched poles and zeros can be used as training data to train the neural networks in the pole-zero-based neuro-TF model. Then, the overall model is fine-tuned in a refinement training process (Zhang et al., 2020).

3.4 Parametric Modeling Using Neuro-TF Model in Pole/Residue Format

3.4.1 *Formulation of the neuro-TF model in pole/residue format*

Another alternative structure for the neuro-TF model is to use the transfer function in pole/residue format. The structure of the neuro-TF model in pole/residue format is illustrated in Fig. 3.5. The neuro-TF model in pole/residue format consists of pole–residue-based transfer functions and neural networks. The outputs of the overall model are the S-parameters of the EM behaviors and the inputs of the model are the geometrical variables of the EM structure and the frequency. The response $H(s)$ of the pole–residue-based transfer function is formulated as follows (Feng et al., 2016b):

$$H(s) = \sum_{i=1}^{N} \frac{r_i}{s - p_i}, \qquad (3.10)$$

where p_i and r_i represent the poles and residues in the transfer function, respectively, and N represents the order of the transfer function.

3.4.2 *Vector fitting for parameter extraction*

To perform the modeling using neuro-TF, the vector fitting (Gustavsen and Semlyen, 1999) is firstly used to obtain a group

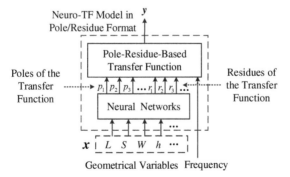

Fig. 3.5. The structure of the neuro-TF model in pole/residue format (based on Feng et al. (2016b)).

of poles $\boldsymbol{p}$ and residues $\boldsymbol{r}$ with the minimum order of the transfer function for every geometrical sample. The given information in the vector fitting process is EM data $\boldsymbol{d}$ (e.g., S-parameters) versus frequency for each geometrical sample. The expected solutions are poles and residues of the transfer function. A frequency scaling and shifting process is set up during the vector fitting to obtain all poles and residues in complex values (Feng et al., 2016b).

Let $\boldsymbol{c}_k$ represent a vector containing the poles/residues of the transfer function for the kth geometrical sample obtained after vector fitting, defined as

$$\boldsymbol{c}_k = \begin{bmatrix} \boldsymbol{p}^{(k)} \\ \boldsymbol{r}^{(k)} \end{bmatrix} = [p_1^{(k)}\ p_2^{(k)} \cdots p_{n_k}^{(k)}\ r_1^{(k)}\ r_2^{(k)} \cdots r_{n_k}^{(k)}]^T, \quad (3.11)$$

where n_k is the transfer function order for the kth geometrical sample. $\boldsymbol{p}^{(k)}$ is a vector of poles for the kth geometrical sample, defined as $\boldsymbol{p}^{(k)} = [p_1^{(k)}\ p_2^{(k)} \cdots p_{n_k}^{(k)}]^T$. $\boldsymbol{r}^{(k)}$ is a vector of residues for the kth geometrical sample, defined as $\boldsymbol{r}^{(k)} = [r_1^{(k)}\ r_2^{(k)} \cdots r_{n_k}^{(k)}]^T$.

Since the relationship between the pole/residues and geometrical variables is nonlinear and unknown, neural networks are used as the mapping function to map the pole/residues onto geometrical variables. Since there are different minimum orders for different values of geometrical parameters, the number of elements in vector $\boldsymbol{c}_k$ will be different for different geometrical samples. This variation of the numbers of pole/residues among different geometrical samples will cause difficulty in training the neural networks. An advanced pole–residue tracking technique is used to handle the variations in transfer function orders.

3.4.3 Pole–residue tracking technique for order-changing

An advanced pole–residue tracking technique is used to solve the order changing problem (Feng et al., 2016b). The main purpose of this technique is to add groups of new pole/residues to bridge the gap of transfer function orders between different geometrical samples while keeping the responses of the transfer functions unchanged. In this way, continuous pole/residues are ultimately obtained with the transfer functions of constant order for all geometrical samples.

Obtaining effective poles and residues is the first step of pole–residue tracking technique. Let n_k represent the order of the transfer function for the kth geometrical sample. Based on the pole–residue-based transfer function in (3.10), each pole has a related residue. Complex pole/residues are used to formulate a systematic pole–residue tracking process. Since each complex pole or residue has a conjugate, all the poles with negative imaginary parts and their corresponding residues can be removed. After this process, the number of pole/residues is reduced by half, and the remaining ones are regarded as effective pole/residues. Let the effective order for the kth geometrical sample be defined as N_k, i.e., $N_k = n_k/2$. Let $N_{\min}$ and $N_{\max}$ represent the minimum and maximum orders among all the geometrical samples, respectively.

Sorting of the effective pole/residues is the second step. Poles are sorted in an ascending sequence according to the values of their imaginary parts, while residues are sorted according to their corresponding poles. The sorting process can guarantee the relative correspondence of pole/residues between different geometrical samples, therefore, ensuring a smooth and continuous model with respect to varying values of geometrical parameters.

Performing the main pole–residue tracking process (Feng *et al.*, 2016b) is the third step. The transfer functions after vector fitting process have different effective orders for different geometrical samples, varying from $N_{\min}$ to $N_{\max}$. The new effective pole/residues need to be added without introducing non-uniqueness problems, such that ultimately the orders for all geometrical samples will be equal to $N_{\max}$.

In order to perform the pole–residue tracking process, every sample is enumerated to find that whose order is equal to $N_{\min}$. For each sample with order $N_{\min}$, say, the kth sample, one pole is selected to be split into two overlapping poles according to the neighboring sample. The neighboring sample (defined as the Lth sample) is the one which has the shortest geometrical distance to the present kth sample among all the samples whose orders are equal to $N_{\min} + 1$ (Feng *et al.*, 2016b).

To determine which particular pole in the pole vector should be selected to be split, $N_{\min}$ trials are performed. In the mth trial, the mth pole is split and the sum of differences between the imaginary parts of the present poles after splitting and the poles in the neighboring sample whose order is $N_{\min} + 1$ is calculated. Assume that the order of the kth geometrical sample is $N_k = N_{\min}$, while the Lth sample is the neighboring sample with order $N_L = N_{\min} + 1$. The formulation of the sum of differences is defined as follows (Feng et al., 2016b):

$$D_m = \sum_{i=1}^{m} \left\| \text{Im}\left(p_i^{(k)}\right) - \text{Im}\left(p_i^{(L)}\right) \right\|^2 + \sum_{i=m}^{N_{\min}} \left\| \text{Im}\left(p_i^{(k)}\right) - \text{Im}\left(p_{i+1}^{(L)}\right) \right\|^2. \tag{3.12}$$

This process is iteratively performed for $N_{\min}$ times with $m = 1, 2, \ldots, N_{\min}$. Because the poles should move continuously when the values of geometrical parameters change slightly, the correct splitting pole should be the one that leads to the minimum sum of differences between the poles of the present sample (present geometrical sample) and the poles of the neighboring sample (whose values of geometrical parameters are slightly different from that of the present sample). Subsequently, the pole to be split is selected according to the ranking of its D_m value. The pole with the minimum D_m value is selected as the splitting pole (Feng et al., 2016b), i.e.,

$$M = \arg \min_{m \in \{1, 2, \ldots, N_{\min}\}} \{D_m\}. \tag{3.13}$$

The Mth pole for the kth training sample is selected for splitting since among the various trials of pole-splitting D_M is the minimum sum of differences. The splitting of pole p_M means that, as the geometrical parameters change, the pole p_M will be split into two new poles, whose values are both equal to p_M at the moment of splitting. Consequently, the Mth residue which is related to the selected pole is also split into two separate residues. Each of the new residues has half value of the original residue in order to remain the transfer function response unchanged. The splitting of the Mth pole/residue is described as follows (Feng et al., 2016b):

$$\frac{r_M}{s - p_M} = \frac{r'_M}{s - p_M} + \frac{r_{N_{\min}+1}}{s - p_{N_{\min}+1}}, \tag{3.14}$$

where

$$p_{N_{\min}+1} = p_M,$$
$$r_{N_{\min}+1} = r'_M = \frac{r_M}{2}. \tag{3.15}$$

The pole–residue tracking is performed iteratively for all the samples with order equal to $N_{\min}$. After one iteration of the pole–residue tracking process is finished, the minimum order $N_{\min}$ of the entire training set increases by one, i.e., $N_{\min} = N_{\min} + 1$. The pole–residue tracking process terminates when the minimum effective order is equal to the maximum effective order, i.e., $N_{\min} = N_{\max}$. By doing this, all the samples now have the same orders. The flowchart of pole–residue tracking algorithm is illustrated in Fig. 3.6.

Let vector C_k represent the poles and residues including both the original and newly added ones after the pole–residue tracking process

$$C_k = \begin{bmatrix} P^{(k)} \\ R^{(k)} \end{bmatrix}$$

$$= \begin{bmatrix} P_1^{(k)} & P_2^{(k)} & \cdots & P_{N_{\max}}^{(k)} & R_1^{(k)} & R_2^{(k)} & \cdots & R_{N_{\max}}^{(k)} \end{bmatrix}^T, \quad (3.16)$$

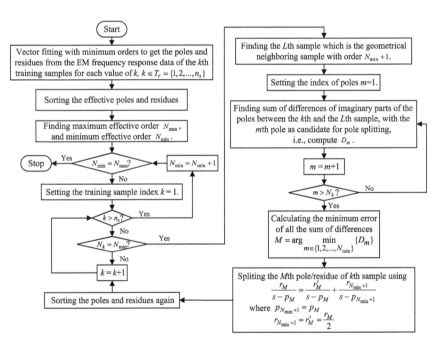

Fig. 3.6. The flowchart of the overall pole–residue tracking algorithm to handle order-changing problem as geometrical parameters change (based on Feng et al. (2016b)).

where $\boldsymbol{P}^{(k)}$ is a vector of poles including both the original and newly added ones of the kth sample, defined as $\boldsymbol{P}^{(k)} = [P_1^{(k)}\ P_2^{(k)} \cdots P_{N_{\max}}^{(k)}]^T$. $\boldsymbol{R}^{(k)}$ is a vector of residues including both the original and newly added ones of the kth sample, defined as $\boldsymbol{R}^{(k)} = [R_1^{(k)}\ R_2^{(k)} \cdots R_{N_{\max}}^{(k)}]^T$. The pole–residue-based transfer function after the order-changing process is rewritten as follows (Feng et al., 2016b):

$$H(\boldsymbol{C}_k, s) = \sum_{i=1}^{N_{\max}} \left(\frac{R_i^{(k)}}{s - P_i^{(k)}} + \frac{R_i^{*(k)}}{s - P_i^{*(k)}} \right), \quad (3.17)$$

where superscript "*" represents complex conjugate. After order-changing, constant order transfer functions are obtained. Those poles and residues (including original and newly added ones) are used to build the neuro-TF model in pole/residue format. The specific training process (Feng et al., 2016b) is used for the training of the neuro-TF model in pole/residue format. After training, the neuro-TF model can well represent the EM responses and can be used in high-level design.

3.5 Sensitivity-Analysis-Based Neuro-TF Modeling Technique

3.5.1 Structure of the sensitivity-analysis-based neuro-TF model

An advanced method for developing neuro-TF models with less training data is the introduction of sensitivity analysis. Figure 3.7 illustrates the sensitivity-analysis-based neuro-TF model. The sensitivity-analysis-based neuro-TF model consists of two sub-models, i.e., the original neuro-TF model and the adjoint neuro-TF model based on sensitivity analysis (Feng et al., 2017). Both sub-models share the same inputs $\boldsymbol{x}$. The derivatives of the outputs $\boldsymbol{y}$ of original neuro-TF model with respect to the inputs $\boldsymbol{x}$, denoted as $d\boldsymbol{y}/d\boldsymbol{x}$, are defined as the outputs of the adjoint neuro-TF model.

The original neural networks are defined as the neural networks in the original neuro-TF model. The adjoint neural networks

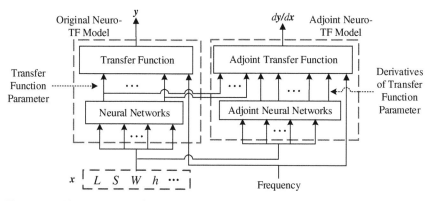

Fig. 3.7. The structure of the sensitivity-analysis-based neuro-TF model (based on Feng et al. (2017)).

(Sadrossadat et al., 2013) are defined as the neural networks in the adjoint neuro-TF model, which maps the relationship between the derivatives of transfer function parameters and the geometrical parameters.

3.5.2 Preliminary training process for the sensitivity-analysis-based neuro-TF model

EM evaluation and sensitivity data are simultaneously used in training the sensitivity-analysis-based neuro-TF model. With same amount of training data, the model trained using EM sensitivity data can be more accurate than that trained without EM sensitivity data. In other words, the same modeling accuracy can be obtained with fewer training data by exploiting EM sensitivities. The sensitivity-analysis-based model is trained through a two-stage training process (Feng et al., 2017), including preliminary training and refinement training.

In the first stage training (i.e., preliminary training), the original and adjoint neural networks are trained simultaneously. Let $\hat{c}_k$ and $\hat{A}_k$ represent the data of transfer function parameters and their derivatives with respect to x, respectively. The training data for the original neural networks are $(x_k, \hat{c}_k)$. The training data for the adjoint neural networks are $(x_k, \hat{A}_k)$. Let $c(x, w)$ be a vector consisting of all the outputs of the neural networks, i.e., transfer function parameters. The error function for the first stage training process is formulated as

$$E_{\text{Pre}}(\boldsymbol{w}) = \frac{1}{2n_s}\sum_{k=1}^{n_s}\|\boldsymbol{c}(\boldsymbol{x}_k,\boldsymbol{w}) - \hat{\boldsymbol{c}}_k\|^2 + \frac{1}{2n_s}\sum_{k=1}^{n_s}\left\|\frac{\partial \boldsymbol{c}(\boldsymbol{x}_k,\boldsymbol{w})}{\partial \boldsymbol{x}^T} - \hat{\boldsymbol{A}}_k\right\|_F^2,$$
(3.18)

where n_s represents the total number of training samples; $\|\cdot\|$ and $\|\cdot\|_F$ represent L_2 norm and Frobenius norm, respectively.

3.5.3 *Refinement training process for the sensitivity-analysis-based neuro-TF model*

After the first-stage training, a second-stage training (i.e., refinement training) is used to refine the overall model (Feng *et al.*, 2017). The training data for both original and adjoint neuro-TF models in the second stage training are $(\boldsymbol{x}_k, \boldsymbol{d}_k)$ and $(\boldsymbol{x}_k, \boldsymbol{d}'_k)$, respectively. During the second-stage training, the original neuro-TF model and the adjoint neuro-TF model are trained simultaneously. The error function of the second-stage training is formulated as

$$E_{\text{Tr}}(\boldsymbol{w}) = E_{\text{orig}}(\boldsymbol{w}) + E_{adj}(\boldsymbol{w}) = \frac{1}{2n_s}\sum_{k=1}^{n_s}\sum_{j=1}^{n_y} a_j \|y_j(\boldsymbol{x}_k,\boldsymbol{w}) - d_{k,j}\|^2$$

$$+ \frac{1}{2n_s}\sum_{k=1}^{n_s}\sum_{j=1}^{n_y}\sum_{i=1}^{n_x} b_{j,i}\left\|\frac{\partial y_j(\boldsymbol{x}_k,\boldsymbol{w})}{\partial x_i} - d'_{k,j,i}\right\|^2,$$
(3.19)

where E_{orig} represents the training error between the EM evaluation data and the original neuro-TF model. E_{adj} represents the training error between the EM sensitivity data and the adjoint neuro-TF model. n_s is the total number of training samples. n_y is the number of elements in $\boldsymbol{y}$. n_x is the number of elements in $\boldsymbol{x}$. $d_{k,j}$ represents the EM evaluation data of the kth sample for the jth output. $d'_{k,j,i}$ is the kth sample's EM sensitivity data of the jth output with respect to the ith input. a_j and $b_{j,i}$ are the weighting parameters for the original neuro-TF model and the adjoint neuro-TF model, respectively. When the total training error is smaller than a user defined error threshold, the overall training process terminates. Although the training process for the sensitivity-analysis-based neuro-TF model is complicated, the usage of the trained model is simple. The adjoint neuro-TF model is only used during the training process. After training, the original neuro-TF model internally has accurate sensitivity information. Therefore, the final model (i.e., original neuro-TF model itself) is simple and can be further used in high-level design.

3.6 Neuro-TF Modeling Examples and RF/Microwave Applications

In this section, three application examples including a three-pole H-plane filter, a microwave junction, and a deplexer, are presented to illustrate the neuro-TF method in pole/zero format in Section 3.3, the neuro-TF method in pole/residue format in Section 3.4, and the sensitivity-analysis-based neuro-TF modeling method in Section 3.5, respectively.

3.6.1 *Neuro-TF modeling in pole/zero format of a three-pole H-plane filter*

In this example, the neuro-TF method in pole/zero format in Section 3.3 is illustrated by a three-pole H-plane filter (Zhang *et al.*, 2020), as shown in Fig. 3.8. This neuro-TF model in pole/zero format has four geometrical variables as inputs, i.e., $x = [L_1\ L_2\ W_1\ W_2]^T$ (mm). The model has one output, i.e., $y = |S_{11}|$. Frequency is an additional input to the model.

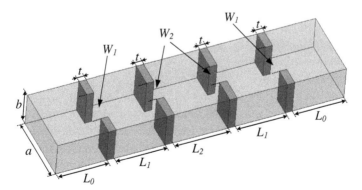

Fig. 3.8. Structure of the three-pole H-plane filter for EM simulation (based on Zhang *et al.* (2020)).

Table 3.1. Definition of training and testing samples for the three-pole H-plane filter example (based on Zhang et al. (2020)).

Geometrical variables	Training samples (49 samples)			Testing samples (36 samples)		
	Min	Max	Step	Min	Max	Step
Case 1 (small range)						
L_1 (mm)	13.7	14.26	0.09	13.75	14.21	0.09
L_2 (mm)	14.9	15.5	0.1	14.95	15.45	0.1
W_1 (mm)	8.82	9.18	0.06	8.85	9.15	0.06
W_2 (mm)	5.88	6.12	0.04	5.9	6.1	0.04
Case 2 (wider range)						
L_1 (mm)	13.28	14.68	0.23	13.4	14.56	0.23
L_2 (mm)	14.44	15.96	0.25	14.57	15.83	0.25
W_1 (mm)	8.55	9.45	0.15	8.63	9.38	0.15
W_2 (mm)	5.7	6.3	0.1	5.75	6.25	0.1

The neuro-TF method in pole/zero format is applied to two different cases, i.e, Case 1 with a smaller geometrical variations and Case 2 with a larger geometrical variations, as defined in Table 3.1. It is noted that Case 2 is more challenging as the steps between two neighboring EM samples are larger than that in Case 1. For the purpose of comparison, three existing modeling methods, i.e, the distance-based method presented in Yue et al. (2018), the distance-based method with refinement training, and the continuation method with refinement training presented in Yue et al. (2019), are also applied to EM modeling of the filter for both cases. The distance-based method matches the poles between two neighboring geometrical samples by minimizing the sum of distances between the two sets of poles in the "γ-space" (Yue et al., 2018). ANNs are used to interpolate the poles and residues in the transfer function. The continuation method matches the poles between two neighboring geometrical samples by predicting the movement of the poles using linear extrapolation (Yue et al., 2019). For both the distance-based method with refinement training and the continuation method with refinement training, ANNs are used to interpolate the poles and zeros in the transfer function and a refinement training process is added to further refine the overall model. Table 3.2 compares these four different

Table 3.2. Comparisons of different modeling methods for the three-pole H-plane filter example (based on Zhang et al. (2020)).

Modeling method	Number of hidden neurons	Average training error (%)	Average testing error (%)
Case 1			
Distance-based method	20	1.37	1.66
Distance-based method with refinement training	10	0.16	0.18
Continuation method with refinement training	10	0.16	0.18
Neuro-TF method in pole/zero format (Zhang et al., 2020)	10	0.16	0.18
Case 2			
Distance-based method	20	35.8	107.57
	40	0.86	81.91
Distance-based method with refinement training	10	1.94	14.09
	20	0.17	26.42
Continuation method with refinement training	10	3.56	12.83
	20	1.16	460.51
Neuro-TF method in pole/zero format (Zhang et al., 2020)	10	0.62	1.56

methods. In Case 1, since the geometrical variations are small, the correct correspondences between the poles/zeros at two neighboring geometrical samples are relatively easy to identify. The training and testing errors of all the methods are comparable and very small. In Case 2, however, the geometrical variations become large and the mismatch patterns of poles and zeros become more complicated. The testing errors of all the three existing methods are large, and only the neuro-TF method in pole/zero format still maintains very small training and testing errors. Figure 3.9 compares the model outputs obtained from the neuro-TF method in pole/zero format, and the model outputs obtained from the three existing methods versus EM

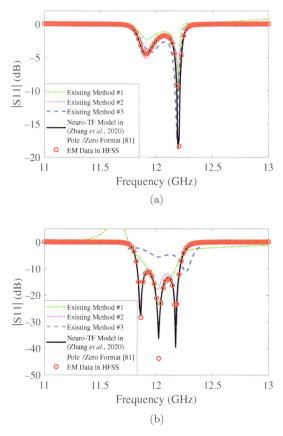

Fig. 3.9. Comparison of the model outputs and the EM data for the three-pole H-plane filter with larger geometrical variations: (a) test geometrical sample #1, and (b) test geometrical sample #2. The three existing methods are: (1) distance-based method, (2) distance-based method with refinement training, and (3) continuation method with refinement training (based on Zhang et al. (2020)).

data for the wider parameter range (i.e., Case 2). Two different test geometrical samples are used to do this comparison.

It can be seen from Fig. 3.9 that, the outputs of the neuro-TF model in pole/zero format match the EM data much better than those of the other three modeling methods. Moreover, the trained neuro-TF model in pole/zero format is accurate at the test geometrical samples even though these samples are not used in training.

3.6.2 Neuro-TF modeling in pole/residue format of a microwave junction

In this example, the neuro-TF method in pole/residue format (Feng et al., 2016b) in Section 3.4 is illustrated by a microwave junction, as shown in Fig. 3.10. g is the gap between two conductive walls; h_c is the height of the tuning cylinder; r_c is the radius of the tuning cylinder; t_w is thickness of the conductive walls; w_s is width of the slot on the upper conductive wall; l_s is the length of the slot on the upper conductive wall; h is the height of the junction cavity. The *CST Studio Suite* software is used for the full-wave EM simulation to generate training and testing data for modeling.

This model has seven input geometrical variables, i.e., $\boldsymbol{x} = [g\ h_c\ r_c\ t_w\ w_s\ l_s\ h]^T$. Frequency is an additional input. The model has eight outputs, i.e., $\boldsymbol{y} = [\text{Re}(S_{11})\ \text{Im}(S_{11})\ \text{Re}(S_{21})\ \text{Im}(S_{21})\ \text{Re}(S_{31})\ \text{Im}(S_{31})\ \text{Re}(S_{41})\ \text{Im}(S_{41})]^T$, which are the real and imaginary parts of S_{11}, S_{21}, S_{31} and S_{41}, respectively.

The neuro-TF modeling method in pole/residue format is applied to two different cases (Feng et al., 2016b), i.e., Case 1 with a small parameter range and Case 2 with a wider parameter range as defined in Table 3.3. For comparison purpose, the brute-force neuro-TF modeling method in pole/residue format which directly uses the maximum orders for all the samples of geometrical parameters and the recent neuro-TF modeling method in bilinear rational format with order-changing technique (Cao et al., 2009) are applied for the two cases. Table 3.4 compares these various modeling methods. In Case 1, since the geometrical parameters vary within a small range and the orders of transfer function change slightly, all methods obtain

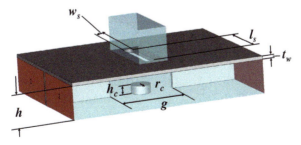

Fig. 3.10. The 3D configuration of the junction for EM simulation (based on Feng et al. (2016b)).

Table 3.3. Definition of training and testing samples for the junction example (based on Feng et al. (2016b)).

Geometrical variables (mm)	Training samples (81 samples)			Testing samples (49 samples)		
	Min	Max	Step	Min	Max	Step
Case 1 (small range)						
g	19	19.8	0.1	19.05	19.75	0.1
h_c	2	2.8	0.1	2.05	2.75	0.1
r_c	3	3.8	0.1	3.05	3.75	0.1
t_w	0.9	1.06	0.02	0.91	1.05	0.02
w_s	3	3.8	0.1	3.05	3.75	0.1
l_s	19.9	20.22	0.04	19.92	20.2	0.04
h	10	10.8	0.1	10.05	10.75	0.1
Case 2 (wider range)						
g	18	22	0.5	18.25	21.25	0.5
h_c	2	3.6	0.2	2.1	3.3	0.2
r_c	2.2	3.8	0.2	2.3	3.5	0.2
t_w	0.84	1.16	0.04	0.86	1.1	0.04
w_s	2.2	3.8	0.2	2.3	3.5	0.2
l_s	19.6	20.4	0.1	19.65	20.25	0.1
h	10	11.6	0.2	10.1	11.3	0.2

comparatively small training and testing errors. In Case 2, when the geometrical parameters vary within a wider range and the orders of transfer function have larger variations, the brute-force neuro-TF modeling method in pole/residue format and the neuro-TF modeling method in bilinear rational format have higher testing errors. The neuro-TF method in pole/residue format (Feng et al., 2016b) can achieve better modeling accuracy than the other two methods in the comparison.

Figure 3.11 shows the outputs of the neuro-TF model in pole/residue format (Feng et al., 2016b) for two different test geometrical samples of junctions, i.e., test geometrical samples #1 and #2, and its comparison with model response using the two different methods and EM data. It is observed that the neuro-TF model in pole/residue format can achieve good accuracy for different geometrical samples even though these samples are never used in training.

Table 3.4. Comparisons of different modeling methods for the junction example (based on Feng et al. (2016b)).

Training method	Coefficient (or pole–residue) continuity?	Coefficient (or pole–residue) sensitivity	Number of hidden neurons	Average training error (%)	Average testing error (%)
Case 1 (small geometry range)					
Brute-force neuro-TF in pole/residue format	Yes	Low	4	0.258	0.273
Neuro-TF in bilinear rational format	Yes	High	5	0.286	0.292
Neuro-TF in pole/residue format (Feng et al., 2016b)	Yes	Low	4	0.254	0.276
Case 2 (wider geometry range)					
Brute-force neuro-TF in pole/residue format	No	Low	5	0.928	1.568
			15	0.528	4.168
Neuro-TF in bilinear rational format	Yes	High	5	0.959	1.237
			15	0.559	2.537
Neuro-TF in pole/residue format (Feng et al., 2016b)	Yes	Low	5	0.516	0.558

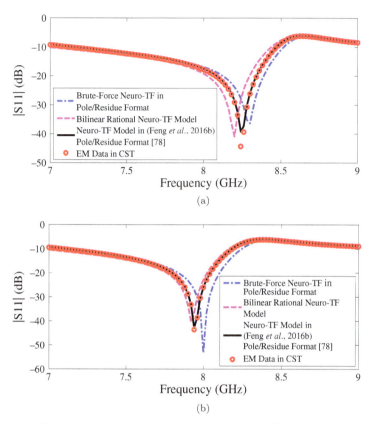

Fig. 3.11. Comparison of the magnitude in decibels of S_{11} of the models developed using different modeling methods and CST EM data: (a) test geometrical sample #1, and (b) test geometrical sample #2 for the junction example. As shown in the figure, the neuro-TF model in pole/residue format matches well with CST EM data even though the testing geometrical samples used in the figures are never used in training (based on Feng et al. (2016b)).

3.6.3 Sensitivity-analysis-based neuro-TF modeling of a diplexer

In this example, a diplexer is used to demonstrate the sensitivity-analysis-based neuro-TF modeling method (Feng et al., 2017) in Section 3.5. The structure of the diplexer is shown in Fig. 3.12. D_1, D_2, D_3, D_4, D_5, and D_6 are the offset distances from the end of each microstrip line to its corresponding fringe.

This model has seven input variables including six geometrical variables and frequency, i.e., $[D_1 \ D_2 \ D_3 \ D_4 \ D_5 \ D_6 \ \omega]^T$. The model

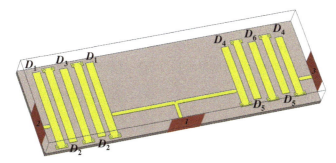

Fig. 3.12. The three-dimensional configuration of the diplexer example for EM simulation (based on Feng *et al.* (2017)).

Table 3.5. Definition of training and testing data for the diplexer example (based on Feng *et al.* (2017)).

	Training samples (25 samples)			Testing samples (64 samples)		
Geometrical variables	Min	Max	Step	Min	Max	Step
Case 1 (small range)						
D_1 (mm)	2	10	2	2.5	9.5	1
D_2 (mm)	−6	2	2	−5.5	1.5	1
D_3 (mm)	−6	2	2	−5.5	1.5	1
D_4 (mm)	2	10	2	2.5	9.5	1
D_5 (mm)	−6	2	2	−5.5	1.5	1
D_6 (mm)	−6	2	2	−5.5	1.5	1
Case 2 (wider range)						
D_1 (mm)	−4	12	4	−3	11	2
D_2 (mm)	−12	4	4	−11	3	2
D_3 (mm)	−12	4	4	−11	3	2
D_4 (mm)	−4	12	4	−3	11	2
D_5 (mm)	−12	4	4	−11	3	2
D_6 (mm)	−12	4	4	−11	3	2

has 42 outputs, including the real/imaginary part of S_{11}, S_{21}, and S_{31}, and their derivatives with respect to six input geometrical variables. EM simulators are used to generate the EM sensitivity data of real/imaginary part of S_{11}, S_{21}, and S_{31} with respect to six sensitivity variables D_1, D_2, D_3, D_4, D_5, and D_6.

Two different cases are used to illustrate the sensitivity-analysis-based neuro-TF modeling in Table 3.5. For comparison purposes, the modeling is also performed using the sensitivity-analysis artificial neural network (SAANN) technique (Sadrossadat et al., 2013) and the existing neuro-TF model without sensitivity (Feng et al., 2016b). Table 3.6 illustrates two cases of comparisons of training and testing errors among the SAANN Model using fewer data (25 training samples), existing neuro-TF model without sensitivity using fewer

Table 3.6. Comparisons of different modeling methods for the diplexer example (based on Feng et al. (2017)).

Model type	Number of hidden neurons	Average training error (%)	Average testing error (%)
Case 1 (small range)			
SAANN model using 25 samples of training data	30	0.906	0.924
Neuro-TF model (Feng et al., 2016b) using 25 samples of training data	10	0.867	0.932
SA neuro-TF model (Feng et al., 2017) using 25 samples of training data	12	0.892	0.915
Case 2 (wider range)			
SAANN model using 25 samples of training data	36	1.165	11.564
Neuro-TF model (Feng et al., 2016b) using 25 samples of training data	12	1.095	9.264
Neuro-TF model (Feng et al., 2016b) using 81 samples of training data	18	1.135	1.368
SA Neuro-TF model (Feng et al., 2017) using 25 samples of training data	12	1.065	1.254

data (25 training samples), existing neuro-TF model without sensitivity using more data (81 training samples), and the sensitivity-analysis-based neuro-TF model (short for SA neuro-TF) using fewer data (25 training samples) (Feng et al., 2017). In Case 1, all methods have relatively low training and testing errors since the geometrical range is relatively small. In Case 2, with wider geometrical range, the SAANN model and neuro-TF model without sensitivity trained with fewer training data cannot achieve good modeling accuracy.

The comparisons of derivatives of real part of S_{11} with respect to sensitivity variables D_1 and D_2 for one training sample by neuro-TF model without sensitivity using fewer data and more data, the sensitivity-analysis-based neuro-TF model using fewer data and CST sensitivity analysis are shown in Fig. 3.13. As shown in Fig. 3.13, sensitivity information from the trained sensitivity-analysis-based neuro-TF model is more accurate than that from the other methods in the comparison.

The comparisons of the outputs of the sensitivity-analysis-based neuro-TF model and the neuro-TF model without sensitivity for two different testing samples #1 and #2 are shown in Fig. 3.14. As shown in Table 3.6 and Fig. 3.14, fewer training data are needed for the sensitivity-analysis-based neuro-TF model (Feng et al., 2017) than that for the neuro-TF model without sensitivity (Feng et al., 2016b) to achieve a good modeling accuracy. Therefore, the sensitivity-analysis-based neuro-TF method can speed up the modeling process by reducing the amount of training data.

3.7 Discussion

This chapter has presented several recent advances in parametric modeling of microwave components using neuro-TF. Here, we discuss the proper usage of different techniques under different practical situations. When the EM responses of the microwave applications can be represented by the transfer function with relatively low order, the neuro-TF modeling method in rational format (Cao et al., 2009) is efficient to use. Compared with neuro-TF method in pole/residue format (Feng et al., 2016b, 2017), the neuro-TF modeling in rational

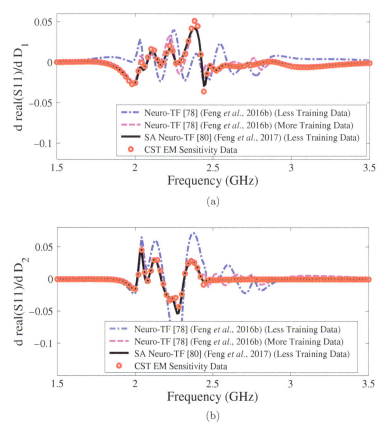

Fig. 3.13. Comparison of the sensitivities of the existing neuro-TF model without sensitivity using fewer and more data, the sensitivity-analysis-based neuro-TF model using fewer data and CST EM sensitivity data for this diplexer example: (a) derivatives of real parts of S_{11} with respect to sensitivity variable D_1, and (b) derivatives of real parts of S_{11} with respect to sensitivity variable D_2 (based on Feng et al. (2017)).

format (Cao et al., 2009) is simpler, because the coefficients of the transfer function in rational format are automatically continuous and do not need to be tracked among different geometrical samples. However, when the order of the transfer function becomes high, the transfer function response becomes very sensitive with respect to the coefficients of the transfer function in rational format, which results in difficult training and non-robustness of the trained neuro-TF model.

Parametric Modeling of Microwave Components 113

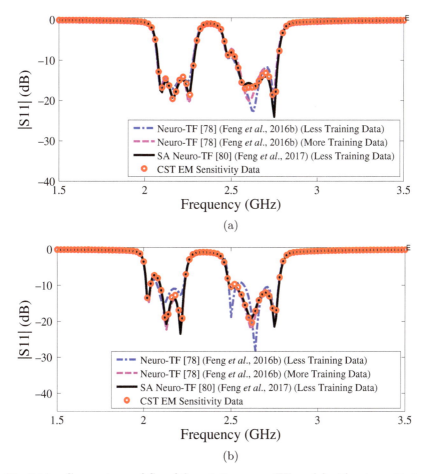

Fig. 3.14. Comparisons of S_{11} of the existing neuro-TF model without sensitivity using fewer and more data, the sensitivity-analysis-based neuro-TF model using fewer data and CST EM data: (a) test geometrical sample #1, and (b) test geometrical sample #2 for this diplexer example (based on Feng *et al.* (2017)).

When the EM responses need to be represented by the transfer function with relatively high order, the modeling method using transfer functions in pole/residue format (Feng *et al.*, 2016b, 2017) can achieve higher accuracy and more robustness. The transfer function response is much less sensitive with respect to the pole–residues than the coefficients of the rational transfer function. When the order of the transfer function becomes higher, this advantage becomes more pronounced. Therefore, the neuro-TF modeling method in pole/residue format (Feng *et al.*, 2016b, 2017) becomes more robust than the neuro-TF modeling in rational format (Cao *et al.*, 2009). However, the neuro-TF modeling method in pole/residue format (Feng *et al.*, 2016b, 2017) needs complex pole tracking process in order to obtain continuous pole/residues among different geometrical samples.

When the poles and zeros in the complex propagation space are provided by the EM solver, the modeling method using transfer functions in pole/zero format (Zhang *et al.*, 2020) is an alternative way to avoid the order-changing problem and to achieve desired modeling accuracy. The poles and zeros in the transfer function are computed by the MPVL algorithm instead of being extracted by the vector fitting technique. The use of MPVL guarantees the consistency of the order of transfer function between different geometrical samples, thereby avoiding the order-changing problem. However, the neuro-TF modeling method in pole/zero format involves more complicated pole- and zero-matching processes to obtain clear correspondences between the poles/zeros at different geometrical samples.

When the neuro-TF model is constructed in a relatively small geometrical range, the neuro-TF modeling method without sensitivity is easier to use (Cao *et al.*, 2009; Feng *et al.*, 2016b). The neuro-TF model in Cao *et al.* (2009); Feng *et al.* (2016b) does not need EM sensitivities which are obtained using extra CPU time during the EM simulations. Additionally, the training structure for the neuro-TF model without sensitivity (Cao *et al.*, 2009; Feng *et al.*, 2016b) is much simpler than the neuro-TF model with sensitivity (Feng *et al.*, 2017), resulting in the faster training process.

However, when the geometrical range of the EM problem becomes relatively large, the neuro-TF modeling method with adjoint sensitivity analysis (Feng *et al.*, 2017) can achieve better accuracy with the same of number of training samples than that without sensitivities (Cao *et al.*, 2009; Feng *et al.*, 2016b). To obtain the same accuracy, the neuro-TF modeling method with adjoint sensitivity analysis (Feng *et al.*, 2017) needs much fewer training samples than that without sensitivities (Cao *et al.*, 2009; Feng *et al.*, 2016b), because EM sensitivities can provide very rich interpolating information for parametric modeling. Furthermore, when the EM structures are very complex resulting in very expensive EM data generation, the neuro-TF modeling method with adjoint sensitivity analysis (Feng et al., 2017) can save a lot of CPU time than that without sensitivities (Cao et al., 2009; Feng et al., 2016b).

The neuro-TF in Cao et al. (2009); Feng et al. (2016b, 2017) are used in frequency-domain analysis. When using neuro-TF models in time-domain analysis, the issue of passivity enforcement needs to be carefully considered (Cao et al., 2013; Guo et al., 2012).

3.8 Conclusion

An overview of parametric modeling of microwave components using neuro-TF has been provided in this chapter. Transfer functions have been used to represent the EM responses of passive components versus frequency. With the help of the transfer function, the nonlinearity of the neural network structure has been significantly decreased. We have first introduced the neuro-TF modeling approach under the rational and pole-zero-based transfer function formulations. Then, we have reviewed the neuro-TF modeling technique in pole/residue format. Lastly, we have discussed the sensitivity analysis-based neuro-TF modeling technique. Using the neuro-TF modeling technique, the trained model can provide rapid and fast prediction of the EM responses with respect to the geometrical variables and can be subsequently used in the high-level circuit and system design.

References

Ayed, R.B., Gong, J., Brisset, S., Gillon, F., Brochet, P. (2012). Three-level output space mapping strategy for electromagnetic design optimization, *IEEE Trans. Magn.*, vol. 48, no. 2, pp. 671–674.

Bakr, M.H., Bandler, J.W., Ismail, M.A., Rayas-Sanchez, J.E., Zhang, Q.J. (2000). Neural space-mapping optimization for EM-based design, *IEEE Trans. Microw. Theory Techn.*, vol. 48, no. 12, pp. 2307–2315.

Bandler, J.W., Ismail, M.A., Rayas-Sanchez, J.E., Zhang, Q.J. (1999). Neuromodeling of microwave circuits exploiting space-mapping technology, *IEEE Trans. Microw. Theory Techn.*, vol. 47, no. 12, pp. 2417–2427.

Bandler, J.W., Ismail, M.A., Rayas-Sanchez, J.E. (2002). Expanded space-mapping EM-based design framework exploiting preassigned parameters, *IEEE Trans. Circuits Syst. I, Fundam. Theory Appl.*, vol. 49, no. 12, pp. 1833–1838.

Bandler, J.W., Cheng, Q.S., Dakroury, S.A., Mohamed, A.S., Bakr, M.H., Madsen, K., Sondergaard, J. (2004a). Space mapping: The state of the art, *IEEE Trans. Microw. Theory and Techn.*, vol. 52, no. 1, pp. 337–361.

Bandler, J.W., Cheng, Q.S., Nikolova, N.K., Ismail, M.A. (2004b). Implicit space mapping optimization exploiting preassigned parameters, *IEEE Trans. Microw. Theory Tech.*, vol. 52, no. 1, pp. 378–385.

Bandler, J.W., Hailu, D.M., Madsen, K., Pedersen, F. (2004c). A space mapping interpolating surrogate algorithm for highly optimized EM-based design of microwave devices, *IEEE Trans. Microw. Theory Techn.*, vol. 52, no. 11, pp. 2593–2600.

Cao, Y., Xu, J.J., Devabhaktuni, V.K., Ding, R.T., Zhang, Q.J. (2003). An adjoint dynamic neural network technique for exact sensitivities in nonlinear transient modeling and high-speed interconnect design, in *IEEE MTT-S Int. Microw. Symp. Dig.*, PA, Philadelphia, pp. 165–168.

Cao, Y., Ding, R.T., Zhang, Q.J. (2004). A new nonlinear transient modeling technique for high-speed integrated circuit applications based on state-space dynamic neural network, in *IEEE MTT-S Int. Microw. Symp. Dig.*, Fort Worth, TX, pp. 1553–1556.

Cao, Y., Ding, R.T., Zhang, Q.J. (2006). State-space dynamic neural network technique for high-speed IC applications: Modeling and stability analysis, *IEEE Trans. Microw. Theory Techn.*, vol. 54, no. 6, pp. 2398–2409.

Cao, Y., Wang, G. (2007). A wideband and scalable model of spiral inductors using space-mapping neural network, *IEEE Trans. Microw. Theory Techn.*, vol. 55, no. 12, pp. 2473–2480.

Cao, Y., Wang, G., Zhang, Q.J. (2009). A new training approach for parametric modeling of microwave passive components using combined neural networks and transfer functions, *IEEE Trans. Microw. Theory Techn.*, vol. 57, no. 11, pp. 2727–2742.

Cao, Y., Wang, G., Gunupudi, P., Zhang, Q.J. (2013). Parametric modeling of microwave passive components using combined neural networks and transfer functions in the time and frequency domains, *Int. J. RF and Microw. CAE*, vol. 23, no. 1, pp. 20–33.

Creech, G.L., Paul, B.J., Lesniak, C.D., Jenkins, T.J., Calcatera, M.C. (1997). Artificial neural networks for fast and accurate EM-CAD of microwave circuits, *IEEE Trans. Microwave Theory Techn.*, vol. 45, no. 5, pp. 794–802.

Devabhaktuni, V.K., Yagoub, M., Zhang, Q.J. (2001). A robust algorithm for automatic development of neural-network models for microwave applications, *IEEE Trans. Microw. Theory Techn.*, vol. 49, no. 12, pp. 2282–2291.

Devabhaktuni, V.K., Xi, C., Wang, F., Zhang, Q.J. (2002). Robust training of microwave neural models, *Int. J. RF Microwave CAE*, vol. 12, no. 1, pp. 109–124.

Devabhaktuni, V.K., Chattaraj, B., Yagoub, M.C.E., Zhang, Q.J. (2003). Advanced microwave modeling framework exploiting automatic model generation, knowledge neural networks, and space mapping, *IEEE Trans. Microw. Theory Techn.*, vol. 51, no. 7, pp. 1822–1833.

Ding, X., Devabhaktuni, V.K., Chattaraj, B., Yagoub, M.C.E., Doe, M., Xu, J.J., Zhang, Q.J. (2004). Neural-network approaches to electromagnetic-based modeling of passive components and their applications to high-frequency and high-speed nonlinear circuit optimization, *IEEE Trans. Microw. Theory Tech.*, vol. 52, no. 1, pp. 436–449.

Fang, Y., Yagoub, M.C.E., Wang, F., Zhang, Q.J. (2000). A new macromodeling approach for nonlinear microwave circuits based on recurrent neural networks, *IEEE Trans. Microw. Theory Techn.*, vol. 48, no. 12, pp. 2335–2344.

Fang, Y.H., Yagoub, M.C.E., Wang, F., Zhang, Q.J. (2000). A new macromodeling approach for nonlinear microwave circuits based on recurrent neural networks, *IEEE Trans. Microw. Theory Tech.*, vol. 48, no. 12, pp. 2335–2344.

Fedi, G., Manetti, S., Pelosi, G., Selleri, S. (2000). Design of cylindrical posts in rectangular waveguide by neural network approach, in *IEEE AP-S Int. Symp. Digest*, Salt Lake City, UT, pp. 1054–1057.

Feng, F., Zhang, C., Gongal-Reddy, V.M.R., Zhang, Q.J. (2014a). Knowledge-based coarse and fine mesh space mapping approach to EM optimization, in *Int. Conf. Numerical Electromagnetic Modeling and Optimization*, Pavia, Italy, pp. 1–4.

Feng, F., Zhang, C., Gongal-Reddy, V.M.R., Zhang, Q.J., Ma, J. (2014b). Parallel space-mapping approach to EM optimization, *IEEE Trans. Microw. Theory Techn.*, vol. 62, no. 5, pp. 1135–1148.

Feng, F., Gongal-Reddy, V.M.R., Zhang, S., Zhang, Q.J. (2015). Recent advances in space mapping approach to EM optimization, in *Proc. Asia-Pacific Microw. Conf.*, Nanjing, China, pp. 1–3.

Feng, F., Zhang, Q.J. (2015a). Neural space mapping optimization for EM design, in *Proc. Asia-Pacific Microw. Conf.*, Nanjing, China, pp. 1–3.

Feng, F., Zhang, Q.J. (2015b). Parametric modeling using sensitivity-based adjoint neuro-transfer functions for microwave passive components, in *IEEE MTT-S Int. Conf. Numerical Electromagnetic and Multiphysics Modeling and Optim.*, Ottawa, Canada, pp. 1–3.

Feng, F., Gongal-Reddy, V.M.R., Zhang, C., Na, W., Zhang, S., Zhang, Q.J. (2016a). Recent advances in parallel EM optimization approaches, in *IEEE MTT-S Int. Conf. Microw. Millimeter Wave Technology*, Beijing, China, pp. 1–3.

Feng, F., Zhang, C., Ma, J., Zhang, Q.J. (2016b). Parametric modeling of EM behavior of microwave components using combined neural networks and pole-residue-based transfer functions, *IEEE Trans. Microw. Theory Techn.*, vol. 64, no. 1, pp. 60–77.

Feng, F., Gongal-Reddy, V.M.R., Zhang, C., Ma, J., Zhang, Q.J. (2017). Parametric modeling of microwave components using adjoint neural networks and pole-residue transfer functions with EM sensitivity analysis, *IEEE Trans. Microw. Theory Techn.*, vol. 65, no. 6, pp. 1955–1975.

Garcia-Lamperez, A., Llorente-Romano, S., Salazar-Palma, M., Sarkar, T.K. (2004). Efficient electromagnetic optimization of microwave filters and multiplexers using rational models, *IEEE Trans. Microw. Theory Techn.*, vol. 52, no. 2, pp. 508–521.

Garcia-Lamperez, A., Salazar-Palma, M. (2016). Multilevel aggressive space mapping applied to coupled-resonator filters, in *IEEE MTT-S Int. Microw. Symp. Dig.*, San Francisco, CA, pp. 1–4.

Gongal-Reddy, V.M.R., Feng, F., Zhang, Q.J. (2015). Parametric modeling of millimeter-wave passive components using combined neural networks and transfer functions, in *Global Symposium On Millimeter Waves (GSMM)*, Montreal, QC, Canada, pp. 1–3.

Gorissen, D., Zhang, L., Zhang, Q.J., Dhaene, T. (2011). Evolutionary neuro-space mapping technique for modeling of nonlinear microwave devices, *IEEE Trans. Microw. Theory Techn.*, vol. 59, no. 2, pp. 213–229.

Gosal, G., Almajali, E., McNamara, D., Yagoub, M. (2015). Transmitarray antenna design using forward and inverse neural network modeling, *IEEE Ant. Wireless Prop. Lett.*, vol. 15, no. 1, pp. 1483–1486.

Guo, Z., Gao, J., Cao, Y., Zhang, Q.J. (2012). Passivity enforcement for passive component modeling subject to variations of geometrical parameters using neural networks, in *IEEE MTT-S Int. Microw. Symp. Dig.*, Montreal, QC, Canada, pp. 1–3.

Gustavsen, B., Semlyen, A. (1999). Rational approximation of frequency domain responses by vector fitting, *IEEE Trans. Power Del.*, vol. 14, no. 3, pp. 1052–1061.

Gutierrez-Ayala, V., Rayas-Sanchez, J.E. (2010). Neural input space mapping optimization based on nonlinear two-layer perceptrons with optimized nonlinearity, *Int. J. RF Microw. Comput.-Aided Eng.*, vol. 20, no. 5, pp. 512–526.

Huang, A.D., Zhong, Z., Wu, W., Guo, Y.X. (2016). An artificial neural network-based electrothermal model for GaN HEMTs with dynamic trapping effects consideration, *IEEE Trans. Microw. Theory Techn.*, vol. 64, no. 8, pp. 2519–2528.

Isaksson, M., Wisell, D., Ronnow, D. (2005). Wide-band dynamic modeling of power amplifiers using radial-basis function neural networks, *IEEE Trans. Microw. Theory Techn.*, vol. 53, no. 11, pp. 3422–3428.

Jin, J., Zhang, C., Feng, F., Na, W., Ma, J., Zhang, Q.J. (2019). Deep neural network technique for high-dimensional microwave modeling and applications to parameter extraction of microwave filters, *IEEE Trans. Microw. Theory Techn.*, vol. 67, no. 10, pp. 4140–4155.

Kabir, H., Zhang, L., Yu, M., Aaen, P.H., Wood, J., Zhang, Q.J. (2010). Smart modeling of microwave device, *IEEE Microw. Mag.*, vol. 11, no. 3, pp. 105–118.

Koziel, S., Bandler, J.W., Madsen, K. (2006). A space mapping framework for engineering optimization: Theory and implementation, *IEEE Trans. Microw. Theory Techn.*, vol. 54, no. 10, pp. 3721–3730.

Koziel, S., Cheng, Q.S., Bandler, J.W. (2008). Space mapping, *IEEE Microw. Mag.*, vol. 9, no. 6, pp. 105–122.

Koziel, S., Bandler, J.W., Madsen, K. (2009). Space mapping with adaptive response correction for microwave design optimization, *IEEE Trans. Microw. Theory Techn.*, vol. 57, no. 2, pp. 478–486.

Koziel, S., Bandler, J.W., Cheng, Q.S. (2011). Tuning space mapping design framework exploiting reduced electromagnetic models, *IET Microw. Antennas Propag.*, vol. 5, no. 10, pp. 1219–1226.

Koziel, S., Ogurtsov, S., Bandler, J.W., Cheng, Q.S. (2013). Reliable space-mapping optimization integrated with EM-based adjoint sensitivities, *IEEE Trans. Microw. Theory Tech.*, vol. 61, no. 10, pp. 3493–3502.

Koziel, S., Cheng, Q.S., Bandler, J.W. (2014). Fast EM modeling exploiting shape-preserving response prediction and space mapping, *IEEE Trans. Microw. Theory Tech.*, vol. 62, no. 3, pp. 399–407.

Liao, S., Kabir, H., Cao, Y., Xu, J., Zhang, Q.J., Ma, J. (2011). Neural-network modeling for 3-D substructures based on spatial EM-field coupling in finite-element method, *IEEE Trans. Microw. Theory Techn.*, vol. 59, no. 1, pp. 21–38.

Liu, T., Boumaiza, S., Ghannouchi, F.M. (2004). Dynamic behavioral modeling of 3G power amplifiers using real-valued time-delay neural networks, *IEEE Trans. Microw. Theory Techn.*, vol. 52, no. 3, pp. 1025–1033.

Meng, J., Koziel, S., Bandler, J.W., Bakr, M.H., Cheng, Q.S. (2008). Tuning space mapping: A novel technique for engineering design optimization, in *IEEE MTT-S Int. Microw. Symp. Dig.*, Atlanta, Georgia, pp. 991–994.

Mkadem F., Boumaiza, S. (2011). Physically inspired neural network model for RF power amplifier behavioral modeling and digital predistortion, *IEEE Trans. Microw. Theory Techn.*, vol. 59, no. 4, pp. 913–923.

Na, W.C., Zhang, Q.J. (2013). Automated parametric modeling of microwave components using combined neural network and interpolation techniques, in *IEEE MTT-S Int. Dig.*, Seattle, WA, USA.

Na, W., Zhang, Q.J. (2014). Automated knowledge-based neural network modeling for microwave applications, *IEEE Microw. Wireless Compon. Lett.*, vol. 24, no. 7, pp. 499–501.

Na, W., Zhang, W., Yan, S., Liu, G. (2019). Automated neural-based modeling of microwave devices using parallel computation and interpolation approaches, *IEEE Access*, vol. 7, pp. 73929–73937.

Nikolova, N.K., Bandler, J.W., Bakr, M.H. (2004). Adjoint techniques for sensitivity analysis in high-frequency structure CAD, *IEEE Trans. Microw. Theory Techn.*, vol. 52, no. 1, pp. 403–419.

O'Brien, B., Dooley, J., Brazil, T.J. (2006). RF power amplifier behavioral modeling using a globally recurrent neural network, in *IEEE MTT-S Int. Microw. Symp. Dig.*, San Francisco, CA, pp. 1089–1092.

Rayas-Sánchez, J.E. (2004). EM-based optimization of microwave circuits using artificial neural networks: The state-of-the-art, *IEEE Trans. Microw. Theory Techn.*, vol. 52, no. 1, pp. 420–435.

Rayas-Sánchez, J.E., Gutierrez-Ayala, V. (2006). EM-based Monte Carlo analysis and yield prediction of microwave circuits using linear-input neural-output space mapping, *IEEE Trans. Microw. Theory Techn.*, vol. 54, no. 12, pp. 4528–4537.

Rizzoli, V., Neri, A., Masotti, D., Lipparini, A. (2002). A new family of neural network-based bidirectional and dispersive behavioral models for nonlinear RF/microwave subsystems, *Int. J. RF Microw. Computer-Aided Eng.*, vol. 12, no. 1, pp. 51–70.

Root, D.E. (2012). Future device modeling trends, *IEEE Microw. Mag.*, vol. 13, no. 7, pp. 45–59.

Sadrossadat, S.A., Cao, Y., Zhang, Q.J. (2013). Parametric modeling of microwave passive components using sensitivity-analysis-based adjoint neural-network technique, *IEEE Trans. Microw. Theory Techn.*, vol. 61, no. 5, pp. 1733–1747.

Sans, M., Selga, J., Velez, P., Rodriguez, A., Bonache, J., Boria, V.E., Martin, F. (2015). Automated design of common-mode suppressed balanced wideband bandpass filters by means of aggressive space mapping, *IEEE Trans. Microw. Theory Techn.*, vol. 63, no. 12, pp. 3896–3908.

Steer, M.B., Bandler, J.W., Snowden, C.M. (2002). Computer-aided design of RF and microwave circuits and systems, *IEEE Trans. Microw. Theory Techn.*, vol. 50, no. 3, pp. 996–1005.

Stievano, I.S., Maio, I.A., Canavero, F.G. (2002). Parametric macromodels of digital I/O ports, *IEEE Trans. Adv. Packag.*, vol. 25, no. 5, pp. 255–264.

Veluswami, A., Nakhla, M.S., Zhang, Q.J. (1997). The application of neural networks to EM-based simulation and optimization of interconnects in highspeed VLSI circuits, *IEEE Trans. Microw. Theory Techn.*, vol. 45, no. 5, pp. 712–723.

Wang, F., Zhang, Q.J. (1997). Knowledge based neural models for microwave design, *IEEE Trans. Microw. Theory Techn.*, vol. 45, no. 12, pp. 2333–2343.

Watson, P.M. Gupta, K.C. (1996). EM-ANN models for microstrip vias and interconnects in multilayer circuits, *IEEE Trans. Microw. Theory Techn.*, vol. 44, no. 12, pp. 2495–2503.

Watson, P.M., Gupta, K.C. (1997). Design and optimization of CPW circuits using EM-ANN models for CPW components, *IEEE Trans. Microw. Theory Techn.*, vol. 45, no. 12, pp. 2515–2523.

Xu, J., Yagoub, M.C.E. Ding, R., Zhang, Q.J. (2002). Neural based dynamic modeling of nonlinear microwave circuits, *IEEE Trans. Microw. Theory Techn.*, vol. 50, no. 12, pp. 2769–2780.

Xu, J., Yagoub, M.C.E. Ding, R., Zhang, Q.J. (2003). Exact adjoint sensitivity analysis for neural-based microwave modeling and design, *IEEE Trans. Microw. Theory Techn.*, vol. 51, no. 1, pp. 226–237.

Yan, S., Zhang, C., Zhang, Q.J. (2014). Recurrrent neural network technique for behavioral modeling of power amplifier with memory effects, *Int. J. RF Microwave CAE*, vol. 25, no. 4, pp. 289–298.

Yan, S., Zhang, Y., Jin, X., Zhang, W., Shi, W. (2018). Multi-physics parametric modeling of microwave passive components using artificial neural networks, *Progress In Electromagnetics Research M*, vol. 72, pp. 79–88.

Yu, H., Chalamalasetty, H., Swaminathan, M. (2019). Modeling of voltage-controlled oscillators including I/O behavior using augmented neural networks, *IEEE Access*, vol. 7, pp. 38973–38982.

Yue, Y., Feng, L., Benner, P. (2018). Interpolation of reduced-order models based on modal analysis, in *IEEE Int. Conf. on Numerical Electromagnetic and Multiphysics Modeling and Optimization (NEMO)*, Reykjavik, Iceland, pp. 1–4.

Yue, Y., Feng, L., Benner, P. (2019). An adaptive method for interpolating reduced-order models based on matching and continuation of poles, in *IEEE Int. Conf. on Numerical Electromagnetic and Multiphysics Modeling and Optimization (NEMO)*, Cambridge, MA, pp. 1–4.

Zhang, Q.J., Nakhla, M.S. (1994). Signal integrity analysis and optimization of VLSI interconnects using neural network models, in *IEEE Int. Circuits Syst. Symp.*, London, England, pp. 459–462.

Zhang, Q.J., Gupta, K.C. (2000). *Neural Networks for RF and Microwave Design*. Norwood, MA, Artech House.

Zhang, Q.J., Gupta, K.C., Devabhaktuni, V.K. (2003). Artificial neural networks for RF and microwave design - From theory to practice, *IEEE Trans. Microw. Theory Techn.*, vol. 51, no. 4, pp. 1339–1350.

Zhang, L., Xu, J., Yagoub, M.C.E., Ding, R., Zhang, Q.J. (2005). Efficient analytical formulation and sensitivity analysis of neuro-space mapping for nonlinear microwave device modeling, *IEEE Trans. Microw. Theory Techn.*, vol. 53, no. 9, pp. 2752–2767.

Zhang, L., Aaen, P.H., Wood, J. (2012). Portable space mapping for efficient statistical modeling of passive components, *IEEE Trans. Microw. Theory Techn.*, vol. 60, no. 3, pp. 441–450.

Zhang, C., Feng, F., Zhang, Q.J. (2013). EM optimization using coarse and fine mesh space mapping, in *Proc. Asia-Pacific Microw. Conf.*, Seoul, Korea, pp. 824–826.

Zhang, J., Ma, K., Feng, F., Zhang, Q.J. (2015). Parallel gradient-based local search accelerating particle swarm optimization for training microwave neural network models, in *IEEE MTT-S Int. Microw. Symp. Dig.*, Phoenix, AZ, pp. 1–3.

Zhang, W., Feng, F., Gongal-Reddy, V.M.R., Zhang, J., Yan, S., Ma, J. Zhang, Q.J. (2018). Space mapping approach to electromagnetic centric multiphysics parametric modeling of microwave components, *IEEE Trans. Microw. Theory Techn.*, vol. 66, no. 7, pp. 3169–3185.

Zhang, J., Feng, F., Zhang, W., Ma, J., Zhang, Q.J. (2020). A novel training approach for parametric modeling of microwave passive components using Padé via Lanczos and EM sensitivities, *IEEE Trans. Microw. Theory Techn.*, vol. 68, no. 6, pp. 2215–2233.

© 2022 World Scientific Publishing Europe Ltd.
https://doi.org/10.1142/9781800610750_0004

Chapter 4

Surrogate Model-Assisted Global Optimization for Antenna Design

Mobayode O. Akinsolu, Peter Excell, and Bo Liu

Abstract

Antenna design automation via optimization continues to attract a lot of interest. This can be mainly attributed to the expected efficiency improvement that numerical optimization methods for antenna design offer in terms of the overall design time compared to traditional antenna design methodologies. Numerical optimization methods for antenna design still present several challenges in terms of their efficiency and optimization ability to handle contemporary antenna designs which are typically complex in terms of topology and performance requirements. Today, parallel computing and machine learning have been demonstrated as proficient ways of addressing the challenges of efficiency and optimization capability. In this chapter, the application of a state-of-the-art antenna optimization method which combines the advantages of parallel computing and machine learning via surrogate-based optimization (SBO) is presented. The method is the parallel surrogate model-assisted hybrid differential evolution for antenna optimization (PSADEA). The efficacy of PSADEA is demonstrated in this chapter through the design and optimization of a compact slotted monopole antenna for ultra-wide band (UWB) body-centric imaging applications and a novel compact quadruple-band indoor base station antenna for 2G/3G/4G/5G systems. The close agreement between the simulated and measured results for the fabricated prototypes of the PSADEA-synthesized designs for the two antennas validate the design solution quality of PSADEA. Further to this, the performance of PSADEA is compared with popular and commercially available Computer Simulation Technology Microwave Studio (CST-MWS) optimizers using the UWB slotted monopole antenna design problem. The outcomes of the comparisons reveal that PSADEA obtains very satisfactory design solutions repeatedly within an affordable computational cost in each run, whereas CST-MWS optimizers fail to obtain an adequate design solution in all runs.

Keywords: Antenna optimization, PSADEA, SADEA, surrogate-based optimization (SBO), surrogate model-assisted optimization

4.1 Introduction

Electromagnetic (EM) devices, particularly microwave antennas, continue to play a leading role as vital components of modern wireless communication systems. From legacy systems and applications (such as radar applications) to more recent systems and applications (such as 5G applications), the optimal performance of antennas in the transmission and reception of radio signals directly impact the overall efficiency and reliability of wireless communication systems (Yoo *et al.*, 2020; Hua *et al.*, 2020). Generally, some experience-based rules of thumb are available for designing antennas to ensure good performance (Weiland *et al.*, 2008; Balanis 2014). Even though these design principles offer a suitable guide for the practical design and implementation of antennas, they are mostly fitted for basic antenna structures (Weiland *et al.*, 2008; Balanis 2014). Additionally, the correct application of some these rules for antenna design can also yield sub-optimal designs (even for basic antenna designs (Grout *et al.*, 2019)).

Present-day antennas are typically complex in terms of topology, material composition and EM characterization. This is mainly because modern antenna applications such as body-centric or on-body wireless communications (Ur-Rehman *et al.*, 2017), multiple-input and multiple-output (MIMO) systems (Alieldin *et al.*, 2018a), and multiband communications (Alieldin *et al.*, 2018b) introduce increasingly stringent design specifications to be met by contemporary antenna structures. As a result, antenna designers and engineers are often saddled with handling a very large number of sensitive design parameters and additional performance requirements per design (Yoo *et al.*, 2020; Hua *et al.*, 2020; Alieldin *et al.*, 2018a). Compared to conventional antenna structures, finding the best designs for contemporary antenna structures could be very challenging due to the inherent complexities and challenges mentioned earlier. To address the above challenges, the design (geometric and/or material) parameters of modern antenna structures are commonly fine-tuned by antenna designers to improve their performances according to the intended applications. Many of the available methods which support this technique largely involve an experience-based parametric study of the performances by sweeping a few design parameters at a time (Hua et al., 2020; Ur-Rehman et al., 2017; Alieldin et al., 2018a; Yang et al., 2018). However, parameter sweeping is often a trial and error process and it is typically time-consuming without any guarantee of successful outcomes. Hence, the need for antenna design automation via optimization has risen.

Optimization of antennas is primarily carried out using local and/or global numerical optimization methods (Grout et al., 2019). Despite their efficiency improvement over experience-based parameter sweeping, these methods also have some drawbacks. The major drawback of local optimization methods is the requirement of a good initial design or starting point to obtain good results for many antenna problems (Grout et al., 2019). When an initial design is available (e.g., based on previous similar designs), efficient local optimization methods are effective to adequately exploit and explore the design space (Koziel and Pietrenko-Dabrowska, 2019, 2020). For example, the optimization of conventional planar antennas for ultra-wide band (UWB) and multi-band applications may be expedited efficiently using available initial designs (Koziel and Pietrenko-Dabrowska, 2019, 2020).

Global optimization methods are more generic. In contrast to local optimization methods, they do not require starting points or initial designs. However, their primary drawback remains the use of a very large (often prohibitive) number of EM simulations to obtain near-optimum designs for many antenna problems (Grout et al., 2019). For the thorough analysis of antenna structures, such computationally expensive numerical technique-based full-wave EM simulations are unavoidable. Whilst a one-time antenna analysis via an expensive EM simulation is clearly not an issue, the substantial amounts of such EM simulations required by global optimization methods often constitute excessive (or even sometimes prohibitive) computational overheads.

In recent times, parallel computing, and artificial intelligence (applied via machine learning techniques) have come to the fore in the expedited design automation of antennas (Richie and Ababei, 2017; Guney and Basbug, 2014; Akinsolu et al., 2020). Modern computational electromagnetic design and computer-aided design tools now exploit the distributed processors and shared memory available on multicore operating systems to allow for faster and concurrent simulation of multiple antenna design models. Also, a priori and/or a posteriori integration of machine learning techniques into the optimization kernel of numerical optimization methods is now a common practice. Amongst the available machine learning techniques for the machine-learning assisted optimization of antennas, surrogate modeling is of interest to many researchers (Koziel and Leifsson, 2013; Koziel and Ogurtsov, 2014). As a matter of fact, it is arguably the foremost machine learning method used for the machine-learning assisted optimization of antennas (Akinsolu et al., 2020).

Primarily, to lower the computational cost of the optimization process via surrogate modeling, computationally expensive EM simulations are replaced with computationally cheap approximation models called surrogate models. Statistical learning techniques are the primary means for the construction of surrogate models (Forrester and Keane, 2009). There are several surrogate modeling techniques available for the expedited optimization of EM designs (Couckuyt et al., 2010), Gaussian process (GP) or kriging (Santner et al., 2003) which is widely used in the antenna design optimization domain (Koziel et al., 2012; Liu et al., 2014a, 2018a). The synergetic running of numerical optimization methods and surrogate modeling techniques in a single optimization framework is referred to as SBO (Koziel and Leifsson, 2013; Koziel and Ogurtsov, 2014). It can, therefore, be said that parallel computing and SBO can significantly expedite the optimization process and overall design time of antennas, severally and conjunctively.

When global numerical optimization methods, specifically evolutionary algorithms (EAs), are employed as the search engine in SBO methods, surrogate model-assisted evolutionary algorithms (SAEAs) are formed (Liu et al., 2017c, 2018b; Akinsolu et al., 2020). There exists a vital trade-off between the quality of the surrogate model and the efficiency (measurable from the required number of full-wave simulations) in SAEAs. Thus, a surrogate model management method is required to find an appropriate trade-off in SAEAs. Due to a multiplicity of surrogate model management procedures, there are various kinds of SAEAs. The surrogate model-aware evolutionary search (SMAS) framework is a state-of-the-art SAEA framework and its efficiency and optimization quality are well established and verified (Liu et al., 2014a, 2014b, 2017b, 2017c, 2018a, 2018b, 2020; Akinsolu et al., 2019, 2020).

In this chapter, the application of a SMAS-based SAEA method is presented for the surrogate model-assisted global optimization of antennas. The method is the parallel surrogate model-assisted hybrid differential evolution for antenna optimization (PSADEA) method (Liu et al., 2018a; Akinsolu et al., 2019). PSADEA harnesses the combined advantages of parallel computing and SBO in a unified SAEA framework to address the challenges of efficiency and robustness in antenna design optimization. It is the state-of-the-art in the surrogate model-assisted differential evolution for antenna synthesis (SADEA) series (Liu et al., 2014a, 2017b, 2018a, 2020; Akinsolu et al., 2019), while the first generation algorithm in the SADEA series (Liu et al., 2014a, 2017a) has already showed clear advantages compared to most global optimization-based antenna design methods (Grout et al., 2019; Liu et al., 2014a).

The optimization capacity of the PSADEA method is demonstrated in this chapter through the global optimization and subsequent physical implementation of two challenging real-world antenna designs. The first antenna is the compact slotted monopole antenna for UWB body-centric imaging applications (Danjuma et al., 2020) and the second antenna is the novel compact quadruple-band indoor base station antenna for 2G/3G/4G/5G systems (Hua et al., 2019). The PSADEA method is also compared to widely used and commercially available antenna optimizers using one of the antenna problems (which is non-complex in terms of structure and implementation) presented in this chapter.

The remainder of this chapter is organized as follows: Section 4.2 presents an overview of the SADEA algorithm family, Section 4.3 discusses the features, implementation and process flow of the PSADEA method, Section 4.4 details the global optimization of the two antenna problems used as examples to demonstrate PSADEA and the concluding remarks are given in Section 4.5.

4.2 Overview of the SADEA Algorithm Family

As mentioned earlier, the SADEA algorithm family is a class of SMAS-based SAEAs purpose-built for the machine learning-assisted optimization of antennas (Liu *et al.*, 2014a, 2017b, 2018a, 2020; Akinsolu *et al.*, 2019). The primary recurring features across all the SADEA generations are differential evolution (DE) (Storn and Price, 1997) global search and Gaussian process (GP) surrogate modeling (Santner *et al.*, 2003; Rasmussen and Williams, 2006). The first-generation SADEA (Liu *et al.*, 2014a; Liu *et al.*, 2017a) is the first global optimization-based online SBO for antenna design to the best of our knowledge. It shows up to 10 times speed optimization time improvement compared to standard global optimization methods with comparable optimization ability (Grout *et al.*, 2019). It also outperforms a few popular SBO methods in the artificial intelligence (AI) domain (Grout *et al.*, 2019). It is suitable for antennas with around 20 or fewer sensitive design variables. The second-generation SADEA (SADEA-II) (Liu *et al.*, 2017b) introduces multi-fidelity EM models, aiming at addressing the discrepancy between them to further improve the efficiency without sacrificing the design quality. The third-generation SADEA (PSADEA) (Liu *et al.*, 2018a; Akinsolu *et al.*, 2019), whose application is the focus of this chapter, developed for expedited antenna design in parallel computing environments where multiple candidate designs can be simulated concurrently using distributed processors and shared memory on multicore operating systems. It aims at satisfying very stringent antenna specifications that most other global optimization methods are not able to. PSADEA employs multiple DE mutation strategies and reinforcement learning techniques to achieve an added 1.5–3 times efficiency improvement and higher design solution quality even in the sequential mode compared to the previous generations of SADEA. The fourth-generation SADEA (training cost reduced-SADEA (TR-SADEA)) is the latest in the SADEA algorithm family (Liu *et al.*, 2020). To the best of our knowledge, TR-SADEA is the first machine learning-assisted optimization method purpose-built for the efficient synthesis of complex antennas with many design variables (e.g., 50) and specifications (e.g., 20) such as 5G base station antennas, for which, the machine learning cost becomes an issue.

The SADEA algorithm family offers high optimization ability and high efficiency without the requirement of specific conditions of use. In other words, no ad hoc processes are required for their successful implementation. The main distinctions across the generations of the SADEA algorithm family are summarized as follows:

(1) In SADEA, a single mutation operator is used to produce the child populations and their respective child populations. The efficiency improvement comes from the SMAS-based global search. More details can be found in Liu *et al.* (2014a, 2017a).
(2) In SADEA-II, a multi-stage optimization involving a coarse model at the first stage, data mining at the second stage and a SADEA-based optimization (enhanced by a local search using the best design from the first stage as the starting point) using a fine model is carried out. More details can be found in Liu *et al.* (2017b).
(3) In PSADEA, a self-adaptive usage of multiple mutation operators and their child solutions enhances the surrogate model quality and better balances the exploration and exploitation of the antenna design space. More details are provided in the subsequent sections.
(4) In TR-SADEA, the cost of building and training multiple GP surrogate models for multiple specifications in higher-dimensional spaces is reduced via a self-adaptive GP surrogate modeling method, as well as a hybrid search method. For additional details see Liu *et al.* (2020).

4.3 The PSADEA Method

4.3.1 *GP surrogate modeling*

As mentioned earlier, GP is the surrogate modeling method used in PSADEA (Santner *et al.*, 2003; Rasmussen and Williams, 2006). The main advantages of GP surrogate modeling include tractability and preciseness (Santner *et al.*, 2003; Rasmussen and Williams, 2006; Buche *et al.*, 2005). GP modeling and prediction primarily works as follows (Akinsolu, 2019).

For a given set of n observations where $x = (x_1, \ldots, x_n)$ and $y = (y_1, \ldots, y_n)$, GP modeling assumes that the objective function

$y(x)$ is a sample of a Gaussian distributed stochastic process with mean μ and variance σ. GP modeling then predicts a new function value $y(x)$ at some design point x by using the available points. If $y(x)$ is a continuous function, the function values of two given points x_i and x_j are close when they are highly correlated. The Gaussian correlation function is defined as

$$\text{Corr}(x_i, x_j) = \exp\left(-\sum_{l=1}^{d} \theta_l |x_i^l - x_j^l|^{pl}\right)$$

$$\theta_l \leq 0, 1 \leq pl \leq 2, \qquad (4.1)$$

where d is the dimension of x and θ_l is the correlation parameter which determines the extent of correlation reduction when x_i^l moves in the l direction.

The smoothness of the correlation function is related to pl with respect to x^l. The hyper-parameters θ_l and pl are determined by maximizing the likelihood function in Equation (4.2).

$$\frac{1}{(2\pi\sigma^2)^{\frac{n}{2}}\sqrt{\det(R)}} \exp\left[-\frac{(y-\mu I)^T R^{-1}(y-\mu I)}{2\sigma^2}\right], \qquad (4.2)$$

where R is a $n \times n$ matrix and I is a $n \times 1$ vector having all its elements as unity.

By maximizing the likelihood function for $y = y^i$ at $x = x^i (i = 1, \ldots, n)$ and the prediction uncertainty based on the best linear unbiased prediction, the function value $y(x^*)$ at a new point x^* can be predicted as

$$\hat{y}(x^*) = \mu + r^T R^{-1}(y - I\mu), \qquad (4.3)$$

where

$$R_{i,j} = \text{Corr}(x_i, x_j); i, j = 1, 2, \ldots, n, \qquad (4.4)$$

$$r = [\text{Corr}(x^*, x_1), \text{Corr}(x^*, x_2), \ldots, \text{Corr}(x^*, x_n)], \qquad (4.5)$$

$$\hat{\mu} = (I^T R^{-1} y)(I^T R^{-1} I)^{-1}. \qquad (4.6)$$

For the GP surrogate model, the mean squared error (MSE) or $\hat{s}^2$ value of the prediction uncertainty used for evaluating the model accuracy is as follows:

$$\hat{s}^2(x^*) = \hat{\sigma}^2[I - r^T R^{-1} r + (I - r^T R^{-1} r)^2 (I^T R^{-1} I)^{-1}], \quad (4.7)$$

where

$$\hat{\sigma}^2 = (y - I\hat{\mu})^T R^{-1}(y - I\hat{\mu}) n^{-1}. \quad (4.8)$$

To evaluate the quality of a candidate design in terms of the predicted value in Equation (1.3) and the prediction uncertainty in Equation (1.7), the lower confidence bound (LCB) method (Emmerich et al., 2006) is used in PSADEA. If the objective function $y(x)$ has a predictive distribution of $N(\hat{y}(x), \hat{s}^2(x))$, an LCB prescreening of $y(x)$ can be defined according to Equation (4.9).

$$y_{\text{lcb}}(x) = \hat{y}(x) - \omega \hat{s}(x) \quad \omega \in [0, 3], \quad (4.9)$$

where ω is a constant, which is often set to 2 to ensure a balance between exploration and exploitation (Emmerich et al., 2006).

In PSADEA, the GP surrogate model is implemented using the ooDACE toolbox (Couckuyt et al., 2012). The computational complexity of GP-based surrogate modeling is $O(N_{it} K^3 d)$, where N_{it} is the number of iterations expended in the optimization of the hyperparameters and K is the number of training data points (Bertsekas, 1999). It should be noted that K is the most critical factor affecting the cost of the training. To build a very reliable surrogate model, for a large d (dimensionality of the search space), very large K will be required to obtain satisfactory results. Yet, considering the dimensionality of typical EM design optimization problems, GP-based surrogate modeling time is often considerably short.

4.3.2 The DE algorithm

DE is recognized to be a very effective algorithm for global optimization over continuous spaces (Storn and Price, 1997; Price et al., 2005). Typically, DE works as follows (Storn and Price, 1997).

Let P be a population with $x = (x_1, \ldots, x_d) \in R^d$ as an individual solution in P. To create a child solution $u = (u_1, \ldots, u_d)$ for x, firstly, mutation occurs to produce a donor vector:

$$v^i = x^{r1} + F \cdot (x^{r2} - x^{r3}), \quad (4.10)$$

where x^{r1}, x^{r2}, and x^{r3} are three mutually exclusive solutions randomly selected from P; v^i is the ith mutant vector; $F \in (0, 2]$ is a control parameter, often called the scaling factor or differential weight. The mutation strategy in equation (4.10) is called DE/rand/1.

The crossover operator is then applied to produce the child solution u:

1. Randomly select a variable index $j_{\text{rand}} \in \{1, \ldots, d\}$.
2. For each $j = 1$ to d, generate a uniformly distributed random number rand from $(0, 1)$ and set:

$$u_j = \begin{cases} v_j, & \text{if (rand} \leq CR) | j = j_{\text{rand}} \\ x_j, & \text{otherwise,} \end{cases} \quad (4.11)$$

where $CR \in [0, 1]$ is a constant called the crossover rate.

For a current generation t, to decide whether or not the child solution or trial vector $u_j(t)$ becomes a member of the next generation $t + 1$, it is compared to the target vector $x_j(t)$ using the greedy criterion. If vector $u_j(t)$ yields a smaller fitness function value than $x_j(t)$, then $x_j(t)$ is set to $u_j(t)$; otherwise, $x_j(t)$ is retained.

DE follows a standard EA flow and it is controlled by only a few parameters (P, CR, and F in most cases). However, there are several DE variants due to the possibility of having several mutation strategies. In the PSADEA algorithmic framework, the DE mutation strategies stated in equation (4.12) to Equation (4.14) are also considered:

(1) *Mutation strategy: DE/best/1*

$$v^i = x^{\text{best}} + F \cdot (x^{r1} - x^{r2}), \quad (4.12)$$

where x^{best} is the best candidate in the current population, P.

(2) *Mutation strategy: DE/current-to-best/1*

$$v^i = x^i + F \cdot (x^{\text{best}} - x^i) + F \cdot (x^{r1} - x^{r2}), \quad (4.13)$$

where x^i is the ith vector in the current population.

(3) *Mutation strategy: DE/rand/2*

$$v^i = x^{r1} + F \cdot (x^{r2} - x^{r3}) + F \cdot (x^{r4} - x^{r5}), \quad (4.14)$$

where x^{r5} is a solution randomly selected from P and is different from x^{r1}, x^{r2}, x^{r3} and x^{r4}.

Arguably, the above mutation strategies are the most widely used DE mutation operators (Mezura-Montes and Velaquez-Reyes, 2006; Das and Suganthan, 2010).

4.3.3 *Implementation of PSADEA*

The flow diagram for the implementation of PSADEA is shown in Fig. 4.1 and PSADEA summarily works as follows:

Step 1: Use the Latin hypercube sampling method (Stein, 1987) to sample a small number of candidates from the design space and simulate these designs to obtain their performances. These designs and their performances form the initial database.

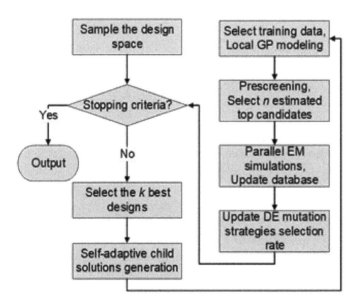

Fig. 4.1. PSADEA flow diagram.

Step 2: If a predetermined stopping criterion such as the maximum number of EM simulations is satisfied, output the best design from the database; otherwise go to Step 3.

Step 3: Choose k top-ranked candidates from the database to form a population.

Step 4: Apply the DE mutation operators using either of equations (4.12)–(4.14) self-adaptively on the population in Step 3 to generate child populations and their solutions.

Step 5: For every candidate design in each population, build a GP surrogate model using the nearest designs based on Euclidean distance from the database and their simulation results as the training data points.

Step 6: Using the GP surrogate models in Step 5 and the LCB prescreening method (Emmerich *et al.*, 2006), prescreen the child solutions in Step 4 and select the best child solutions based on the lower confidence bound values.

Step 7: Evaluate (via simulation) the estimated best child solutions from Step 6 in parallel. Add them and their simulation results to the database. Go back to Step 2.

4.4 Case Study

PSADEA has been used to address quite a few challenging real-world antenna design problems, such as wearable (or body-centric) antennas, multi-band antennas, and reconfigurable antenna arrays (Danjuma *et al.*, 2020; Hua *et al.*, 2019; Zhang *et al.*, 2021). In this section, the compact slotted monopole antenna for UWB body-centric imaging applications (Danjuma *et al.*, 2020) and the novel compact quadruple-band indoor base station antenna for 2G/3G/4G/5G systems (Hua *et al.*, 2019) are used to demonstrate PSADEA. Also, comparisons between PSADEA and commercially available antenna optimizers are made for the slotted monopole antenna example. The reference methods are Computer Simulation Technology–Microwave Studio (CST-MWS) trust region framework (TRF) and CST-MWS particle swarm optimization (PSO), which are popular for antenna design optimization (Grout *et al.*, 2019; Dassault Systemes, 2020). The parameter settings adopted for PSADEA are described in Liu *et al.* (2018a) and Akinsolu *et al.* (2019), while the default settings in CST-MWS are used for the algorithmic parameters of the reference methods. These default settings are adopted due to the general assumption that they are well verified to be fitting for most cases (Grout *et al.*, 2019). However, to ensure a fair comparison, a population size of 50 and a swarm size of 50 are used for PSADEA and CST-MWS PSO, respectively, and the sigma value for CST-MWS TRF is set to unity to improve exploration. Note that for CST-MWS TRF, the initial design used as the starting point for each run is randomly generated using the Latin hypercube sampling method (Stein, 1987). The time consumption reported for all tests is wall clock time.

4.4.1 *Example one*

The first example is the compact slotted monopole antenna for UWB body-centric applications. The layout of its basic/primitive design is shown in Fig. 4.2. The slotted monopole antenna structure mainly consists of a circular patch radiator driven by a 50Ω microstrip line, with four T-slots conjoined at the head and centrally etched on the radiator to have a quasi-cross slot configuration. Two uniform rectangular planes which are abreast and separated by the microstrip line constitute the co-planar partial ground of the monopole antenna. It is implemented on an FR-4 substrate having a thickness of 0.8 mm, loss tangent ($\tan(\delta)$) of 0.025 and a relative permittivity (ε_r) of 4.3. The slotted monopole antenna is modeled and discretized in CST-MWS using the time-domain finite integration technique (FIT) with an accuracy of −40 dB and a maximum cell density of 20 cells per wavelength resulting in about 300,000 hexahedral mesh cells in total. Each simulation costs about 2 min on the average on a workstation with Intel 4-core i7 CPU and 24 GB RAM.

The design of the compact slotted monopole antenna for UWB body-centric applications could be very challenging, particularly the determination of the optimal structure, mainly due to the following reasons:

(1) The monopole antenna is to be compact to ensure its correct placement (physical) and integration (as a sensor for most typical body-centric applications) with compact active and/or passive circuit components on the same PCB.

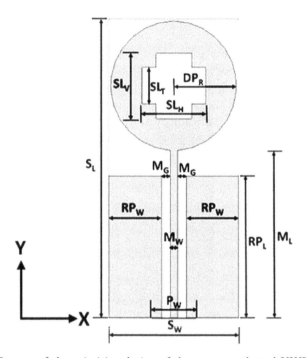

Fig. 4.2. Layout of the primitive design of the compact slotted UWB monopole antenna.

(2) The monopole antenna is required to have a large bandwidth in both free space and close to the human body to ensure greater accuracy in UWB body-centric short-range wireless communications. In other words, the antenna is to exhibit good impedance matching and limited pulse distortion without the loss of radiation efficiency over the frequency band 3.1–10.6 GHz.
(3) For microwave imaging applications, the monopole antenna will typically act as a sensor (transceiver) node within the microwave imaging system setup. As such, its beam should be broad and flat (i.e., not steered in any specific direction) for easy detection by other connected antennas and/or devices from various angles. Consequently, the antenna's bore-sight gain must be controlled to be within a moderate range (i.e., not too low, and not too high) over the UWB spectrum.

The above objectives cannot be easily achieved using manual design techniques; thus, the need for design automation via optimization. Typically, the geometry of the slots etched on planar antennas determines the direction of the current flow and nature of their surface current distributions (Yeboah-Akowuah et al., 2017). As a result, the radiation of the slotted monopole antenna can be varied by adjusting the dimensions of its four T-slots. According to Fig. 4.2, the configuration of the slot can evolve from quasi-cross to rectangular configuration (X-axis dominant or Y-axis dominant) according to the depths and widths of the heads and throats of the conjoined four T-slots (i.e., SL_H, SL_V, and SL_T).

For the synthesis of the monopole antenna, the design parameters, their given search ranges, and the geometric constraints in Table 4.1 are considered. The optimization goal is the minimization of the fitness function (F_{uwb}) in (4.15) to satisfy the design specifications in Table 4.2. Note that the geometric constraints and search bounds

Table 4.1. Search ranges of the design variables and a typical design by PSADEA (all sizes in mm) (example 1).

Design parameters	Lower bound	Upper bound	PSADEA-optimum
Substrate width (S_W)	$2 \times DP_R$	$3 \times DP_R$	14.90
Microstrip length (M_L)	RP_L	50.00	18.52
Microstrip width (M_W)	0.50	7.50	0.84
Microstrip gap (M_G)	>0.00	21.5	0.12
Circular patch radius (DP_R)	2.00	25.00	7.21
Width of slot throat (SL_T)	>0.00	$2 \times DP_R$	5.98
Vertical slots' depth (SL_V)	>0.00	$2 \times DP_R$	12.06
Horizontal slots' depth (SL_H)	>0.00	$2 \times DP_R$	5.98
Partial ground plane length (RP_L)	DP_R	M_L	18.05
Feed guide width (P_W)	$6 \times M_W$	$10 \times M_W$	2.56
Substrate length (S_L) = $M_L + (2 \times DP_R) + 0.2$			
Partial ground plane width (RP_W) = $(S_W - (2 \times M_G) - M_W) \div 2$			

Table 4.2. Performance specifications (example 1).

Item	Specification	PSADE-optimum
Maximum return loss (S_{11})	$\leq -10\,\text{dB}$	$-10.61\,\text{dB}$
Minimum bore-sight gain ($G_{\min}$)	$\geq 2\,\text{dBi}$	$2.17\,\text{dBi}$
Maximum bore-sight gain ($G_{\max}$)	$\leq 5\,\text{dBi}$	$4.74\,\text{dBi}$

provided in Table 4.1 have been carefully defined by the designer to maintain a low profile for the antenna in the optimization process. For example, the microstrip length (M_L) cannot be greater than 50 mm and the partial ground plane length (RP_L) cannot be larger than M_L according to Table 4.1.

$$F_{\text{uwb}} = \max(S_{11}) + w \times \max([2dBi - G_{\min}, 0]) \\ + \ldots w \times \max([G_{\max} - 5dBi, 0]), \qquad (4.15)$$

where w is the penalty coefficient and it is set to 50. By setting w to be equal to 50, the specifications for the bore-sight gain in Table 4.2 (i.e., $G_{\min}$ and $G_{\max}$) are satisfied first, by largely penalizing F_{uwb} if they are violated. Then, meeting the S_{11} requirement becomes the primary focus of the optimization procedure as soon as the requirements for $G_{\min}$ and $G_{\max}$ are met.

For the experiments, the computing budgets are as follows: 500 parallel simulations (i.e., 1,500 EM simulations in total considering three simulations are carried out in parallel) over 10 runs are used for PSADEA, while 5000 simulations over three runs and 3,000 simulations over five runs are used for CST-MWS PSO and CST-MWS TRF, respectively. Note that for CST-MWS PSO and CST-MWS TRF, it takes several days to complete a single run and additional runs are not affordable.

A typical optimal design obtained by PSADEA after 50 hours of optimization is shown in Table 1.1 and the performance for this design is shown in Table 4.2. The physical implementation of this design is shown in Fig. 4.3 and the overall size is 33.14 mm × 14.90 mm × 0.8 mm, which corresponds to $0.34\,\lambda_o \times 0.15\lambda_o \times 0.008\lambda_o$, where λ_o is the free space wavelength at 3.1 GHz. The size (in linear dimensions) of this prototype is about half the size of a similar state-of-the-art design (Yeboah-Akowuah et al., 2017). The simulated and measured results for the return loss (free-space, on-phantom and on-body) of the PSADEA-synthesized design are shown in Fig. 4.4.

Figure 4.4 shows that the simulated and measured results are in reasonable agreement except that in free space a new resonance was observed for the measured results around 5.5 GHz. This acceptable discrepancy can be mainly attributed to fabrication tolerances and process variations during the measurements carried out using an Agilent Technologies N5242A vector network analyzer (VNA). Figure 4.4

Fig. 4.3. Physical implementation of the PSADEA-synthesized UWB slotted monopole antenna.

also demonstrates that the slotted monopole antenna maintains its UWB operation in free space and it is not overly detuned in the presence of a human phantom modeled according to Beard *et al.* (2006) and Lin *et al.* (2012) and the human body.

In terms of quality of results, all methods satisfy the bore-sight gain requirements over all runs and the following can be observed for the simulated return loss from Table 4.3:

(1) In all 10 runs, PSADEA satisfies the return loss requirement in Table 1.2. Even for the worst case, the $\max(S_{11})$ performance obtained by PSADEA is still very good.
(2) PSADEA shows very good robustness because the standard deviation is low.
(3) In all five runs, CST-MWS TRF obtained designs with geometric incongruities and did not satisfy the return loss requirement in Table 4.2. Thus, all the results for CST-MWS TRF are termed as not applicable (N/A). This is because designs with geometric incongruities are impractical and cannot be prototyped for end-use.
(4) In all three runs, CST-MWS PSO did not obtain designs that meet the return loss requirement in Table 4.2.

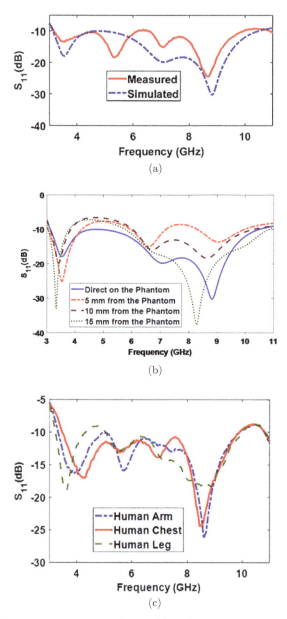

Fig. 4.4. Frequency response for PSADEA-synthesized antenna design (example 1). (a) free space return loss, (b) on-phantom return loss, (c) on-body return loss.

Table 4.3. Statistics of the best max(S_{11}) using different methods (example 1).

Method	Best	Worst	Mean	Median	Std.
PSADEA (10 runs)	−10.61 dB	−10.05 dB	−10.38 dB	−10.39 dB	0.1815
CST-MWS TRF (5 runs)	N/A	N/A	N/A	N/A	N/A
CST-MWS PSO (3 runs)	−5.92 dB	−5.02 dB	−5.61 dB	−5.90 dB	0.5113

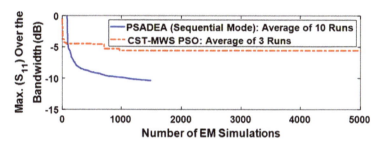

Fig. 4.5. Convergence trends of PSADEA and CST-MWS PSO on example 1.

In terms of efficiency, the convergence trends can be seen in Fig. 4.5. Figure 4.5 shows that PSADEA used 1090 EM simulations on average over 10 runs to satisfy all the specifications in Table 4.2 and then obtains an average max(S_{11}) of −10.38 dB upon convergence after 1450 EM simulations. Also, from Fig. 4.5, CST-MWS PSO used 972 EM simulations on average over 3 runs to obtain an average max(S_{11}) of −5.61 dB and it did not show any further improvement over 5000 EM simulations. PSADEA needs 127 EM simulations on average to obtain the result of CST-MWS PSO. Thus, it can be inferred that PSADEA is 7.7 times faster than CST-MWS PSO on the average for this example. Owing to geometric incongruities earlier stated, the convergence trend CST-MWS TRF is not N/A and it is not shown in Fig. 4.5.

4.4.2 Example two

The second example is the compact quadruple-band indoor base station antenna for 2G/3G/4G/5G systems whose layout is shown in

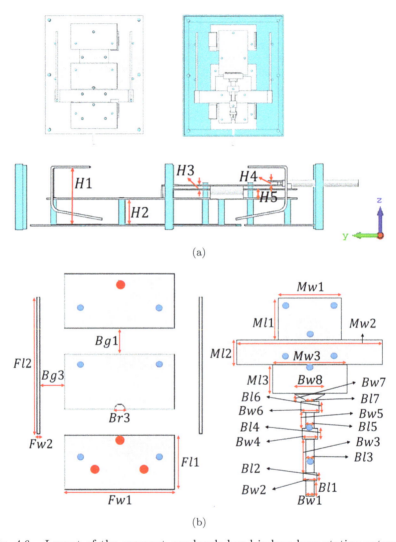

Fig. 4.6. Layout of the compact quadruple-band indoor base station antenna. (a) 3D geometry, (b) first and second layers, (c) third and fourth layers, (d) metal strips and reflector base.

Fig. 4.6. The antenna structure mainly comprises a reflector base, four bent metal strips and a total of four layers. The resonant frequencies are introduced, and the bandwidths are enhanced for the lower frequency bands (i.e., 0.69–0.96 GHz and 1.71–2.7 GHz)

Surrogate Model-Assisted Global Optimization for Antenna Design 143

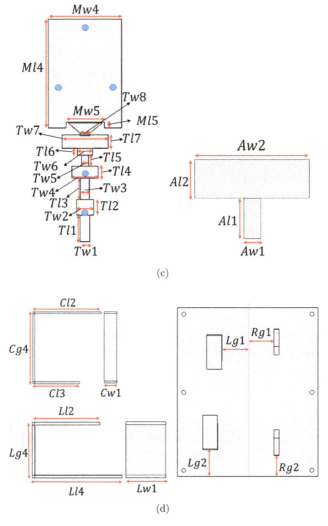

Fig. 4.6. (*Continued*)

via three coupled metal patches and two parasitic metal strips in the first layer. Three additional metal patches and a feeding structure are implemented in the second layer. A single patch and a single feeding structure are implemented in the third layer. A small T-shaped patch is implemented in the fourth layer to have resonant frequencies in the higher frequency bands.

The indoor base station antenna is driven by a 50Ω coaxial cable connected to the feeding structure. It is modeled and discretized in CST-MWS using the time-domain FIT method with an accuracy of −40 dB and a maximum cell density of 12 cells per wavelength resulting in about 5,200,000 mesh cells in total. The cost of each EM simulation is about 12 minutes on average on a workstation with Intel 8-core i7 5960×3 GHz CPU, 64 GB RAM and 16 GB NVIDIA Quadro GP100 GPU. Due to the computational cost of each EM simulation, comparisons with the reference methods are not made for this example. More particularly, non-satisfactory results were obtained for a typical run of CST-MWS TRF and a prohibitive computational overhead without any tendency of successful outcomes is envisaged to be incurred for CST-MWS PSO.

The determination of the optimal structure for the compact quadruple-band indoor base station antenna could be challenging due to the following reasons:

(1) A light weight, a low geometric profile, and a simple feeding structure are required to ensure low maintenance.
(2) Stable radiation patterns and high polarization purity are desired within the 2G/3G/4G/5G frequency bands.
(3) Multi-antenna system requirements in terms of reflection coefficient and gain (directivity) to have an integration of conventional 2G, 3G and 4G antenna systems' specifications with modern 5G antenna specifications to cover a quadruple-band (i.e., 2G, 3G, 4G and 5G bands).
(4) A large fractional bandwidth is required in the new sub-6 GHz spectrum stipulated for 5G systems.

In design of the indoor base station antenna, the design parameters described in Fig. 4.6 along with their search bounds provided in Table 4.4 are considered. Note that this base station antenna has over 45 sensitive design parameters in total (see Fig. 4.6) and design knowledge has been used to select the most critical ones (i.e., 19 design parameters) for the optimization. The optimization goal is

Table 4.4. Search ranges of the design variables and PSADEA-optimum (all sizes in mm) (example 2).

Design parameters	Lower bound	Upper bound	PSADEA-optimum
$Bw1$	70	90	86.18
$Bl1$	33.5	50	41.71
$Bg1$	13	33	22.70
$Bg2$	39	59	48.90
$Bg3$	12.50	32.50	31.94
$Bl2$	104	124	106.74
$Bw2$	1	10	9.25
$Rg1$	27.8	47.8	47.51
$Rw1$	1	10	9.25
$Rg2$	7.5	27.5	27.06
$Rl2$	14	34	15.71
$Rg4$	4	17	14.22
$Rl3$	18	28	26.21
$Lg1$	30	40	30.63
$Lw1$	3.8	23.8	23.55
$Lg2$	7.5	27.5	25.92
$Ll2$	14	34	17.63
$Ll4$	22	42	39.53
$Lg4$	4	15	12.11

Table 4.5. Performance specifications (example 2).

Item	Specification (dB)	PSADEA-optimum (dB)
Maximum return loss (S_{11}) (0.69–0.96 GHz)	≤ -10	-11.33
Maximum return loss (S_{11}) (1.71–2.7 GHz)	≤ -10	-10.15
Maximum return loss (S_{11}) (3.3–3.8 GHz)	≤ -10	-11.16
Maximum return loss (S_{11}) (4.8–5.0 GHz)	≤ -10	-10.20

the minimization of the fitness function (F_{indoor}) in (4.16) to satisfy the design specifications in Table 4.5.

$$F_{\text{indoor}} = \max(S_{11}^{i=1}) + w \times \sum_{i=2}^{4} \max([S_{11}^{i} + 10\,dB, 0]), \quad (4.16)$$

where i is the index for the corresponding frequency band out of the four frequency bands (from low to high), and w is the penalty coefficient set to 50. This setting ensures that S_{11}^i is firstly satisfied in the upper three frequency bands by largely penalizing F_indoor if it is not satisfied and then the optimization focuses on satisfying S_{11}^i in the first band as soon as the specifications in the other bands are met. Also note that the fitness function in (4.16) is based on design knowledge which suggests that the meeting the return loss specifications in the first lower band could be quite challenging due to a noticeable trade-off with having a good impedance matching in the other bands.

The PSADEA-optimum design obtained after 3 days of optimization is shown in Table 4.4. The performance of this design is shown in Table 4.5. Though comparisons are not made with the other methods for this example (for reasons earlier stated above), it is worth mentioning that PSADEA obtained its optimized design using 552 EM simulations. The physical implementation of this design is shown in Fig. 4.7 and the overall size is 204 mm × 175 mm × 39 mm, which is about $0.47\lambda_o \times 0.40\lambda_o \times 0.09\lambda_o$, where λ_o is the free space wavelength at 0.69 GHz. The simulated and measured results for the return loss and realized gain of the PSADEA-synthesized design are shown in Fig. 4.8. From Fig. 4.8, the simulated and measured results are in reasonable agreement.

Fig. 4.7. Physical implementation of the PSADEA-synthesized compact quadruple-band indoor base station antenna.

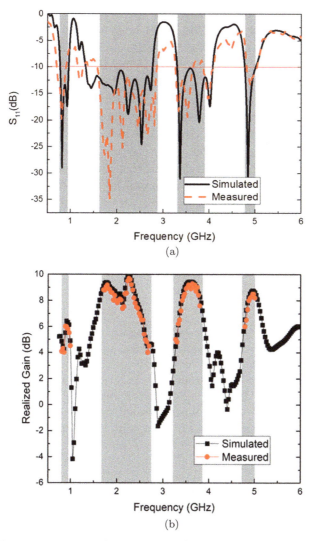

Fig. 4.8. Frequency response for the PSADEA-synthesized antenna (example 2). (a) return loss, (b) realized gain.

4.5 Conclusions

In this chapter, surrogate model-assisted global optimization for antenna design is demonstrated through the application of the authors' PSADEA method to real-world antenna design problems.

The antenna design problems are the optimization of a compact slotted monopole antenna for UWB body-centric applications and a novel compact quadruple band antenna for 2G/3G/4G/5G systems. For both antenna design problems, PSADEA obtains high-quality design solutions. The selected optimum designs have been fabricated and the measured and simulated results for each prototype have been found to be in close agreement. Hence, the design solution quality of PSADEA was validated. PSADEA is also compared to widely used antenna optimizers available in the popular CST-MWS product for the UWB slotted monopole antenna design problem. In all cases and over repeated runs, PSADEA obtained high-quality design solutions for this example. The other optimizers (i.e., CST-MWS TRF and CST-MWS PSO) show high failure rates by obtaining design solutions that do not meet the desired specifications and/or have geometric incongruities.

References

Akinsolu, M.O. (2019). Efficient surrogate model-assisted evolutionary algorithm for electromagnetic design automation with applications, University of Chester, Chester, UK.

Akinsolu, M.O., Liu, B., Grout, V., Lazaridis, P.I., Mognaschi, M.E., P. Di Barba. (2019). A parallel surrogate model assisted evolutionary algorithm for electromagnetic design optimization, *IEEE Transactions on Emerging Topics in Computational Intelligence*, vol. 3, no. 2, pp. 93–105.

Akinsolu, M.O., Liu, B., Lazaridis, P.I., Mistry, K.K., Mognashchi, M.E., Di Barba, P., Zaharis, Z. D. (2020). Efficient design optimization of high-performance MEMS based on a surrogate-assisted self-adaptive differential evolution, *IEEE Access*, vol. 8, pp. 80256–80268.

Akinsolu, M.O., Mistry, K.K., Liu, B., Lazaridis, P.I., Excell, P. (2020). Machine learning-assisted antenna design optimization: A review and the state-of-the-art, *Proc. 14th European Conference on Antennas and Propagation*, EuCAP, pp. 1–5 (in English).

Alieldin, A., Huang, Y., Boyes, S.J., Stanley, M., Joseph, S.D., Al-juborri, B. (2018a). A dual-broadband dual-polarized fylfot-shaped antenna for mobile base stations using MIMO over-lapped antenna subarrays, *IEEE Access*, vol. 6, pp. 50260–50271.

Alieldin, A., Huang, Y., Boyes, S.J., Stanley, M., Joseph, S.D., Hua, Q., Lei, D. (2018b). A triple-band dual-polarized indoor base Station antenna for 2G, 3G, 4G and sub-6 GHz 5G applications, *IEEE Access*, vol. 6, pp. 49209–49216.

Balanis, C.A. (2014). *Antenna Theory: Analysis and Design*, 4th edn., John Wiley and Sons Inc., New Jersey, USA.

Beard, B.B., Kainz, W., Onishi, T., Iyama, T., Watanabe, S., Fujiwara, O., Wang, J., Bit-Babik, G., Faraone, A., Wiart, J., Christ, A., Kuster, N., Lee, A-K., Kroeze, H., Siegbahn, M., Keshvari, J., Abrishamkar, H., Simon, W., Manteuffel, D., Nikoloski, N. (2006). Comparisons of computed mobile phone induced SAR in the SAM phantom to that in anatomically correct models of the human head, *IEEE Transactions on Electromagnetic Compatibility*, vol. 48, no. 2, pp. 397–407.

Bertsekas, D.P. (1999). *Nonlinear Programming*, 2nd edn., Athena Scientific, Belmont, USA.

Buche, D., Schraudolph, N.N., Koumoutsakos, P. (2005). Accelerating evolutionary algorithms with gaussian process fitness function models, *IEEE Transactions on Systems, Man, and Cybernetics*, vol. 35, no. 2, pp. 183–194.

Couckuyt, I., Declercq, F., Dhaene, T., Rogier, H., Knockaert, L. (2010). Surrogate-based infill optimization applied to electromagnetic problems. *International Journal of RF and Microwave Computer-Aided Engineering*, vol. 20, no. 5, pp. 492–501.

Couckuyt, I., Forrester, A., Gorissen, D., Turck, F.D., Dhaene, T. (2012). Blind kriging: implementation and performance analysis, *Advances in Engineering Software*, vol. 49, pp. 1–13.

Danjuma, I.M., Akinsolu, M.O., See, C.H., Liu, B., Abd-Alhameed, R.A. (2020). Design and optimization of a slotted monopole antenna for ultra-wide band body centric imaging applications, IEEE Journal of Electromagnetics, *RF and Microwaves in Medicine and Biology*, vol. 4, no. 2, pp. 140–147.

Das, S., Suganthan, P.N. (2010). Differential evolution: a survey of the state-of-the-art, *IEEE Transactions on Evolutionary Computation*, vol. 15, no. 1, pp. 4–31.

Dassault Systemes. Automatic Optimization: CST Studio Suite. (2020). [Online]. Available: https://www.3ds.com/products-services/simulia/products/cst-studio-suite/optimization/.

Emmerich, M.T.M., Giannakoglou, K.C., Naujoks, B. (2006). Single-and multiobjective evolutionary optimization assisted by gaussian random field metamodels, *IEEE Transactions on Evolutionary Computation*, vol. 10, no. 4, pp. 421–439.

Forrester, A.I., Keane, A.J. (2009). Recent advances in surrogate-based optimization, *Progress in Aerospace Sciences*, vol. 45, no. 1–3, pp. 50–79.

Grout, V., Akinsolu, M.O., Liu, B., Lazaridis, P.I., Mistry, K.K., Zaharis, Z.D. (2019). Software solutions for antenna design exploration: A comparison of packages, tools, techniques, and algorithms for various design challenges, *IEEE Antennas and Propagation Magazine*, vol. 61, no. 3, pp. 48–59.

Guney, K., Basbug, S. (2014). A parallel implementation of seeker optimization algorithm for designing circular and concentric circular antenna arrays, *Applied Soft Computing*, vol. 22, pp. 287–296.

Hua, Q., Song, C., Huang, Y., Akinsolu, M.O., Liu, B., Jia, T. Xu, Q., Alieldin, A. (2019). A novel compact quadruple-band indoor base station antenna for 2G/3G/4G/5G systems, *IEEE Access*, vol. 7, pp. 151350–151358.

Hua, Q., Huang, Y., Alieldin, A., Song, C., Jia, T., Zhu, X. (2020). A dual-band dual-polarized base station antenna using a novel feeding structure for 5G communications, *IEEE Access*, vol. 8, pp. 63710–6317.

Koziel, S., Ogurtsov, S., Couckuyt, I., Dhaene, T. (2012). Variable-fidelity electromagnetic simulations and co-kriging for accurate modeling of antennas, *IEEE Transactions on Antennas and Propagation*, vol. 61, no. 3, pp. 1301–1308.

Koziel, S., Leifsson, L. (2013). *Surrogate-based Modeling and Optimization*. New York, USA, Springer-Verlag.

Koziel, S., Ogurtsov, S. (2014). *Antenna Design by Simulation-Driven Optimization*, Chapter 2 "Surrogate-based optimization", Springer, Cham, Switzerland, pp. 13–24.

Koziel, S., Pietrenko-Dabrowska, A. (2019). Reduced-cost electro-magnetic-driven optimisation of antenna structures by means of trust-region gradient-search with sparse Jacobian updates, *IET Microwaves, Antennas and Propagation*, vol. 13, no. 10, pp. 1646–1652.

Koziel, S., Pietrenko-Dabrowska, A. (2020). Expedited feature-based quasi-global optimization of multi-band antenna input characteristics with Jacobian variability tracking, *IEEE Access*, vol. 8, pp. 83907–83915.

Lin, C.H., Saito, K., Takahashi, M., Ito, K. (2012). A compact planar inverted-F antenna for 2.45 GHz on-body communications, *IEEE Transactions*, vol. 60, no. 9, pp. 4422–4426.

Liu, B., Aliakbarian, H., Ma, Z., Vandenbosch, G.A., Gielen, G., Excell, P. (2014a). An efficient method for antenna design optimization based on evolutionary computation and machine learning techniques, *IEEE Transactions on Antennas and Propagation*, vol. 62, no. 1, pp. 7–18.

Liu, B., Zhang, Q., Gielen, G.G.E. (2014b). A Gaussian process surrogate model assisted evolutionary algorithm for medium scale expensive optimization problems, *IEEE Transactions on Evolutionary Computation*, vol. 18, no. 2, pp. 180–192.

Liu, B., Irvine, A., Akinsolu, M.O., Arabi, O., Grout, V., Ali, N. (2017a). GUI design exploration software for microwave antennas, *Journal of Computational Design and Engineering*, vol. 4, no. 4, pp. 274–281.

Liu, B., Koziel, S., Ali, N. (2017b). SADEA-II: A generalized method for efficient global optimization of antenna design, *Journal of Computational Design and Engineering*, vol. 4, no. 2, pp. 86–97.

Liu, B., Yang, H., Lancaster, M.J. (2017c). Global optimization of microwave filters based on a surrogate model-assisted evolutionary algorithm, *IEEE Transactions on Microwave Theory and Techniques*, vol. 65, no. 6, pp. 1976–1985.

Liu, B., Akinsolu, M.O., Ali, N., Abd-Alhameed, R. (2018a). Efficient global optimisation of microwave antennas based on a parallel surrogate model-assisted evolutionary algorithm, *IET Microwaves, Antennas and Propagation*, vol. 13, no. 2, pp. 149–155.

Liu, B., Grout, V., Nikolaeva, A. (2018b). Efficient global optimization of actuator based on a surrogate model assisted hybrid algorithm, *IEEE Transactions on Industrial Electronics*, vol. 65, no. 7, pp. 5712–5721.

Liu, B., Akinsolu, M.O., Song, C., Hua, Q., Huang, Y., Excell, P., Imran, M., Xu, Q. (2021). An efficient method for complex antenna synthesis based on a self adaptive surrogate model assisted optimization technique, *IEEE Transactions on Antennas and Propagation*, vol. 69, no. 4, pp. 2302–2315.

Mezura-Montes, E., Velaquez-Reyes, J. (2006). A comparative study of differential evolution variants for global optimization, *Proc. Genetic and Evolutionary Computation Conference*, GECCO, pp. 1–8 (in English).

Price, K., Storn, R. and Lampien, J.A. (2005). *Differential Evolution: A Practical Approach to Global Optimization*, 1st edn., Springer-Verlag, Berlin Heidelberg, Germany.

Rasmussen, C.E., Williams, C.K.I. (2006). *Gaussian Processes for Machine Learning*, MIT Press, Cambridge, USA.

Richie, J.E., Ababei, C. (2017). Optimization of patch antennas via multithreaded simulated annealing based design exploration, *Journal of Computational Design and Engineering*, vol. 4, no. 4, pp. 249–255.

Santner, T.J., Williams, B.J., Notz, W.I. (2003). *The Design and Analysis of Computer Experiments*, Springer-Verlag, New York, USA.

Stein, M. (1987). Large sample properties of simulations using latin hypercube sampling, *Technometrics*, pp. 143–151.

Storn, R., Price, K. (1997). Differential evolution–a simple and efficient heuristic for global optimization over continuous spaces, *Journal of Global Optimization*, vol. 11, no. 4, pp. 341–359.

Ur-Rehman, M., Malik, N.A., Yang, X., Abbasi, Q.H., Zhang, Z., Zhao, N. (2017). A low profile antenna for millimeter-wave body-centric applications, *IEEE Transactions on Antennas and Propagation*, vol. 65, no. 12, pp. 6329–6337.

Weiland, T., Timm, M., Munteanu, I. (2008). A practical guide to 3-D simulation, *IEEE Microwave Magazine*, vol. 9, no. 6, pp. 62–75.

Yeboah-Akowuah, B., Kosmas, P., Chen, Y. (2017). A Q-slot monopole for UWB body-centric wireless communications, *IEEE Transactions on Antennas and Propagation*, vol. 65, no. 10, pp. 5069–5075.

Yang, X., Qin, P., Liu, Y., Yin, Y., Guo, Y.J. (2018). Analysis and design of a broadband multifeed tightly coupled patch array antenna, *IEEE Antennas and Wireless Propagation Letters*, vol. 17, no. 2, pp. 217–220.

Yoo, S., Milyakh, Y., Kim, H., Hong, C., Choo, H. (2020). Patch array antenna using a dual coupled feeding structure for 79 GHz automotive radar applications, *IEEE Antennas and Wireless Propagation Letters*, vol. 19, no. 4, pp. 676–679.

Zhang, J., Liu, B., Vandenbosch, G.A.E. (2021). Automatic AI-driven design of mutual coupling reducing topologies for frequency reconfigurable antenna array, *IEEE Transactions on Antennas and Propagation*, vol. 69, no. 3, pp. 1831–1836.

© 2022 World Scientific Publishing Europe Ltd.
https://doi.org/10.1142/9781800610750_0005

Chapter 5

Surrogate-Based Modeling and Design Optimization Techniques for Signal Integrity in High-Performance Computer Platforms

Francisco E. Rangel-Patiño and José E. Rayas-Sánchez

Abstract

As microprocessor design scales to nanometric technology, traditional post-silicon validation techniques are inappropriate to get a full system functional coverage. Physical complexity and extreme technology process variations introduce design challenges to guarantee performance over different process, voltage, and temperature (PVT) conditions. In addition, there is an increasingly higher number of mixed-signal circuits within microprocessors; many of them employed in high-speed input/output (HSIO) links. Improvements in signaling methods, circuits, and process technology have allowed HSIO data rates to scale beyond 10 Gb/s, where undesired effects can create multiple signal integrity problems. With all of these elements, post-silicon validation of HSIO links is tough and time-consuming. One of the major challenges in post-silicon electrical validation of HSIO links lies in the physical layer (PHY) tuning process, where equalization techniques are used to cancel these undesired signal integrity effects. Typical current industrial practices for PHY tuning require massive lab measurements, since they are based on exhaustive enumeration methods. In this chapter, dedicated surrogate-based modeling and optimization methods, including space mapping, are described to efficiently tune transmitter and receiver equalizers of HSIO links. The proposed methodologies are validated by laboratory measurements on realistic industrial post-silicon validation platforms.

Keywords: Equalization, eye diagram, high-speed links, jitter, kriging, neural networks, orthogonal arrays, physical layer tuning, post-silicon validation, SATA, space mapping, Sobol, support vector machines, surrogate-based optimization, system margining, USB

5.1 Introduction

Technology scaling and advanced silicon packaging have allowed an increasingly higher density of integration. However, as process technologies scale down to nanometric dimensions, traditional integrated circuit (IC) design methods are challenged by the problem of increased silicon process variation. The combined effects of increased product complexity, performance requirements, and time-to-market (TTM) commitments have added tremendous pressure on post-silicon validation (Keshava et al., 2010).

A significant portion of the circuits to be validated in modern microprocessors corresponds to high-speed input/output (HSIO) links. Undesired effects such as jitter, inter-symbol interference (ISI), crosstalk and others can create multiple signal integrity problems in HSIO circuits, making maximum bus speeds difficult to achieve in practice. This problem is aggravated by the fact that channel speeds keep increasing from one generation bus technology to the next one. This is of particular concern for current industrial HSIO interfaces, such as Peripheral Component Interconnect Express (PCIe), Serial Advanced Technology Attachment (SATA), Universal Serial Bus (USB), and Ethernet.

Physical layer (PHY) tuning is a key strategy to face the above challenges in modern HSIO links. Digital PHY tuning coefficients are usually embedded in the I/O links and can be digitally tuned to appropriate values. Considering the large die-to-die process variations, as well as the typical fluctuations in operating conditions, board impedance, channel loss, and different add-in cards, the performance of HSIO links can exhibit large variation. PHY tuning provides a way to reconfigure I/O links to cancel various fluctuations. However, it is usually unknown in pre-silicon (before fabrication of the first prototypes) which configuration the overall best performance gives. Hence, it becomes necessary to search for an "optimal" configuration of PHY tuning coefficients. Most current industrial practices to perform PHY tuning consist of exhaustive enumeration methods, turning them into the most time-consuming processes in industrial post-silicon validation (Rangel-Patiño et al., 2020).

This chapter starts by briefly describing the main components of post-silicon validation in modern computer platforms. Next, we describe several surrogate modeling approaches for PHY tuning of HSIO links, all of them based on actual measurements on industrial server post-silicon validation platforms. We compare several surrogate modeling methods combined with different design of experiments (DoE) techniques to find the best approach to efficiently simulate the PHY tuning process, verifying the accuracy of the resultant surrogate models by comparing with actual measurements. We next propose applying surrogate-based optimization (SBO), exploiting the best surrogate model found so that the optimal PHY tuning receiver (Rx) equalizer settings can be obtained. We validate our approach by measuring the actual functional eye diagram on the real system using the optimal settings predicted by the best surrogate. Finally, we also describe in this chapter how the space mapping technique can be used to efficiently optimize the PHY tuning Rx equalizer settings by exploiting a rough kriging surrogate as the coarse model, and a measurement-based post-silicon validation platform as the fine model. The overall results, confirmed by realistic industrial validation platform measurements, demonstrate the adequacy of the proposed methods.

5.2 Post-Silicon Validation

The complexity of new computing embedded systems has grown to an amazing level. Today's most advanced processors and systems on chip (SoC) incorporate billions of transistors, and must be compatible with dozens of operating systems, hundreds of component platforms, and thousands of hardware devices and software applications. To ensure leading performance, reliability, and compatibility in this complex environment, silicon companies invest millions of dollars annually in validation platforms and rigorous testing procedures.

This complexity creates a huge number of possible testing scenarios and paramount validation challenges. Each processor or SoC is intended to be used in an enormous variety of platform configurations, and it will have to interoperate with many different motherboards, components, peripherals, operating systems, and applications. It will also face various extremes of temperature, voltage, and clock frequency. This variability of configurations and operating environments can dramatically increase the number of necessary tests.

The combined effects of increased product complexity, performance requirements and TTM commitments have added tremendous pressure on post-silicon electrical validation, which is usually the last step prior to volume manufacturing. This challenges the validation teams to continuously assess their methods and processes, and look for opportunities to make validation faster and cheaper (Keshava et al., 2010).

The standard industrial interfaces coming out from the new microprocessors, SoC, and chipsets are now much faster and more diverse. Additionally, the increase in the number of cores and threads and the inclusion of advanced power management and virtualization drive additional complexity, as well as new boundary conditions and bus requirements. Therefore, post-silicon validation is undertaking fundamental transformations from radically changing environmental conditions (Keshava et al., 2010; Patra, 2007).

Post-silicon validation techniques employ hundreds of manufactured silicon samples in actual application environments to validate correct behavior across specified operating conditions and industrial standards. Among the different post-silicon validation disciplines, electrical validation (EV) focuses on assessing electrical parameters/behaviors mainly associated with high-speed interconnects, phase-locked loop (PLLs), power grids, clock networks, analog/mixed-signal circuits, and PHY tuning settings. A large portion of EV is devoted to the validation of HSIO links, including industry-standard interfaces such as PCIe (PCI SIG Org, 2016), SATA (SATA Org, 2016), USB (USB Org, 2016), and Ethernet (10GEA Org, 2016).

The goal of HSIO post-silicon electrical validation is to statistically predict the I/O behavior in a real system environment over lots of dies and operating conditions. Such a prediction leads to a reasonable production release qualification (PRQ) decision. This goal itself poses the greatest challenge for EV since, in practice, only a limited number of experiments is finished within the post-silicon validation time-frame. On top of that, complexity in EV is aggravated by the need to compensate for circuit non-idealities to reach the increasing speeds. This complexity translates to increased compensation and adaptation loops (Gu, 2012).

Similarly, functional verification of modern computing embedded systems has become a great challenge due to this increasing system complexity (Wile et al., 2005). It requires tens or hundreds of person-years and needs the computing power of thousands of workstations. Despite this tremendous effort, it is virtually impossible to detect and fix all bugs in the design before it is fabricated. Given the highly time-sensitive nature of the semiconductor industry, waiting for exhaustive tests is clearly a failing economic approach, even without considering the expense associated with EV (Goodenough and Aitken, 2010).

5.2.1 *Post-silicon tuning*

Modern nanometric transistor technologies exhibit larger die-to-die process variations. Different techniques based on statistical design exist to maximize the parametric yield for high-performance computing platforms. These techniques usually fall into two categories: design-time optimization and post-silicon tuning (Yao et al., 2009).

Design-time optimization techniques explore the design space at system-level and device-level to maximize the yield. However, accurate simulations for nonlinear circuits, which are dominant in digital CMOS circuits, are computationally very expensive given the complexity of the system involved, especially if interconnect high-frequency effects are considered. On the other hand, tunable elements are proposed to adjust the analog circuit performance (Huang and Lee, 2001; Miller et al., 2003). The adoption of circuit tuning, however, introduces great challenges at the design-time level. A tunable circuit may contain a large number of adjustable parameters, making extremely expensive to repeatedly run a large number of highly accurate simulations over all process variations and environmental corners to validate a given design during pre-silicon validation (Wang et al., 2016), making it necessary to perform tuning at post-silicon stage based on physical measurements. Therefore, post-silicon tuning in analog design has been widely adopted to confront the silicon process variation and other undesired effects (Rangel-Patiño et al., 2020).

During post-silicon electrical validation, we may observe a failure that requires debugging. Ideally, we root-cause the bug and fix it by redesigning. However, this approach is inefficient and costly. In practice, instead of redesigning the circuit, many link failures can be fixed by modifying the PHY tuning settings, which are usually embedded in the I/O links and can be digitally adjusted. They provide a way to reconfigure I/O links in post-silicon to cancel out undesired effects, including those of the system channels' variability. Typical PHY tuning settings include: parameters of an equalizer at the transmitter (Tx), Rx, or both; the clock and data recovery (CDR) circuit settings; the variable gain amplifier (VGA) parameters, and others. Finding the optimal PHY tuning settings that guarantee the bit error ratio (BER) required by an industrial specification is called PHY tuning. More specifically, PHY tuning aims at finding the combination of settings that maximizes the functional eye diagram and simultaneously passes the jitter tolerance (JTOL) measurements mask under the specified BER, considering the system variability caused by multiple factors, such as signal integrity phenomena, technology process variation, power supply noise, thermal effects, etc.

A typical post-silicon validation platform may have hundreds of combinations of just equalization parameter values. Sweeping all possible combinations of all PHY tuning settings across different PVT conditions is practically prohibitive for current TTM commitments (Gu, 2012). Typical industrial practices for PHY tuning in HSIO links are very time consuming since they are mostly based on exhaustive enumeration methods requiring massive lab measurements (Rangel-Patino *et al.*, 2016), resulting in an extremely high cost. This empirical approach heavily depends on the expert knowledge of the validation engineers.

5.2.2 *System margining*

System margin validation (SMV) (Rangel-Patino *et al.*, 2016) is a methodology for verifying the signal integrity of a high-performance computing platform and assessing how much margin is in the design relative to semiconductor technology processes and operating conditions, including voltage, temperature, frequency, humidity, and even component aging, among other factors. The intent of SMV is to ensure the silicon parts meet the industry specifications, and their operation is robust across frequency, voltage, temperature, and manufacturing process variations.

The fundamental process behind SMV consists of systematically adjusting the corner conditions under which the validation platform operates, then measure the Rx functional eye opening by using on-die design for test (DFT) features. The DFT circuitry is able to send data traffic through the device under test (DUT) link while monitoring reception errors. The next step is to sweep margin parameters to verify the system stability. Some examples of sweep margin parameters include: signal amplitude, timing parameters of the Rx circuit, I/O buffer impedance, and I/O buffer voltage references.

During SMV, the on-die DFT circuitry varies the voltage threshold and the sampling point relative to time (phase interpolation) while errors are checked at each margin parameter setting. The resulting pass/fail matrix yields the solution space in which the system can operate without failure, as illustrated in Fig. 5.1. The resulting graph, known as the functional eye diagram, resembles the classical time-domain eye diagram seen on a traditional oscilloscope, only that it provides more comprehensive data regarding reliability margins. The horizontal axis in Fig. 5.1 provides the timing margin (eye-width), while the vertical axis indicates the voltage margin (eye height). We aim at maximizing the area of this functional

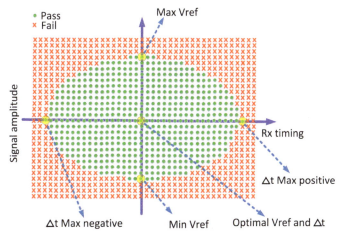

Fig. 5.1. Example of a system margin validation (SMV) functional eye diagram.

eye diagram by using equalization techniques to achieve optimum electrical margining of the corresponding HSIO interconnects.

5.2.3 Jitter tolerance testing

Typical jitter sources in HSIO links that contribute to the overall accumulated jitter at the Rx are illustrated in Fig. 5.2. A non-ideal clock synthesizer within the Tx block induces two kinds of jitter: random jitter (J_R) and deterministic jitter (J_D) (Wang *et al.*, 2008). ISI, losses, reflections, and crosstalk also degrade the signal integrity depending on the quality of the physical channel. At the Rx side, non-ideal equalizers and PLL-inherent phase noise on the clock and data recovery (CDR) circuitry will additionally induce jitter (Agilent Tech. *et al.*, 2003; Casper and O'Mahony, 2009; Li and Jitter, 2007; Li, 2009). Despite these conditions, the Rx must be capable of decoding the incoming signals for correct link communication.

One of the most common ways to measure the performance of a HSIO link is by measuring the BER through the HSIO link (Hong and Cheng, 2008). The fewer the errors measured, the better the performance of the link. BER measurement is typically used to characterize the Rx JTOL performance in order to determine compliance with the industry-standard specifications such as XAUI (10GEA Org, 2016), PCIe (PCI SIG Org, 2016), USB (USB Org, 2016), and SATA (SATA Org, 2016). Most HSIO protocols require a BER in the range from 10^{-12} to 10^{-15}.

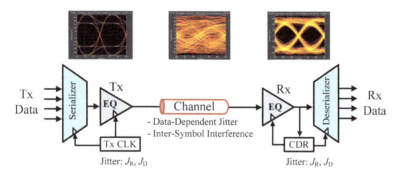

Fig. 5.2. Common jitter sources in a HSIO link.

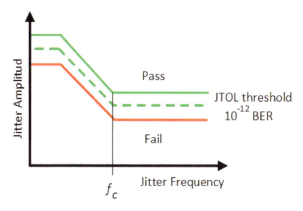

Fig. 5.3. Typical Rx jitter tolerance (JTOL) mask for testing.

The goal for Rx JTOL testing is to verify that it can operate at a target BER when operating under worst-case signaling conditions. JTOL testing consists of verifying that the measured Rx clock-recovery tolerance across frequencies is above the target threshold. If the measured JTOL curve is above the threshold curve, it indicates a passing result; thus, the Rx can tolerate some more jitter. On the other hand, if the measured curve is below the threshold curve, it indicates a failing result. Both scenarios are shown in Fig. 5.3 (Wang et al., 2008), where f_c is the corner frequency in the jitter mask that defines two different areas: the in-band and the out-band jitter. The single inflection point at f_c in the curve comes from the dominant pole in the Rx's CDR PLL loop filter implementation (Wang et al., 2008).

JTOL for HSIO links requires validating the specified BER following rigorous industrial standards. JTOL testing time depends on the link data rate and the required BER. Performing JTOL testing at a 10^{-12} BER target is very time-consuming since it takes several minutes to perform a single BER test (Kossel and Schmatz, 2004; Agilent Tech., 2005). In addition, JTOL is usually measured across a range of frequencies to determine the frequency response of the device (see Fig. 5.3). The measurement time for a complete set of JTOL values can take more than 70 h in a typical industrial post-silicon validation platform (Viveros-Wacher et al., 2018).

5.3 PHY Surrogate Modeling

Surrogate modeling is a scalable mathematical approach to generate parameterized approximations of the input–output relationship of a complex system within a design space of interest (Yelten *et al.*, 2012; Garitselov *et al.*, 2012). Surrogate models can be constructed using data from highly-accurate but computationally expensive models. They can also be built from complex but reliable laboratory measurements. Once the surrogate model is available, it provides fast approximations of the original complex system at new design points within the region of interest (Queipo *et al.*, 2005). They can further be conveniently exploited for efficient design optimization and statistical analysis.

Response surface approximations, neural network techniques, splines, and kriging are examples of core methods used to generate surrogates for simulation and optimization of complex problems (Mack *et al.*, 2007a). The major benefit of surrogate models is the ability to quickly obtain any number of additional function evaluations without resorting to more expensive numerical models or lab measurements.

Surrogate models can also be considered as metamodels (Yelten *et al.*, 2012) when they emulate the system behavior over a user-defined design space following a "black-box" approach. The problem of modeling in post-silicon validation can be mapped to a mathematical problem of function estimation in the presence of noisy data points.

In this section, we revisit several surrogate modeling techniques to efficiently approximate the PHY equalizer circuitry of the Rx of a HSIO link within an actual post-silicon validation platform. These surrogate modeling techniques include polynomial-based surrogate modeling (PSM), support vector machines (SVM), kriging, generalized regression neural networks (GRNN), and three-layer perceptron neural networks (3LP ANN). We generate these surrogate models by using reduced sets of training and testing data exploiting several design of experiments (DoE) techniques, namely: Box Behnken (BB); orthogonal arrays (OA), and Sobol. Finally, we evaluate the models' performance by comparing with actual measured responses on an industrial server HSIO link. We aim at assessing several surrogate modeling and DoE techniques to identify the best approach to address a highly relevant industrial challenge: optimizing a receiver equalizer in a measurement-based server HSIO validation platform, focusing on system margining performance.

5.3.1 Design of experiments

Design of experiments (DoE) is a set of statistical methods for allocating points in the design space with the objective to maximize the amount of useful information. When sampling the points, there is a clear trade-off between the number of points used and the amount of information that can be extracted from these points. The samples should be suitably spread apart in order to capture global trends in the design space (Koziel and Yang, 2011).

A large amount of training and testing data is usually needed to ensure acceptable surrogate model accuracy. However, generating large amounts of data is very expensive in the post-silicon validation environment. DoE can be exploited to reduce the size of these data sets, ensuring adequate parameter coverage (Mack *et al.*, 2007a). Then surrogate models can be constructed using the measurement data and provide fast approximations of the original system at new design points (Queipo *et al.*, 2005).

Here we use DoE to sample the complete design space in an efficient manner by selecting a relatively small number of base points. For instance, with k variables and three levels for each variable, a full factorial space search requires 3^k experimental runs. We employ three different DoE techniques to explore the desired solution space with a much smaller number of runs, namely: Box Behnken (BB), Orthogonal Arrays (OA), and Sobol. For each technique, we use five input variables that represent Rx PHY parameters ($n = 5$), which are settings used in three main Rx circuitry blocks (CTLE, VGA, and CDR), and then we retrieve the eye measurements from the system under test. The samples collected are subsequently used as the training and testing data required for surrogate modeling.

5.3.1.1 Box Behnken

Response surface modeling (RSM) is a collection of techniques commonly used to obtain the model of a system under study whose response is affected by a set of input variables (Montgomery, 2014). RSM helps to find first- and second-order effects of n variables on the measured outputs. First-order effects are easily obtained through two-level full or fractional designs, whereas second-order effects are usually captured by spherical designs such as the central composite design (Viveros-Wacher and Rayas-Sanchez, 2016) that requires up to five levels for each variable (the center points, plus factors ± 1 and $\pm \alpha$, where $\alpha = n^{1/2}$).

Box Behnken (BB) is a type of second order RSM design that combines factorial designs with balanced incomplete blocks designs (Wu and Hamada, 2000). This characteristic is particularly helpful for variables that are not able to take $n^{1/2}$ factorial values, such as digitally controlled variables, as in our system under test. In this manner, we use only three levels for each of the five variables, yielding a total number of $N = 46$ experiments defined by $2n(n-1) + c_p$, where c_p is the number of central points. We denote this DoE as BB.

5.3.1.2 Orthogonal arrays

Orthogonal arrays (OA) are experimental designs identified by $L_N(s^n)$, where N is the number of experimental runs, s is the number of states (or levels) for each variable and n is the number of variables (Chang et al., 2005). Their most important feature is that for each variable, all possible levels appear equally often. OA help to reduce the number of experiments while maintaining the ability to measure the effect of each variable on the output without the need to test all possible combinations.

When $s = 2$, the resulting OA allows for observing linear effects. By increasing the value of s, nonlinear effects can be assessed. We use an $L_{27}(3^5)$ OA in our work in order to capture nonlinear effects in the objective function by only running 27 experiments. We denote this DoE as OA27.

5.3.1.3 Sobol sequence

The most commonly used stochastic sampling algorithm is Monte Carlo. However, Monte Carlo sampling suffers from a number of disadvantages: it usually leads to unnecessary samples and it tends to generate clusters of points, leaving gaps in the solution space. One approach to overcome these issues is to use quasi-Monte Carlo methods such as low-discrepancy sequences (Cheng and Druzdzel, 2000); here, the measure of non-uniformity of a sequence of points is known as the discrepancy.

We select the Sobol (Sobol, 1967) low-discrepancy sequence as the third DoE option to sample the solution space. It is generated from a set of binary fractions u_i^j of length w bits, where $i = 1, 2, \ldots, w$ and $j = 1, 2, \ldots, d$ is the dimension. The Sobol sequence x_n^j in dimension j is generated by

$$x_n^j = b_1 u_1^j \oplus b_2 u_2^j \oplus \cdots \oplus b_w u_w^j, \quad (5.1)$$

where

$$n = \sum_{i=0}^{w} b_i 2^i, \quad (5.2)$$

and operator $\oplus$ denotes the bitwise XOR operation, and $b_i \in \{0, 1\}$.

The solution space is better explored as the number of samples increases, at the expense of increasing test time on the real system. Therefore, we use three different Sobol DoEs, denoted as Sobol50, Sobol100 and Sobol150, with 50, 100, and 150 samples, respectively.

5.3.2 *Polynomial-based surrogate modeling*

PSM is a straightforward technique to approximate computationally expensive models. It has been used to efficiently model microwave circuits in the frequency domain (Chávez-Hurtado and Rayas-Sánchez, 2016), as well as microwave structures subject to multi-physics variations (Chávez-Hurtado *et al.*, 2016). It can also be used for EM-based design optimization (Rayas-Sánchez *et al.*, 2017). PSM is implemented by exploiting the multinomial theorem, which allows the algorithm to raise a polynomial to an arbitrary power with all cross-terms and no redundancies. A polynomial function is used to represent the behavior of the response around a reference design. The order of the polynomial function is increased until generalization performance deteriorates at each frequency point. A particular advantage of this surrogate model technique relies on the fact that weighting factors are calculated in closed form, achieving a global minimum in the least-squares sense and exhibiting a very cheap computational cost. In this chapter, we employ the mathematical formulation of PSM implemented in Matlab that has been described in Chávez-Hurtado and Rayas-Sánchez (2016).

5.3.3 Generalized regression neural networks

GRNN is a special type of artificial neural network (ANN) that does not require an iterative training procedure (Mahouti et al., 2014). Moreover, the number of neurons in the hidden layers is equal to the number of learning samples (Specht, 1991). As the number of samples becomes sufficiently large, this technique exhibits fast learning and convergence to the optimal regression surface (Panda et al., 2014). GRNN uses a special kind of radial basis functions; a detailed formulation of GRNN is provided in Goulermas et al. (2007b). In Goulermas et al. (2007a), the authors demonstrate the functional estimation capability of GRNN. In this chapter, GRNN was implemented using available toolboxes in Matlab with default parameters.

5.3.4 Supported vector machines

SVM are trained by using the structural risk minimization principle, instead of the empirical risk minimization principle used by ANN models. This allows SVM models to exhibit a good trade-off between model complexity and generalization capability (Tokan and Günes, 2008). During training, the SVM technique solves a constrained quadratic optimization problem, finding a global optimum for the model parameters. The optimization problem is feasible due to the use of specific kernel functions, being the radial basis function the most employed kernel (Angiulli et al., 2007; Xia et al., 2006). A detail formulation of SVM is in Angiulli et al. (2007). Our implementation exploits the SVM regression available in Matlab, with Gaussian kernel functions and sequential minimal optimization solver (default).

5.3.5 Kriging

Kriging is a surrogate modeling technique originally developed to approximate deterministic data. It has proven to be very useful for tasks such as optimization (Jones et al., 1998), design space exploration, visualization, prototyping, and sensitivity analysis (Wang and Shan, 2006). A detailed mathematical description of kriging is given in Santner et al. (2003) and Forrester et al. (2008). The popularity of kriging has generated research in many areas, including several extensions of kriging to handle different problem settings, e.g., by adding gradient information in the prediction (Morris et al., 1993), or by adaptations to approximate stochastic simulations (Staum, 2009).

It is named after the pioneering work of D. G. Krige (a South African mining engineer), and was formally developed in Matheron (1963). Its popularity has been originated by the works in Sacks et al. (1989a, 1989b) in the context of the modeling and optimization of deterministic functions. The kriging method in its basic formulation estimates the value of a function at some unsampled location as the sum of two components: the linear model (e.g., polynomial trend) and a systematic departure representing low (large scale) and high frequency (small scale) variation components, respectively.

Kriging considers both the distance and the degree of variation between known data points when estimating values in unknown areas. A kriging estimate is a weighted linear combination of the known sample values around the point to be estimated. Applied properly, kriging allows deriving the weights that result in optimal and unbiased estimates. It attempts to minimize the error variance and aims at setting the mean of the prediction errors to zero so that there are no over- or under-estimates. A unique feature of kriging is that it provides an estimation of the error at each interpolated point, providing a measure of confidence in the modeled surface.

In this chapter, we use ordinary kriging (Queipo et al., 2005), which estimates a deterministic function f as

$$f(\boldsymbol{x}) = \mu + \varepsilon(\boldsymbol{x}), \tag{5.3}$$

$$E(\varepsilon) = 0, \tag{5.4}$$

$$\mathrm{cov}(\varepsilon(\boldsymbol{x}^i), \varepsilon(\boldsymbol{x}^j)) \neq 0 \; \forall i, j, \tag{5.5}$$

where μ is the mean of the response at base points $(\boldsymbol{x})$, and ε is the error with zero expected value, and with a correlation structure being a function of a generalized distance between the base points.

A possible correlation structure (Sacks et al., 1989a) is given by

$$\mathrm{cov}(\varepsilon(\boldsymbol{x}^i), \varepsilon(\boldsymbol{x}^j)) = \sigma^2 \boldsymbol{R}, \tag{5.6}$$

where σ^2 is the variance, and $\boldsymbol{R}$ is the correlation $N \times N$ matrix between the base points $\boldsymbol{x}^1, \boldsymbol{x}^2, \ldots, \boldsymbol{x}^N$,

$$\boldsymbol{R} = \begin{bmatrix} R(\boldsymbol{x}^1, \boldsymbol{x}^1) & R(\boldsymbol{x}^1, \boldsymbol{x}^2) & R(\boldsymbol{x}^1, \boldsymbol{x}^N) \\ R(\boldsymbol{x}^2, \boldsymbol{x}^1) & R(\boldsymbol{x}^2, \boldsymbol{x}^2) & R(\boldsymbol{x}^1, \boldsymbol{x}^N) \\ \vdots & \vdots & \vdots \\ R(\boldsymbol{x}^N, \boldsymbol{x}^1) & R(\boldsymbol{x}^N, \boldsymbol{x}^1) & R(\boldsymbol{x}^N, \boldsymbol{x}^N) \end{bmatrix}. \tag{5.7}$$

We use a Gaussian correlation function of the form

$$R(\boldsymbol{x}^i, \boldsymbol{x}^j) = \exp\left[-\sum_{k=1}^{N} \theta_k |x_k^i - x_k^j|^2\right], \tag{5.8}$$

where θ_k are unknown correlation parameters used to fit the model, while $\boldsymbol{x}_k^i$ and $\boldsymbol{x}_k^j$ are the kth components of the base points $\boldsymbol{x}^i$ and $\boldsymbol{x}^j$, and N denotes the number of dimensions in the set of design variables $\boldsymbol{x}$, σ identifies the standard deviation of the response at sampled design points.

The kriging-based surrogate model $\boldsymbol{R}_s$ is defined as

$$\boldsymbol{R}_s(\boldsymbol{x}) = [R_{s.1}(\boldsymbol{x}) \cdots R_{s.m}(\boldsymbol{x})]^{\mathrm{T}}. \tag{5.9}$$

The model estimates the responses at unsampled points by the kriging predictor (Sacks et al., 1989a) as

$$R_{s.j}(\boldsymbol{x}) = \overline{\mu}_j + \boldsymbol{r}^{\mathrm{T}}(\boldsymbol{x})\boldsymbol{R}^{-1}(\boldsymbol{f}_j - \boldsymbol{1}\overline{\mu}_j), \tag{5.10}$$

where $\boldsymbol{1}$ denotes an N-vector of ones, and

$$\boldsymbol{f}_j = [R_{sm.j}(\boldsymbol{x}^1) \cdots R_{sm.j}(\boldsymbol{x}^N)]^{\mathrm{T}}. \tag{5.11}$$

The bar above the variables in (5.10) denotes estimates, $\boldsymbol{r}$ identifies the correlation vector between the set of prediction points $\boldsymbol{x}$ and the base points,

$$\boldsymbol{r}^{\mathrm{T}}(\boldsymbol{x}) = [R(\boldsymbol{x},\ \boldsymbol{x}^1) \cdots R(\boldsymbol{x}, \boldsymbol{x}^N)]^{\mathrm{T}}. \tag{5.12}$$

The mean $\overline{\mu}_j$ is the estimated value of μ_j and can be calculated using

$$\overline{\mu}_j = \frac{\boldsymbol{1}^{\mathrm{T}}\boldsymbol{R}^{-1}\boldsymbol{f}_j}{\boldsymbol{1}^{\mathrm{T}}\boldsymbol{R}^{-1}\boldsymbol{1}}. \tag{5.13}$$

The unknown parameter θ_k, for the kriging model, can be estimated by maximizing the following likelihood function $g(\boldsymbol{R})$ given by Queipo *et al.* (2005)

$$g(\boldsymbol{R}) = -\frac{N}{2}\ln(\overline{\sigma}^2) - \frac{1}{2}\ln|\boldsymbol{R}|, \quad (5.14)$$

in which the variance

$$\overline{\sigma}_{j^2} = \frac{1}{N}(\boldsymbol{f}_j - \boldsymbol{1}\overline{\mu}_j)^{\mathrm{T}}\boldsymbol{R}^{-1}(\boldsymbol{f}_j - \boldsymbol{1}\overline{\mu}_j), \quad (5.15)$$

and $|\boldsymbol{R}|$ are both functions of θ_k.

5.3.6 *Artificial neural networks*

ANN are particularly suitable to approximate high-dimensional and highly nonlinear relationships, in contrast to more conventional methods such as numerical curve-fitting, empirical or analytical modeling, or response surface approximations (Vicario *et al.*, 2016). ANNs have been used in many areas of applications, including RF and microwave circuits (Zhang and Gupta, 2000), EM-based design optimization (Rayas-Sánchez, 2004), control process, telecommunications, biomedical, remote sensing, pattern recognition, and manufacturing, just to mention a few (Haykin, 1999). Recently, ANNs have been used for HSIO simulations, as in Bistola (2015) and Liu and Tsai (2015), by modeling the nonlinear relationships between channel parameters and system performance to speed up system simulations. In Goay and Goh (2017), authors proposed ANNs for eye diagram modeling based on simulations and they use an adaptive sampling method for data collection process.

Once trained, an ANN model can provide a fast way to perform a large number of I/O links and channel simulations that take into account the die-to-die process variations, board impedances, channel losses, add-in cards, end-point devices, and operating conditions (Beyene, 2007). ANN modeling involves two inter-related processes: (a) neural network model development — that includes the selection of representative training data, network topology, and training algorithms, and (b) neural model validation — the neural network model is tested and validated according to its generalization performance in a given region of interest. A large amount of training data is usually needed to ensure model accuracy, and this could be very expensive in the post-silicon validation environment. An alternative to reduce the dimension of the learning set is to properly select the learning base points by using DoE to ensure adequate design space parameter coverage (Mack et al., 2007b).

5.3.6.1 ANN topology

Multilayer perceptrons are feedforward networks widely used as the preferred ANN topology (Rayas-Sánchez, 2004). Since a three-layer perceptron (3LP) is, in principle, sufficient to achieve an arbitrarily defined accuracy (Hornik et al., 1989), we use a 3LP to implement our neuromodel, with n inputs (equal to the number of Rx equalization coefficients), h hidden neurons, and m outputs (number of system responses of interest), as shown in Fig. 5.4. The required complexity of the ANN, determined by h, depends on the generalization performance for a given set of training and testing data (Rayas-Sánchez, 2001). Following Rayas-Sánchez and Gutiérrez-Ayala (2006), we gradually increase h during training as a regularization scheme.

5.3.6.2 ANN modeling and training

Let $\boldsymbol{R}_\mathrm{f} \in \Re^m$ represent the actual electrical margining system response, denoted as a fine model response, which consists of the eye

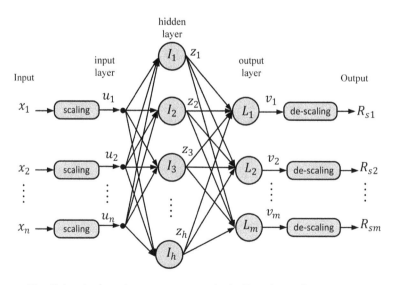

Fig. 5.4. A three-layer perceptron including the scaling process.

width $e_\mathrm{w} \in \Re$ and eye height $e_\mathrm{h} \in \Re$ of the measured eye diagram,

$$\boldsymbol{R}_\mathrm{f}(\boldsymbol{x},\boldsymbol{\psi},\boldsymbol{\delta}) = \begin{bmatrix} e_\mathrm{w}(\boldsymbol{x},\boldsymbol{\psi},\boldsymbol{\delta}) & e_\mathrm{h}(\boldsymbol{x},\boldsymbol{\psi},\boldsymbol{\delta}) \end{bmatrix}^\mathrm{T}. \qquad (5.16)$$

The electrical margining system response depends on the Rx equalization settings $\boldsymbol{x} \in \Re^n$, the operating conditions $\boldsymbol{\psi}$ (voltage and temperature), and the devices $\boldsymbol{\delta}$ connected to the system. The ANN is trained to find an optimal vector of weighting factors $\boldsymbol{w}$, such that the ANN response, denoted as $\boldsymbol{R}_\mathrm{s}$, is as close as possible to the fine model response $\boldsymbol{R}_\mathrm{f}$ for all $\boldsymbol{x}$, $\boldsymbol{\psi}$, $\boldsymbol{\delta}$ in the region of interest,

$$\boldsymbol{R}_\mathrm{s}(\boldsymbol{x},\boldsymbol{\psi},\boldsymbol{\delta},\boldsymbol{w}) \approx \boldsymbol{R}_\mathrm{f}(\boldsymbol{x},\boldsymbol{\psi},\boldsymbol{\delta}). \qquad (5.17)$$

The ANN main input–output relationship is denoted as

$$\boldsymbol{R}_\mathrm{s} = \boldsymbol{f}(\boldsymbol{x}). \qquad (5.18)$$

We aim to develop a fast and accurate ANN model for $\boldsymbol{f}$ by training the ANN with a set of measured learning data. The learning data are pairs of $(\boldsymbol{x}_L, \boldsymbol{t}_L)$, with $L = 1, 2, \ldots, l$, where $\boldsymbol{t}_L$ contains the desired outputs or targets (obtained from the measurements) for the ANN model at the $\boldsymbol{x}_L$ inputs, and l is the total number of learning samples. During training, we keep fixed the system at voltage/temperature (VT) nominal conditions and without changing the external device. Under these conditions, $\boldsymbol{\psi}$ and $\boldsymbol{\delta}$ remain constant. Therefore, the ANN model during training is treated as

$$\boldsymbol{R}_{\mathrm{s}L} = \boldsymbol{R}_\mathrm{s}(\boldsymbol{x}_L, \boldsymbol{w}), \qquad (5.19)$$

where $\boldsymbol{w}$ is the vector containing the ANN weighting factors. The ANN performance during training is evaluated by computing the difference between ANN outputs and the targets for all the learning samples,

$$\boldsymbol{E}_L(\boldsymbol{w}) = \boldsymbol{R}_{\mathrm{s}L}(\boldsymbol{x}_L, \boldsymbol{w}) - \boldsymbol{t}_L, \qquad (5.20)$$

where $\boldsymbol{E}_L$ is the learning error matrix.

Following Rayas-Sánchez and Gutiérrez-Ayala (2006), the problem of training the ANN is formulated as

$$\boldsymbol{w}^* = \arg\min_{\boldsymbol{w}} \|\boldsymbol{E}_L(\boldsymbol{w})\|_\mathrm{F}. \qquad (5.21)$$

To control the generalization performance while solving (5.21), we use T testing base points $(\boldsymbol{x}_T)$ not used during training. The scalar learning and testing errors are given by

$$\varepsilon_L = \|\boldsymbol{R}_{\mathrm{s}L}(\boldsymbol{x}_L, \boldsymbol{w}) - \boldsymbol{R}_{\mathrm{f}L}\|_{\mathrm{F}}, \tag{5.22}$$

$$\varepsilon_T = \|\boldsymbol{R}_{\mathrm{s}T}(\boldsymbol{x}_T, \boldsymbol{w}) - \boldsymbol{R}_{\mathrm{f}T}\|_{\mathrm{F}}, \tag{5.23}$$

where $\boldsymbol{R}_{\mathrm{f}T}$ and $\boldsymbol{R}_{\mathrm{s}T}$ are the output matrices of the fine model and ANN model, respectively, at the T testing base points, and $\boldsymbol{R}_{\mathrm{f}L}$ is the fine model response at the L learning base points.

In this chapter, the 3LP is trained by using the Bayesian regularization (MacKay, 1992) method available in Matlab Neural Network Toolbox. The algorithm for training the ANN is shown in Fig. 5.5. We first define the learning ratio to split the pairs of inputs and targets into the learning and testing data sets. The learning process often begins by initializing the ANN weights with arbitrary values using a random number generator (Thimm and Ra, 1997), however, in our case, we use a decoupling network process with an initial set of inputs and outputs to compute initial weighting factors $\boldsymbol{w}^{(0)}$ and corresponding initial error $\varepsilon_{T\mathrm{old}}$. Then, we start training the 3LP with just one hidden neuron ($h = 1$), and calculate the corresponding learning and testing errors. We keep increasing the complexity of the ANN (h) until the current testing error is larger than the previous one, and the current learning error is smaller than the current testing error, as in Rayas-Sánchez and Gutiérrez-Ayala (2006) (see Fig. 5.5).

5.3.7 PHY surrogate modeling results and comparisons

The system under test is a server post-silicon validation platform, comprised mainly of a CPU and a platform controller hub (PCH). The PCH is a family of Intel microchips, which integrates a range of common I/O blocks required in many market segments and these include USB, PCIe, SATA, SD/SDIO/MMC, and Gigabit Ethernet MAC, as well as general embedded-interfaces such as SPI, I2C, UART, and GPIO. The PCH also provides control data paths with the Intel CPU through direct media interface (DMI), as shown in Fig. 5.6. The automation mechanism to read the Rx eye diagram parameters (functional eye width and eye height) is shown in Fig. 5.7.

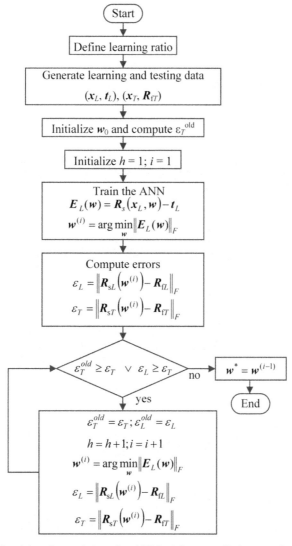

Fig. 5.5. Algorithm for training the ANN with controlled generalization performance.

Within the PCH, the methodology was tested on two different HSIO links: USB3 Super-speed Gen 1 and SATA Gen 3.

The measurement system is based on the SMV process as described in Section 5.2.2, which is a methodology to verify the signal integrity of a circuit board and assess how much margin is in

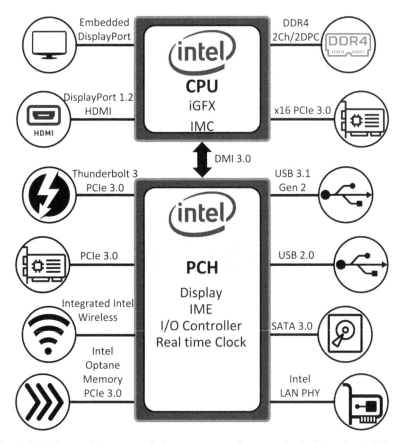

Fig. 5.6. The architecture of the system under test, including the PCH and CPU, as well as the USB, SATA, and PCIe HSIO links.

the design relative to silicon characteristics and processes. The SMV methodology consists of measuring the Rx functional eye width and eye height by using on-die DFT features until the eye opening has been shrunk to a point where the Rx detects errors, or the system fails (Rangel-Patiño et al., 2017).

We employ three different DoE techniques as described in Section 5.3.1 to explore the desired solution space with a reduced number of test cases, and in the case of Sobol, we use three different Sobol DoEs, denoted as Sobol50, Sobol100, and Sobol150, with 50, 100, and 150 samples, respectively. Also, as described in Section 5.3.1, we use five input variables that represent Rx PHY parameters, and then we

Surrogate-Based Modeling and Design Optimization Techniques 175

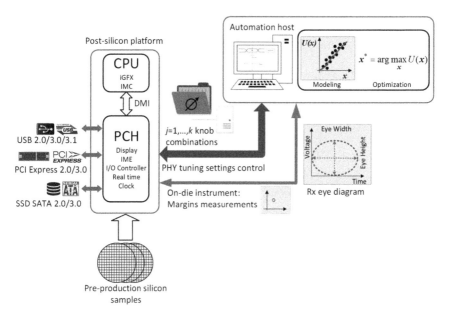

Fig. 5.7. HSIO server post-silicon hardware configuration for Rx modeling.

retrieve the eye measurements from the system under test. The samples are subsequently used as the training and testing data required for surrogate modeling.

System margining testing is very time-consuming when running many test cases for PHY tuning. A single test case with three repetitions can take up 20 min, and then running a Sobol150, for example, can take up 50 h of testing for a single VT corner. The objective of comparing several DoEs is to find a suitable sampling strategy that provides adequate model performance with the least amount of testing time.

We first evaluate the accuracy of the obtained surrogate models by comparing them with actual measured responses from a HSIO link. The average relative error ε for eye height and eye width at testing base points $(\boldsymbol{x}_\mathrm{T})$ not seen during training is calculated as

$$\varepsilon = \frac{\|\boldsymbol{R}_\mathrm{f}(\boldsymbol{x}_\mathrm{T}) - \boldsymbol{R}_\mathrm{s}(\boldsymbol{x}_\mathrm{T})\|_2}{\|\boldsymbol{R}_\mathrm{f}(\boldsymbol{x}_\mathrm{T})\|_2}. \tag{5.24}$$

In all the modeling cases, we use randomly distributed testing base points not seen during training to measure the generalization performance. Norms in (5.24) are calculated using these testing base points.

5.3.7.1 *SATA Gen3 HSIO link PHY modeling*

Tables 5.1 and 5.2 show a summary of the generalization performance, obtained from (5.24), for the eye height and eye width, comparing the five surrogate models using the five DoEs: (a) OAL27, (b) BB, (c) Sobol50, (d) Sobol100, and (e) Sobol150. It can be concluded from the results given in these tables that, overall, the PSM technique yields the lowest testing average relative errors for both eye measurements when using Sobol150, which is the DoE technique yielding the best generalization performance.

Figures 5.8 and 5.9 show the absolute error at all testing samples for both eye height and eye width for the five surrogate models using the five DoE techniques. When using OAL27 DoE, it is observed that the SVM model shows the best accuracy for eye height (Fig. 5.8(a), while the 3LP-ANN model provides the best accuracy for eye width (Fig. 5.9(a)). When using the BB DoE, the PSM shows the best

Table 5.1. Surrogate models generalization error for eye height (SATA).

Model	BB(%)	OAL27(%)	Sobol50(%)	Sobol100(%)	Sobol150(%)
PSM	2.77	8.90	2.68	2.05	0.42
SVM	6.35	6.70	6.69	6.79	6.77
Kriging	3.10	7.01	2.74	1.89	1.45
GRNN	7.47	9.27	2.86	2.15	1.58
3LPANN	3.33	7.14	2.49	1.96	1.15

Table 5.2. Surrogate models generalization error for eye width (SATA).

Model	BB(%)	OAL27(%)	Sobol50(%)	Sobol100(%)	Sobol150(%)
PSM	1.66	2.79	1.37	1.23	0.11
SVM	3.27	4.32	3.43	3.48	3.49
Kriging	2.71	5.36	1.23	1.28	0.55
GRNN	3.82	4.33	1.14	1.04	0.53
3LPANN	2.96	2.59	1.71	1.27	0.56

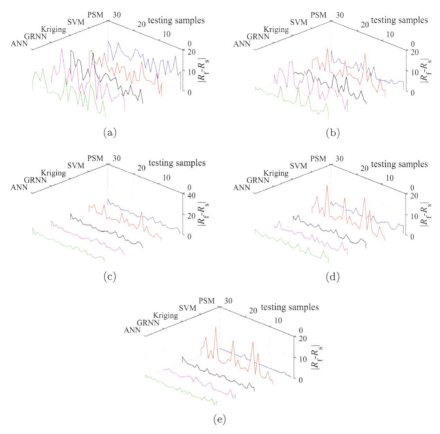

Fig. 5.8. Surrogate models absolute testing errors for eye height, using: (a) OAL27, (b) BB, (c) Sobol50, (d) Sobol100, and (e) Sobol150.

accuracy for both eye height and width (Figs. 5.8(b) and 5.9(b)). 3LP-ANN and GRNN models exhibit the best performance when using Sobol50 DoE (Figs. 5.8(c) and 5.9(c)). When the surrogate models are developed using Sobol100 DoE, the best performance is achieved by kriging and GRNN models (Figs. 5.8(d) and 5.9(d)). Finally, it is observed that the PSM technique with Sobol150 DoE yields the best generalization performance (Figs. 5.8(e) and 5.9(e)), with the lowest average relative testing errors, as confirmed by the results provided in Tables 5.1 and 5.2.

The actual responses from the best surrogate model found are compared in Figs. 5.10 and 5.11 with the fine model response (real

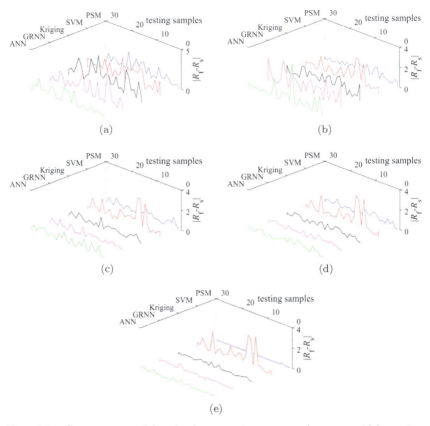

Fig. 5.9. Surrogate models absolute testing errors for eye width, using: (a) OAL27, (b) BB, (c) Sobol50, (d) Sobol100, and (e) Sobol150.

measurements), at the same 30 randomly distributed testing base points. It is observed that the PSM model effectively simulates the fine model.

Figure 5.12 shows the generalization error of the already trained neural model (at w^*), comparing the different DoEs for SATA. It is seen that the best performance is achieved with Sobol150. The three Sobol cases provide the best generalization performance, as confirmed in Fig. 5.12. However, Sobol50 is able to achieve acceptable accuracy with only 50 samples.

Figure 5.13 shows the learning performance of the neural training algorithm for SATA. The best performance is achieved with $h = 3$ for the eye width ANN, achieving a maximum relative learning error

Surrogate-Based Modeling and Design Optimization Techniques   179

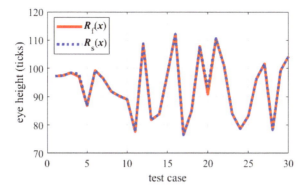

Fig. 5.10. Comparison between fine model responses and polynomial surrogate model responses at testing base points for the eye height.

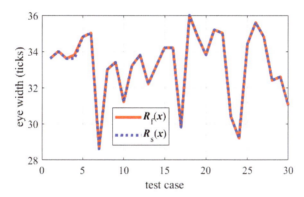

Fig. 5.11. Comparison between fine model responses and polynomial surrogate model responses at testing base points for the eye width.

of 3.65%, and 7.63% for the relative testing error. For the eye height ANN, the best performance is achieved with $h = 4$, yielding 7.98% of learning error and 6.75% of testing error. Thus, the metamodels are able to reach above 90% of accuracy for these initial sampling points.

The neural model response at $\boldsymbol{w}^*$ and $h = 3$ for e_w and $h = 4$ for e_h from Sobol50 is compared in Figs. 5.14(a) and 5.14(b), respectively, with the fine model (real measurements), by using 30 testing base points not used during training, in order to test the generalization performance. It is observed that the neural model effectively simulates the actual physical measurements with a total relative error

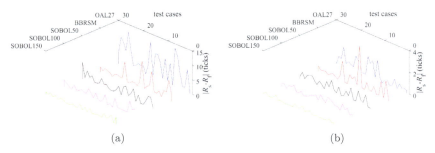

Fig. 5.12. Comparison of SATA neural model generalization performance for different DoE techniques: (a) eye height error, and (b) eye width error.

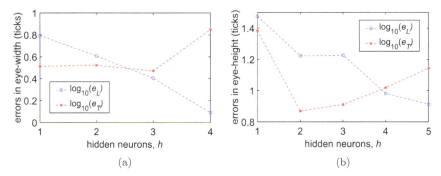

Fig. 5.13. Learning and testing errors during SATA neural training using Sobol50 for: (a) eye width, and (b) eye height.

of 1.7% for the e_w response and 2.5% for the e_h response. In other words, the ANN metamodel is able to predict margins with up to 95% of accuracy when using equalization values not used during the ANN training.

5.3.7.2 USB3.1 Gen1 HSIO link PHY modeling

Tables 5.3 and 5.4 show a summary of the generalization performance for the eye height and eye width, comparing the five surrogate models using four DoEs: (a) OAL27, (b) BB, (c) Sobol50, and (d) Sobol100. The results provided in these tables indicate, overall, the PSM and the 3LP ANN techniques yield the lowest testing average relative errors for both eye measurements when using Sobol100, which is the DoE technique yielding the best generalization performance at the

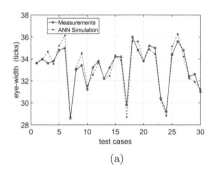 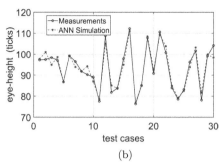

Fig. 5.14. Neural model generalization performance using Sobol50, for: (a) SATA eye width, and (b) SATA eye height.

Table 5.3. Surrogate models generalization error for eye height (USB).

Model	BB(%)	OAL27(%)	Sobol50(%)	Sobol100(%)
PSM	6.07	29.60	4.73	5.70
SVM	5.47	42.18	5.09	4.65
Kriging	5.91	41.77	5.42	4.63
GRNN	14.77	45.44	14.22	10.13
3LPANN	7.32	40.77	5.74	6.19

Table 5.4. Surrogate models generalization error for eye width (USB).

Model	BB(%)	OAL27(%)	Sobol50(%)	Sobol100(%)
PSM	11.61	21.56	18.97	8.24
SVM	20.15	21.65	21.00	19.04
Kriging	16.48	21.06	15.83	13.61
GRNN	20.46	26.79	22.06	18.04
3LPANN	12.28	19.13	6.73	7.35

expense of a relatively high number of measurements on the real system.

Testing the Rx functional eye diagram on a USB3.1 HSIO link is also a time-consuming task, taking around 38 min for a single test case. Therefore, collecting data to build a model from a Sobol100 or BB can take up 76 h, which is excessive in the post-silicon validation

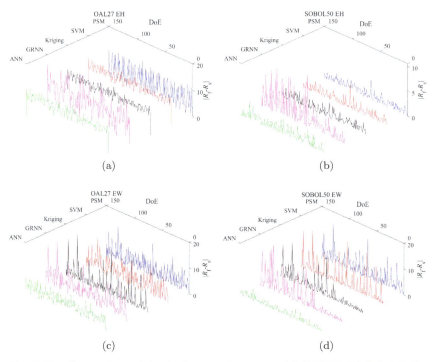

Fig. 5.15. Surrogate models absolute testing errors: (a) OAL27, (b) Sobol50, for eye height, (c) OAL27, and (d) Sobol50, for eye width.

environment. On the other hand, Tables 5.3 and 5.4 indicate that Sobol50 and OAL27 can yield good-enough surrogate models.

Figure 5.15 shows the absolute error at all testing samples for both eye height and eye width for the five surrogate models using OAL27 and Sobol50 DoE techniques. When using OAL27 DoE, it is observed that the PSM model shows good accuracy (Figs. 5.15(a) and 5.15(c)), while the 3LPANN model with Sobol50 also provides a good accuracy (Figs. 5.15(b) and 5.15(d)).

For the neural model response, similar results to those for SATA were obtained for the case of USB3 Super-speed Gen 1, where we use ten input variables ($n = 10$) that represent the corresponding Rx equalization coefficients, which again are settings used in the three main Rx circuitry blocks. Figure 5.16 shows the final results. It is seen that for USB, the resultant neural model also effectively simulates the fine model (physical platform), finding a total relative

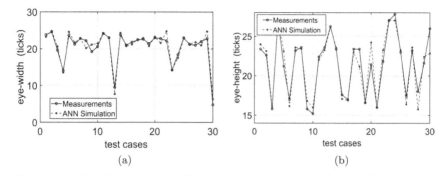

Fig. 5.16. Neural model generalization performance using Sobol50 for: (a) USB eye width, and (b) USB eye height.

error of 6.7% for the e_w response, as shown in Fig. 5.16(a), and a 5.7% relative error for the e_h response, as shown in Fig. 5.16(b). This metamodel performance was achieved also using a Sobol50 DoE.

5.4 Receiver Equalization Surrogate-Based Optimization

A direct optimization algorithm is now applied using the best surrogate model to find the optimal parameters of the PHY tuning settings for system margining. We are looking to maximize the corresponding eye diagram by maximizing a suitable objective function (Rangel-Patiño et al., 2017). Once a set of optimal parameters is found, it is tested on the actual electrical margining system to corroborate the resultant eye diagram area.

5.4.1 Objective function for system margining

Let $\boldsymbol{R}_m \in \Re^2$ denote the electrical margining system response, which consists of the eye width e_w and eye height e_h of the functional eye diagram,

$$\boldsymbol{R}_m = \boldsymbol{R}_m(\boldsymbol{x}, \boldsymbol{\psi}, \boldsymbol{\delta}) = \begin{bmatrix} e_w(\boldsymbol{x}, \boldsymbol{\psi}, \boldsymbol{\delta}) \\ e_h(\boldsymbol{x}, \boldsymbol{\psi}, \boldsymbol{\delta}) \end{bmatrix}. \quad (5.25)$$

This electrical margining system response depends on the PHY tuning settings $\boldsymbol{x}$, also known as equalization (EQ) coefficients, the operating conditions ψ (voltage and temperature), and the devices δ (silicon skew and external devices).

The functional eye diagram area is defined by

$$e_a = (e_{\mathrm{w}})(e_{\mathrm{h}}), \tag{5.26}$$

where $e_a \in \Re$ is the eye area, $e_{\mathrm{w}} \in \Re$ and $e_{\mathrm{h}} \in \Re$ are the eye width and height, respectively, obtained from measured parameters,

$$e_{\mathrm{w}}(\boldsymbol{x}, \boldsymbol{\psi}, \boldsymbol{\delta}) = e_{\mathrm{wr}}(\boldsymbol{x}, \boldsymbol{\psi}, \boldsymbol{\delta}) + e_{\mathrm{wl}}(\boldsymbol{x}, \boldsymbol{\psi}, \boldsymbol{\delta}), \tag{5.27}$$

$$e_{\mathrm{h}}(\boldsymbol{x}, \boldsymbol{\psi}, \boldsymbol{\delta}) = e_{\mathrm{hh}}(\boldsymbol{x}, \boldsymbol{\psi}, \boldsymbol{\delta}) + e_{\mathrm{hl}}(\boldsymbol{x}, \boldsymbol{\psi}, \boldsymbol{\delta}), \tag{5.28}$$

where $e_{\mathrm{wr}} \in \Re$ and $e_{\mathrm{wl}} \in \Re$ are the eye width-right and eye width-left measured parameters, respectively, and $e_{\mathrm{hh}} \in \Re$ and $e_{\mathrm{hl}} \in \Re$ are the eye height-high and eye height-low measured parameters, respectively, as illustrated in Fig. 5.17(a).

Since the objective is to maximize the area of the functional eye diagram, the initial objective function is given by

$$u(\boldsymbol{x}) = [e_{\mathrm{w}}(\boldsymbol{x}, \boldsymbol{\psi}, \boldsymbol{\delta})][e_{\mathrm{h}}(\boldsymbol{x}, \boldsymbol{\psi}, \boldsymbol{\delta})]. \tag{5.29}$$

Based on the operating conditions and devices, the eye diagram can be decentered with respect to the eye-width, eye-height, or both. A well centered Rx eye diagram is required to have a proper sampling on the CDR. The better Rx data is aligned, the easier the

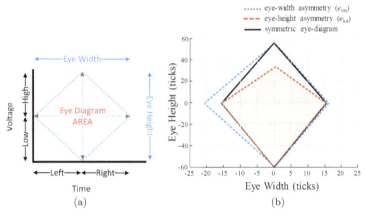

Fig. 5.17. Eye diagram: (a) graphical representation of the objective function based on the functional eye diagram area, and (b) asymmetries of the functional eye diagram.

phase interpolator circuitry will track for edges on the recovered data. Therefore, the objective function must consider the asymmetries of the eye diagram, as illustrated in Fig. 5.17(b).

Let e_{wa} and e_{ha} be the eye-width asymmetry and eye-height asymmetry, respectively. They are defined as

$$e_{wa}(x, \psi, \delta) = |e_{wr}(x, \psi, \delta) - e_{wl}(x, \psi, \delta)|, \quad (5.30)$$

$$e_{ha}(x, \psi, \delta) = |e_{hh}(x, \psi, \delta) - e_{hl}(x, \psi, \delta)|. \quad (5.31)$$

The area of the eye diagram and the asymmetries should be scaled by weighting factors $w_1, w_2, w_3 \in \Re$ such that they become comparable. The values of these weighting factors depend on the operating conditions and devices, and they can be selected by using initial e_w and e_h measurements.

Therefore, the objective function to be maximized is now defined as

$$u(x) = w_1[e_w(x, \psi, \delta)][e_h(x, \psi, \delta)] - w_2[e_{wa}(x, \psi, \delta)]$$
$$- w_3[e_{ha}(x, \psi, \delta)], \quad (5.32)$$

with w_1, w_2, and w_3 calculated from

$$w_1 = \frac{3}{\frac{1}{n}\sum_{i=1}^{n} u1_i(x)}, \quad (5.33)$$

$$w_2 = \frac{1}{\frac{1}{n}\sum_{i=1}^{n} u2_i(x)}, \quad (5.34)$$

$$w_3 = \frac{1}{\frac{1}{n}\sum_{i=1}^{n} u3_i(x)}, \quad (5.35)$$

where $u1$, $u2$, and $u3$ are vectors containing initial e_w and e_h measurements.

From all the above, the optimization problem for system margining is

$$x^* = \arg\max_{x} u(x), \quad (5.36)$$

with $u(x)$ defined by (5.32).

5.4.2 Surrogate-based optimization for system margining

An SBO algorithm is applied to the selected surrogate model to find the optimal parameters for the PHY tuning Rx settings. We are looking to maximize the corresponding eye diagram by solving (5.36) using the Nelder-Mead simplex-based method (Lagarias *et al.*, 1998). Notice that solving (5.36) is computationally very efficient since $u(\boldsymbol{x})$ in (5.32) is evaluated from the already available surrogate. During optimization, both ψ and δ are kept fixed.

We validate the SBO results by measuring the USB3.1 link Rx inner eye height/width at $\boldsymbol{x}^*$ on the real validation platform with a commercial USB device, and compare the SBO results with the exhaustive enumeration method (see Section 5.2). The results shown in Fig. 5.18, indicate that our approach yields better results in terms of eye diagram area and development time than the exhaustive method. Differences in the eye diagrams between PSM with OAL27 and 3LPANN with Sobol50 are not very significant. However, there is almost 50% of time savings (around 16 hours) by using OAL27

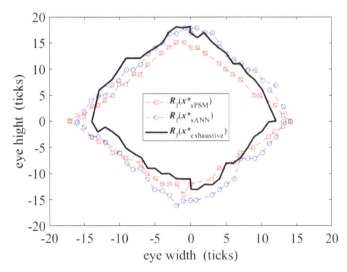

Fig. 5.18. Comparison between the USB3.1 link fine model responses after SBO using PSM with OAL27, and 3LPANN with Sobol50 surrogate models vs. exhaustive enumeration method.

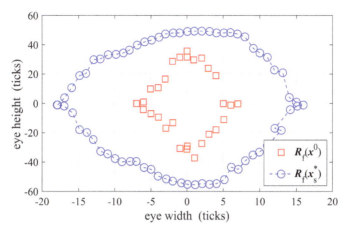

Fig. 5.19. Comparison between the SATA link fine model responses before and after surrogate-based optimization (square and circle marks, respectively).

as compared against Sobol50. Therefore, using PSM modeling technique with OAL27 as the DoE approach yields an efficient coarse model of the system.

The same procedure is applied to obtain the optimal PHY tuning Rx equalizer settings for SATA. We validate the SBO results by measuring the SATA link Rx inner eye height/width at x^* on the real validation platform with a commercial SATA device. The results shown in Fig. 5.19, indicate an improvement of 400% on eye diagram area as compared to the initial PHY tuning settings, demonstrating the high effectiveness of our approach.

The methodology outlined in this section can be applied to any high-performance computer server that has a similar receiver circuitry with adjustable EQ parameters.

5.4.3 Objective function for system margining and jitter tolerance

Here, we describe a holistic optimization approach that merges system margining and jitter tolerance measurements to optimize the Rx analog circuitry during industrial post-silicon validation. The methodology concurrently optimizes Rx system margins and JTOL, by defining an objective function that combines both types of measurements, and by using a kriging surrogate-based modeling approach to efficiently perform optimization. The method can deliver optimal system margins and guarantee jitter tolerance compliance while substantially decreasing the typical post-silicon validation time.

The optimization problem is now modified such that the optimal set of EQ coefficients simultaneously maximizes the eye diagram and exceeds the JTOL mask.

The JTOL system response is denoted as vector $\boldsymbol{R}_\text{J}$ and consists of the measurements of the sinusoidal jitter amplitude $\boldsymbol{S}_\text{JA}$ over a frequency range of interest (see Fig. 5.3),

$$\boldsymbol{R}_\text{J} = \boldsymbol{R}_\text{J}(\boldsymbol{x}, \boldsymbol{\psi}) = \boldsymbol{S}_\text{JA}(\boldsymbol{x}, \boldsymbol{\psi}). \tag{5.37}$$

The new optimization problem can be defined through a constrained formulation,

$$\boldsymbol{x}^* = \arg\max_{\boldsymbol{x}} u(\boldsymbol{x}) \quad \text{subject to } \boldsymbol{g}(\boldsymbol{x}) \leq \boldsymbol{0}, \tag{5.38}$$

with $u(\boldsymbol{x})$ defined by (5.32) and

$$\boldsymbol{g}(\boldsymbol{x}) = \boldsymbol{S}_\text{JAspec} - \boldsymbol{S}_\text{JA}, \tag{5.39}$$

where $\boldsymbol{S}_\text{JAspec}$ is the JTOL specification mask in the frequency range of interest.

A more convenient unconstrained formulation can be defined by adding a penalty term, as

$$U(\boldsymbol{x}) = u(\boldsymbol{x}) - r_0^g \|\boldsymbol{G}(\boldsymbol{x})\|_2^2, \tag{5.40}$$

where $\boldsymbol{G}(\boldsymbol{x})$ is the JTOL penalty function defined as

$$\boldsymbol{G} = \max\{\boldsymbol{0}, \boldsymbol{g}(\boldsymbol{x})\}. \tag{5.41}$$

The optimal solution depends on the value of $r_0^g \in \Re$, which is a penalty coefficient. We define r_0^g as

$$r_0^g = \frac{|u(\boldsymbol{x}_0)|}{\|\boldsymbol{g}(\boldsymbol{x}_0)\|_2^2}, \tag{5.42}$$

where $\boldsymbol{x}_0$ is the starting point for optimization.

In summary, the holistic objective function to optimize system margining and meet the JTOL mask is

$$\boldsymbol{x}^* = \arg\max_{\boldsymbol{x}} U(\boldsymbol{x}), \tag{5.43}$$

with

$$\begin{aligned}U(\boldsymbol{x}) = {} & w_1 [e_\text{w}(\boldsymbol{x}, \boldsymbol{\psi}, \boldsymbol{\delta})][e_\text{h}(\boldsymbol{x}, \boldsymbol{\psi}, \boldsymbol{\delta})] - w_2 [e_\text{wa}(\boldsymbol{x}, \boldsymbol{\psi}, \boldsymbol{\delta})] \\ & - w_3 [e_\text{ha}(\boldsymbol{x}, \boldsymbol{\psi}, \boldsymbol{\delta})] - r_0^g \|\boldsymbol{G}(\boldsymbol{x})\|_2^2.\end{aligned} \tag{5.44}$$

5.4.4 Surrogate-based optimization for system margining and jitter tolerance

In order to solve (5.43), we follow an SBO optimization strategy. As the underlying modeling technique, we select kriging (Sacks et al., 1989b) as described in Section 5.3.5. Our implementation exploits the Matlab kriging toolbox DACE (Lophaven et al., 2002). To enhance the efficiency of our approach, we apply a DoE for selecting fitting points. For optimization, we use the DACE toolbox that employs the *fmincon* optimization routine from the Matlab optimization toolbox.

The proposed SBO method was applied on a single random unit for each interface, at nominal voltage and temperature conditions. However, the general validity of the obtained optimal parameter values was later verified following the traditional coverage in analog post-silicon validation, that includes a large volume of silicon units considering process variations as well as full-range voltage and temperature conditions.

As mentioned before, the complexity of HSIO buses used on Intel server platforms are exacerbated by the interaction of channel components characteristics, such as packages, PCB, input/output density, connectors, cables and devices, as well as its intrinsic elements such as insertion loss (IL), signal-to-noise ratio (SNR), high volume manufacturing (HVM) variations, temperature and humidity impact on IL, etc. Several channel optimizations, including flexible routing (Lopez Miralrio et al., 2013), hybrid PCB stack-up (Mendez Ruiz et al., 2013), crosstalk cancellation (Ye et al., 2012), and impedance mismatch (Ye et al., 2011), are utilized to mitigate the aforementioned complexity. Despite the use of these techniques during platform design, the Rx of each interface still needs to be tuned for optimal performance during post-silicon validation time frame.

5.4.4.1 Test case 1: USB3

In the case of USB3, the channel topology is comprised of the Tx driver, the Tx based board transmission lines (TL), several via transitions, an I/O card connector, an internal cable that attaches a daughter card, followed by an external cable at which another connector is attached at the other end for the Rx I/O card, followed by another set of TL, and DC blocking capacitors at the Rx side of the device.

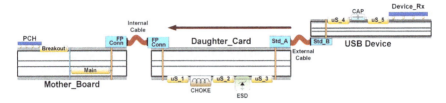

Fig. 5.20. USB3 Rx channel topology.

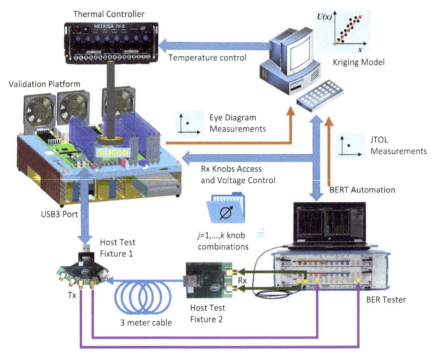

Fig. 5.21. The holistic methodology test setup for USB3 system margining and JTOL optimization.

Its simplified topology is illustrated in Fig. 5.20. The bandwidth limitations and inherent non-idealities of this system essentially arise from a large number of interconnects. Hence, we are looking to optimize the Rx equalization coefficients as a way to compensate for the channel limitations.

Figure 5.21 illustrates the test setup specific for USB3. We stress the Rx with a BER tester, sending a compliant pattern including all jitter impairments as per specification. The host computer is capable

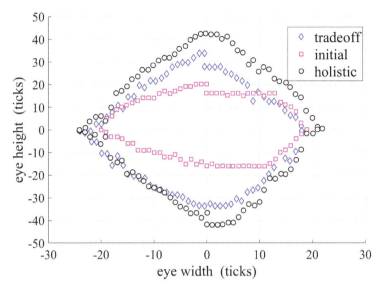

Fig. 5.22. USB3 eye width versus eye height results: comparing the proposed methodology against the initial design and the trade-off approach.

of modifying Rx EQ coefficients and DFT circuitry of the DUT as well as sending commands to the BER tester to sweep the injected jitter amplitude and frequencies. Then, we measure system margins and JTOL and record results for each set of Rx EQ coefficients.

Following the surrogate-based optimization methodology described in Sections 5.3.4 and 5.4.4, an optimal set of EQ coefficient values was found. The obtained set of values were verified by measuring the Rx inner eye height/width as well as JTOL using a commercial device. The EQ settings obtained with our method show an improvement of 125% on eye diagram area as compared to the initial EQ settings, and a 32% improvement as compared with the traditional (trade-off) approach, as shown in Fig. 5.22. Similarly, the JTOL results show a substantial improvement with margins well above the specification limit template, as seen in Fig. 5.23. The efficiency of this approach was also demonstrated by a significant time reduction on post-silicon validation: while the traditional process requires days for a complete empirical "optimization", the method proposed here can be completed in a few hours.

192   *Surrogate Modeling for High-Frequency Design*

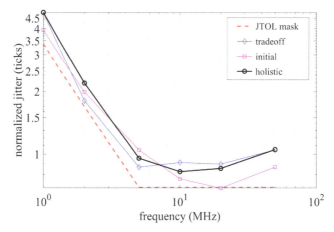

Fig. 5.23. USB3 JTOL testing results: comparing the proposed methodology against the initial design and the trade-off approach.

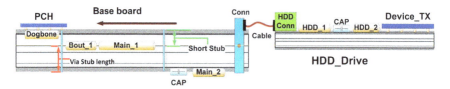

Fig. 5.24. SATA3 Rx channel topology.

5.4.4.2 *Test case 2: SATA3*

A similar scenario to the USB3 topology is found in the SATA3 channel, which also includes board interconnects, several via transitions and I/O card connectors, however, a 1 m SATA cable is used to connect the base board to the device I/O card (HDD drive), as illustrated in Fig. 5.24. Figure 5.21 also applies in general to the test setup used for SATA3, with the exception of replacing the respective test fixtures and switching the 3 m cable for a SATA3 compliance interconnect channel. The eye diagram area measured when using the EQ coefficients obtained through our holistic methodology show a 182% improvement against the initial values, as depicted in Fig. 5.25.

The trade-off approach derives a slightly larger eye area than the one obtained with our proposal. However, the JTOL results from the trade-off approach fall below the spec mask at 33 MHz, as seen in Fig. 5.26, rendering a compliance failure. Thus, it is clear in the

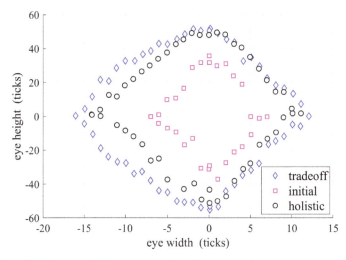

Fig. 5.25. SATA eye width versus eye height: comparing the proposed methodology against the initial design and the trade-off approach.

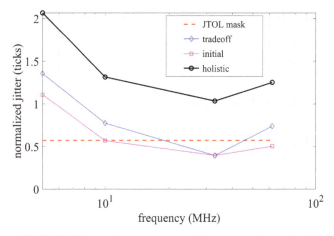

Fig. 5.26. SATA JTOL results: comparing the proposed methodology against the initial design and the trade-off approach.

SATA3 case that with our holistic approach, both the eye diagram and the JTOL margins are optimized. Furthermore, the execution following our method took less than 30% of the time required for the trade-off approach to reach a passing solution for both types of measurements.

5.4.4.3 *Test case 3: PCIe*

Figure 5.27 shows the PCIe channel topology implemented. As in the previous test cases, the PCIe topology includes the Tx driver, the base board interconnects and via transitions. It also includes a slot connector and an add-in card in which interconnects and other internal devices are found. In the PCIe test setup, a compliance load board is used to connect the measuring instruments, as seen in Fig. 5.27, instead of the test fixtures and cables shown in Fig. 5.21 for USB3 and SATA3.

Results from the PCIe test case provide a clear example of the role of asymmetries in the objective function discussed in Section 5.4.1. Figure 5.28 shows the PCIe functional eye diagram results, where the smallest area is obtained with the initial Rx EQ coefficients. The area obtained with our holistic approach is 14% larger than the area measured with the initial Rx EQ settings. The center of both of these eyes is located near the 0-tick value in both axes, thus the width and height asymmetry values are low. The area obtained from the trade-off approach is the largest from the three eyes, however, the

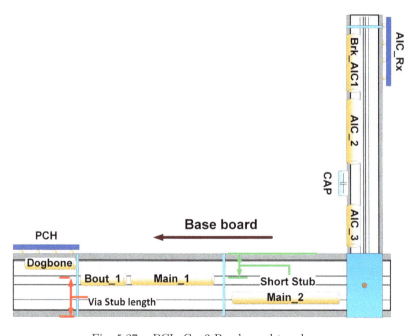

Fig. 5.27. PCIe Gen3 Rx channel topology.

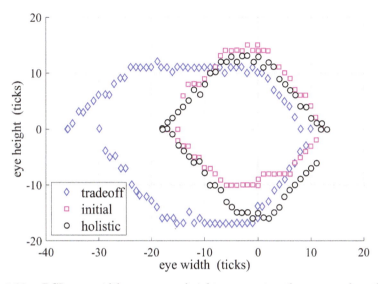

Fig. 5.28. PCIe eye width versus eye height: comparing the proposed methodology against the initial design and the trade-off approach.

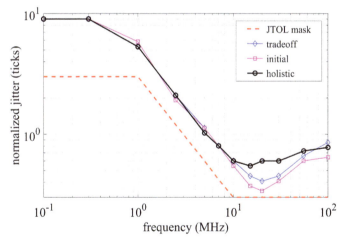

Fig. 5.29. PCIe JTOL results: comparing the proposed methodology against the initial design and the trade-off approach.

large asymmetry seen on the horizontal axis could eventually lead to system failures (see Fig. 5.28). Additionally, the JTOL results obtained with our holistic approach show the largest margins with respect to the specification limits, as shown in Fig. 5.29. As with the other two test cases, validation time was significantly decreased using our holistic approach, in this test case, by up to 70% with respect to the traditional trade-off approach.

5.5 Space Mapping Optimization for PHY Tuning

While an accurate surrogate model is desirable for direct SBO, it can still be computationally expensive to develop. By combining an adequate modeling technique with a suitable DoE approach, a rough or coarse surrogate model can be efficiently developed with a very reduced set of data, as in Rangel-Patino *et al.* (2017, 2019). Once this coarse surrogate model is available, obtained from a reduced number of experiments at post-silicon validation, SM techniques (Bandler *et al.*, 1994; Cheng *et al.*, 2008; Koziel *et al.*, 2008) can be exploited to obtain the optimal PHY settings.

In this section, the Broyden-based input SM algorithm, better known as aggressive SM (ASM) (Bandler *et al.*, 1995; Rayas-Sánchez, 2016), is used in HSIO PHY tuning optimization. Our SM approach takes advantage of a coarse surrogate model developed following Rangel-Patino *et al.* (2017). In our case, the fine model is a measurement-based post-silicon validation industrial platform. Our approach is illustrated by optimizing the PHY tuning receiver (Rx) equalizer settings for a SATA Gen 3 channel topology, which allows to accelerate tuning from several days to a few hours.

5.5.1 *Broyden-based input space mapping*

SM optimization methods belong to the general class of surrogate-based optimization algorithms (Booker *et al.*, 1999). They are specialized in the optimization of computationally expensive models. The most widely used SM approach to efficient design optimization is the ASM or Broyden-based input SM algorithm (Rayas-Sánchez, 2016). ASM efficiently finds an approximation of the optimal design of a computationally expensive model (fine model) by exploiting a fast but inaccurate surrogate representation (coarse model) (Rayas-Sánchez, 2016). ASM aims at quickly finding a solution that makes the fine-model response close enough to the desired response.

5.5.2 Fine model

As mentioned above, our fine model is an Intel server post-silicon validation platform in an industrial environment. Within the PCH, our methodology is applied to a HSIO link SATA Gen3 (SATA Org, 2016). As described in Section 5.4.4.2, the SATA channel topology is comprised of the Tx driver, the Tx base board transmission lines, several via transitions, an I/O card connector, and 1 m SATA cable used to connect the base board to the device I/O card, as illustrated in Fig. 5.24. The measurement system shown in Fig. 5.30 is based on an Intel process called SMV, as described in Section 5.2.2. The fundamental process behind the SMV consists of systematically adjusting the corner conditions under which the validation platform operates, then measure the Rx functional eye opening by using on-die design for test (DFT) features until the eye opening has been shrunk to a

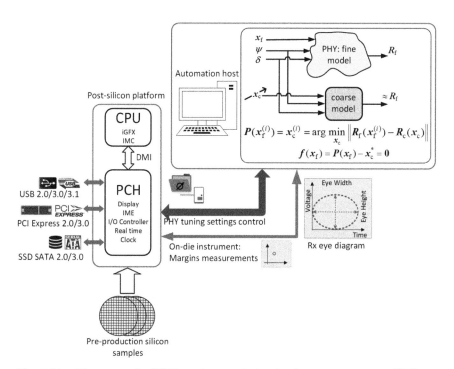

Fig. 5.30. Test setup for PHY settings optimization by space mapping (SM): an Intel server post-silicon validation platform is used as the fine model and a rough surrogate as the coarse model.

point where the Rx detects errors or the system fails (Rangel-Patiño et al., 2017).

Let $\boldsymbol{R}_\text{f} \in \Re^m$ represent the actual (measured) electrical margining system response, denoted as a fine model response, which consists of the eye width $e_\text{w} \in \Re$ and eye height $e_\text{h} \in \Re$ of the measured eye diagram,

$$\boldsymbol{R}_\text{f}(\boldsymbol{x}, \boldsymbol{\psi}, \boldsymbol{\delta}) = \begin{bmatrix} e_\text{W}(\boldsymbol{x}, \boldsymbol{\psi}, \boldsymbol{\delta}) & e_\text{h}(\boldsymbol{x}, \boldsymbol{\psi}, \boldsymbol{\delta}) \end{bmatrix}^\text{T}. \qquad (5.45)$$

This electrical margining system response depends on the PHY tuning settings $\boldsymbol{x}$ (EQ coefficients), the operating conditions $\boldsymbol{\psi}$ (voltage and temperature), and the devices $\boldsymbol{\delta}$ (silicon skew and external devices). In this case, we use five input variables that represent the SATA Rx PHY tuning coefficients, which are the settings used in three main Rx circuitry blocks (CTLE, VGA, and CDR). The functional eye width and height are obtained from measured parameters,

$$e_\text{w}(\boldsymbol{x}, \boldsymbol{\psi}, \boldsymbol{\delta}) = e_\text{wr}(\boldsymbol{x}, \boldsymbol{\psi}, \boldsymbol{\delta}) + e_\text{wl}(\boldsymbol{x}, \boldsymbol{\psi}, \boldsymbol{\delta}), \qquad (5.46)$$

$$e_\text{h}(\boldsymbol{x}, \boldsymbol{\psi}, \boldsymbol{\delta}) = e_\text{hh}(\boldsymbol{x}, \boldsymbol{\psi}, \boldsymbol{\delta}) + e_\text{hl}(\boldsymbol{x}, \boldsymbol{\psi}, \boldsymbol{\delta}), \qquad (5.47)$$

where $e_\text{wr} \in \Re$ and $e_\text{wl} \in \Re$ are the eye width-right and eye width-left measured parameters, respectively, and $e_\text{hh} \in \Re$ and $e_\text{hl} \in \Re$ are the eye height-high and eye height-low measured parameters, respectively. During SM optimization, both ψ and δ are kept fixed when evaluating (5.45).

5.5.3 Coarse model

As mentioned in Section 5.3, surrogate models can be constructed using data from high-reliability simulation models or from measurements, and provide fast approximations of the original system or component at new design points (Queipo et al., 2005). In Section 5.3, we analyze several surrogate models trained with different DoE techniques to find a good coarse model that is able to approximate a HSIO link with a very reduced amount of measurements, selecting the best combination of surrogate modeling technique and DoE in terms of accuracy and development time. Here, as the coarse model for SM we use a rough surrogate of the HSIO link SATA Gen3: a kriging model with a Sobol DoE approach using only 50 samples.

5.5.4 *ASM optimization*

ASM starts by finding the optimal coarse model design x_c^* from direct numerical optimization, that yields the optimal coarse model response, $R_c(x_c^*) = R_c^*$. ASM takes R_c^* as the target response for the fine model, aiming to find a fine model design, x_f^{SM} (also known as the space-mapped solution) that makes the fine model response $R_f(x_f^{SM})$ as close as possible to the target response R_c^*.

The central part of the ASM algorithm is the parameter extraction process (Rayas-Sánchez, 2016), which can be considered as a vector function P representing the mapping between both design parameter spaces at the i-th iteration, $x_c^{(i)} = P(x_f^{(i)})$. If the current extracted parameters $x_c^{(i)}$ correspond approximately to x_c^*, then the current fine model response approximates the desired response, $R_f(x_f^{(i)}) \approx R_c^*$. To find x_f^{SM}, the ASM algorithm solves a system of nonlinear equations defined as

$$f(x_f) = P(x_f) - x_c^*. \quad (5.48)$$

The parameter extraction process consists of finding, for the i-th fine model design $x_f^{(i)}$, the coarse model design $x_c^{(i)}$ whose corresponding response $R_c(x_c^{(i)})$ is as close as possible to $R_f(x_f^{(i)})$. This can be realized by solving

$$P(x_f^{(i)}) = x_c^{(i)} = \arg\min_{x_c} \|R_f(x_f^{(i)}) - R_c(x_c)\|_2^2. \quad (5.49)$$

The system of equations $f(x_f)$ is directly solved by using Broyden's updating formula (Broyden, 1965). Notice that solving $f(x_f)$ is equivalent to solving the mapping equation $P(x_f) = x_c^*$, which means that a solution to the system is found when the extracted parameters $x_c^{(i)}$ are equal to the optimal coarse model design x_c^*, also implying that the fine model response is sufficiently close to the target response.

The next iterate in the algorithm is predicted by

$$x_f^{(i+1)} = x_f^{(i)} + h^{(i)}, \quad (5.50)$$

where $h^{(i)}$ solves the linear system defined as

$$B^{(i)} h^{(i)} = -f(x_f^{(i)}) = -f^{(i)}, \quad (5.51)$$

where Broyden matrix B is an approximation of the Jacobian of f with respect to x_f at the current iterate i. The matrix B is first initialized by the identity matrix and updated by using Broyden's formula,

$$B^{(i+1)} = B^{(i)} + \frac{f^{(i)} h^{(i)^\text{T}}}{h^{(i)^\text{T}} h^{(i)}}. \qquad (5.52)$$

In this chapter, the pseudo-code of Rayas-Sánchez and Gutiérrez-Ayala (2006) shown in Fig. 5.31 was used for a constrained Broyden-based input SM. The stopping criteria considered in this work include four possibilities: when a root of the nonlinear system is found; when the relative change in the fine-model design parameters is small enough; when the maximum relative error in the fine-model response with respect to the target response is small enough; or when

---

Begin
    find x_c^*, define x_f^min and x_f^max
    $i = 0, x_f^{(i)} = x_c^*, B^{(i)} = I, \delta = 0.3$
    $f^{(i)} = P(x_f^{(i)}) - x_c^*$ using (49)
    repeat until StoppingCriteria
        solve $B^{(i)} h^{(i)} = -f^{(i)}$ for $h^{(i)}$
        $x_f^{(test)} = x_f^{(i)} + h^{(i)}$
    while $x_f^{(test)} \prec x_f^\text{min} \vee x_f^{(test)} \succ x_f^\text{max}$
        $h^{(i)} = \delta h^{(i)}$
        $x_f^{(test)} = x_f^{(i)} + h^{(i)}$
    end
    $x_f^{(i+1)} = x_f^{(test)}$
    $f^{(i+1)} = P(x_f^{(i+1)}) - x_c^*$ using (49)
    $B^{(i+1)} = B^{(i)} + \dfrac{f^{(i+1)} h^{(i)^T}}{h^{(i)^T} h^{(i)}}, \; i = i + 1$
end

Fig. 5.31. Pseudo-code for a constrained Broyden-based input space mapping optimization.

a maximum number of iterations is reached; as follows:

$$\|f(x_\mathrm{f}^{(i)})\|_\infty < \varepsilon_1 \vee, \tag{5.53}$$

$$\|x_\mathrm{f}^{(i+1)} - x_\mathrm{f}^{(i)}\|_2 \leq \varepsilon_2 \left(\varepsilon_2 + \|x_\mathrm{f}^{(i)}\|_2\right) \vee, \tag{5.54}$$

$$\|R_\mathrm{f}(x_\mathrm{f}^{(i)}) - R_\mathrm{c}(x_\mathrm{c}^*)\|_\infty \leq \varepsilon_3 \left(\varepsilon_3 + \|R_\mathrm{c}(x_\mathrm{c}^*)\|_\infty\right) \vee, \tag{5.55}$$

$$i > i_\mathrm{max}, \tag{5.56}$$

where ε_1, ε_2, and ε_3 are arbitrary small positive scalars.

5.5.5 *Optimization results*

After applying the Broyden-based input SM algorithm (Rayas-Sánchez, 2016) shown in Fig. 5.31, a space-mapped solution x^SM is found in just six iterations (or fine model evaluations), as shown in Fig. 5.32. The set of Rx EQ coefficients contained in x^SM makes the measured SATA Rx inner eye height and width of the PCH as open as that one predicted by the optimized coarse surrogate model. Since we want to indirectly find the optimal set of PHY tuning settings x that maximize the functional eye diagram area, we plot in

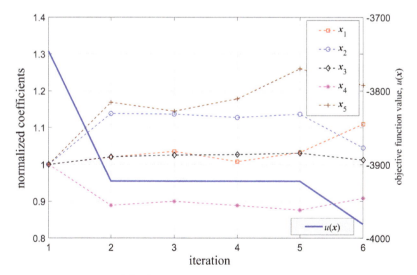

Fig. 5.32. Normalized SATA coefficients and objective function values across SM optimization iterations.

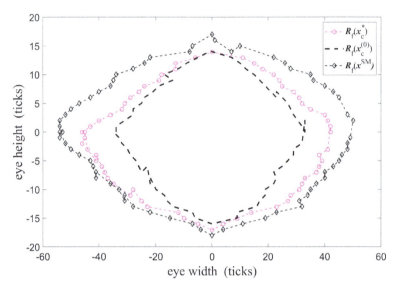

Fig. 5.33. Comparison between the system fine model responses at the initial Rx EQ coefficients, $x_c^{(0)}$, at the optimal coarse model solution, x_c^*, and at the space-mapped solution found, x^{SM}.

Fig. 5.32 the negative of the objective function defined in (5.32). The SM solution (x^{SM}) makes an improvement of 85% on the fine model eye diagram area as compared to that one with the initial settings ($x_c^{(0)}$), and a 33% improvement as compared to that one with the optimal coarse model solution (x_c^*), as shown in Fig. 5.33.

The efficiency of this approach is also demonstrated by a very significant time reduction in post-Si validation and PHY tuning Rx equalization. While the traditional industrial process requires days for a complete empirical optimization (based on engineering expertise), the method proposed here can be completed in just a few hours. This SM technique can easily be applied to other interfaces such as USB and PCIe.

5.6 Discussion and Conclusion

As microprocessor design scales down to the 10-nm technology and bellow, traditional pre- and post-silicon validation techniques are unsuitable to get a full system functional coverage. Technology scaling and advanced silicon packaging techniques are allowing a continuously higher density of integration. Physical complexity and extreme technology process variations severely limit the effectiveness and reliability of pre-silicon validation techniques. To ensure leading performance, reliability, and compatibility in this complex environment, companies invest over hundreds of millions of dollars annually in component and platform validation.

This scenario imposes the need for sophisticated post-silicon validation approaches to consider complex electromagnetic phenomena and large manufacturing fluctuations observed in actual physical platforms. However, product complexity, performance requirements, and TTM commitments further pressure on engineering teams to search for opportunities to make validation faster and cheaper.

The empirical optimization of receiver analog circuitry in modern HSIO links is a very time-consuming post-silicon validation process. Current industrial practices are based on exhaustive enumeration methods to improve either the system margins or the jitter tolerance compliance test.

In this chapter, we discussed surrogate modeling and surrogate-based optimization methods, including space mapping, based on suitable objective functions to efficiently tune the Tx and Rx equalizers coefficients.

We presented several surrogate modeling techniques to find the best approach to simulate the Rx equalization circuitry in an industrial HSIO link. Several surrogate models were trained with different DoE techniques to choose the best sampling approach. All surrogate models were evaluated by comparing with actual measured responses on a real server HSIO link. We selected the best combination of surrogate modeling technique and DoE in terms of accuracy and generalization performance, and applied a SBO to maximize the eye diagram area. The values obtained through our SBO procedure were evaluated by measuring the real functional eye diagram of the physical system, showing a great improvement as compared with the initial performance.

A holistic optimization approach that merges system margining and jitter tolerance measurements for PHY tuning was also demonstrated. The experimental results confirm its efficiency to deliver optimal margins while ensuring jitter tolerance compliance, showing a substantial improvement for both system margins and jitter tolerance as compared with the current industrial practice, and dramatically accelerating the typical time required for PHY tuning.

A metamodeling technique based on artificial neural networks was also presented to efficiently simulate the effects of the Rx EQ circuitry in industrial HSIO links. Through the proposed neural modeling procedure, an efficient surrogate model is found that approximates the system with a reduced set of testing and training data.

Finally, it was also described how the Broyden-based input space mapping algorithm can be used to efficiently optimize the PHY tuning Rx equalizer settings by exploiting a low-cost low-accuracy kriging surrogate as the coarse model, and a measurement-based post-silicon validation platform as the fine model. The experimental results, based on a real industrial validation platform, demonstrated the efficiency of the method, showing a substantial performance improvement and a dramatic acceleration of the typical required time for PHY tuning.

The proposed optimization techniques, along with the objective functions, are not limited to server PHY tuning validation platforms. The same procedures could be applied to other computer platforms for post-silicon validation, such as the client market segment to include laptops and desktops, and devices segment to include cell phones and tablets.

References

10GEA Org. (2016). *XAUI Interface* [Online]. Available: http://www.10gea.org/whitepapers/xaui-interface/.

Agilent Tech. (2003). *Jitter Analysis Techniques for High Data Rates* [Online]. Available: www.agilent.com.

Agilent Tech (2005). *Total Jitter Measurement at Low Probability Levels, using Optimized BERT Scan Methods* [Online]. Available: www.agilent.com.

Angiulli, G., Cacciola, M., Versaci, M. (2007). Microwave devices and antennas modelling by support vector regression machines, *IEEE Trans. Magn.*, vol. 43, no. 4, pp. 1589–1592.

Bandler, J.W., Biernacki, R., Chen, S.H., Grobelny, P.A., Hemmers, R.H. (1994). Space mapping technique for electromagnetic optimization, *IEEE Trans. Microwave Theory Tech.*, vol. 42, no. 12, pp. 2536–2544.

Bandler, J.W., Biernacki, R.M., Chen, S.H., Hemmers, R.H., Madsen, K. (1995). Electromagnetic optimization exploiting aggressive space mapping, *IEEE Trans. Microw. Theory Techn.*, vol. 41, no. 12, pp. 2874–2882.

Beyene, W.T. (2007). Application of artificial neural networks to statistical analysis and nonlinear modeling of high-speed interconnect systems, *IEEE Trans. Computer-Aided Design of Integrated Circuits Systems*, vol. 26, no. 1, pp. 166–176.

Bistola, S. (2015). High-speed interconnect simulation using artificial neural networks, in *Intel Design & Test Technology Conf. (DTTC)*, Oregon, CA.

Booker, A.J., Dennis Jr., J.E., Frank, P.D., Serafini, D.B., Torczon, V., Trosset, M.W. (1999). A rigorous framework for optimization of expensive functions by surrogates, *Struct. Optim.*, vol. 17, no. 1, pp. 1–13.

Broyden, C.G. (1965). A class of methods for solving non-linear simultaneous equations, *Math. Comp.*, vol. 19, no. 92, pp. 577–593.

Casper, B., O'Mahony, F. (2009). Clocking analysis, implementation and measurement techniques for high-speed data links — A tutorial, *IEEE Trans. Circuits Sys. I*, vol. 56, no. 1, pp. 17–39.

Chang et al., Y.P. (2005). Design of discrete-value passive harmonic filters using sequential neural-network approximation and orthogonal array, in *IEEE PES Transmission & Distribution Conf. & Expo.: Asia and Pacific*, Dalian, China, pp. 1–6.

Chávez-Hurtado, J.L., Rayas-Sánchez, J.E. (2016). Polynomial-based surrogate modeling of RF and microwave circuits in frequency domain exploiting the multinomial theorem, *IEEE Trans. Microw. Theory Techn.*, vol. 64, no. 12, pp. 4371–4381.

Chávez-Hurtado, J.L., Rayas-Sánchez, J.E., Brito-Brito, Z. (2016). Multiphysics polynomial-based surrogate modeling of microwave structures in frequency domain, in *IEEE MTT-S Latin America Microw. Conf. (LAMC)*, Puerto Vallarta, Mexico, pp. 1–3.

Cheng, J., Druzdzel, M.J. (2000). Computational Investigation of Low-Discrepancy Sequences in Simulation Algorithms for Bayesian Networks, in *Proc. Conf. Uncertainty in Artificial Intelligence*, San Francisco, CA, pp. 72–81.

Cheng, Q.S., Bandler, J.W., Koziel, S. (2008). Combining coarse and fine models for optimal design, *IEEE Microw. Magaz.*, vol. 9, no. 1, pp 79–88.

Forrester, A., Sobester, A., Keane, A. (2008). *Engineering Design Via Surrogate Modelling: A Practical Guide*, Hoboken, NJ, Wiley.

Garitselov, O., Mohanty, S.P., Kougianos, E. (2012). A comparative study of metamodels for fast and accurate simulation of nano-CMOS circuits, *IEEE Trans. Semicond. Manuf.*, vol. 25, no.1, pp. 26–36.

Goodenough, J., Aitken, R. (2010). Post-silicon is too late avoiding the $50 million paperweight starts with validated designs, in *Proc. Design Automation Conf. (DAC), 47th ACM/IEEE*, Anaheim,CA, pp. 8–11.

Goulermas, J.Y., Liatsis, P., Xiao-Jun, Z., Cook, P. (2007a). Density-driven generalized regression neural networks (DD-GRNN) for function approximation, *IEEE Trans. Neural Netw.*, vol. 18, no. 6, pp. 1683–1696.

Goulermas, J.Y., Zeng, X.J., Liatsis, P., Ralph, J.F. (2007b). Generalized regression neural networks with multiple-bandwidth sharing and hybrid optimization, *IEEE Trans. Syst., Man., and Cyb.*, vol. 37, no. 6, pp. 1434–1445.

Goay, C.H., Goh, P. (2017). Neural networks for eye height and eye width prediction with an improved adaptive sampling algorithm, in *Asian Simulation Conf. (AsiaSim)*, Melaka, Malaysia, pp. 189–201.

Gu, C. (2012). Challenges in post-Silicon validation of high-speed I/O links, in *Proc. Int. Conf. on Computer-Aided Design (ICCAD), IEEE/ACM*, San Jose, CA, pp. 547–550.

Haykin, S. (1999). *Neural Networks: A Comprehensive Foundation*, New Jersey, MA, Prentice Hall.

Hong, D., Cheng, K.T. (2008). Bit-error rate estimation for bang-bang clock and data recovery circuit, in *26th IEEE VLSI Test Symp.*, San Diego, CA, pp. 17–22.

Hornik, K., Stinchcombe, M., White, H. (1989). Multilayer feedforward networks are universal approximators, *Neural Networks*, vol. 2, no. 5, pp. 359–366.

Huang, H., Lee, E.K.F. (2001). Design of low-voltage CMOS continuous time filter with on-chip automatic tuning, *IEEE J. Solid-State Circuits*, vol. 36, no. 8, pp. 1168–1177.

Keshava, J., Hakim, N., Prudvi, C. (2010). Post-silicon validation challenges: How EDA and academia can help, in *Proc. Design Automation Conf. (DAC)*, Anaheim, CA, pp. 3–7.

Kossel, M., Schmatz, M. (2004). Jitter measurements of high-speed serial links, *IEEE Design & Test Comput.*, vol. 21, no. 6, pp. 536–543.

Koziel, S., Cheng, Q.S., Bandler, J.W. (2008). Space Mapping, *IEEE Microwave Magaz.*, vol. 9, no. 6, pp 105–122.

Koziel, S., Yang, X.S. (2011). *Computational Optimization, Methods and Algorithms*, Heidelberg, Germany, Springer.

Jones, D.R., Schonlau, M., Welch, W.J. (1998). Efficient global optimization of expensive black-box functions, *J. Global Optimiz.*, vol. 13, pp. 455–492.

Lagarias, J.C., Reeds, J.A., Wright, M.H., Wright, P.E. (1998). Convergence properties of the Nelder-Mead simplex method in low dimensions, *SIAM J. Optim.*, vol. 9, no. 1, pp. 112–147.
Li, M.P., Jitter (2007). *Noise, and Signal Integrity at High-Speed*, Boston, MA, Prentice Hall.
Li, M. (2009). Jitter challenges and reduction techniques at 10 Gb/s and beyond, *IEEE Trans. Adv. Packaging*, vol. 32, no. 2, pp. 290–297.
Liu, M., Tsai, J.H. (2015). USB3.1 silicon and channel design optimization using artificial neural network modeling, in *IEEE Electromagnetic Compatibility and Signal Integrity Symp.*, Santa Clara, CA, pp. 289–293.
Lophaven, S.N., Nielsen, H.B., Søndergaard, J. (2002). Aspects of the Matlab toolbox DACE, in *Tech. Report, Informatics and Mathematical Modelling, Technical University of Denmark*, DTU, Lyngby.
Lopez Miralrio, E., Mendez Ruiz, C., Lin, T., Sytwu, J., Hsu, J., Su, T., Ye, v, Ye, X. (2013). Signal integrity design of via with extra routing stub for device routing flexibility, in *Proc. IEEE Electrical Design of Advanced Packaging & Systems Symp.*, Nara, Japan, pp. 189–192.
Mack, Y., Goel, T., Shyy, W., Haftka, R. (2007a). Surrogate model-based optimization framework: A case study in aerospace design, *Stud. Comput. Intell.*, vol. 51, no. 2, pp. 323–342.
Mack, Y., Goel, T., Shyy, W., Haftka, R. (2007b). Surrogate model-based optimization framework: A case study in aerospace design, in Yang, S., YS. Ong, Jin, Y. (ed.), *Evolutionary Computation in Dynamic and Uncertain Environments. Studies in Computational Intelligence*, Berlin: Springer, vol. 51, pp. 323–342.
MacKay, D.J.C. (1992). Bayesian interpolation, *Neural Computation*, vol. 4, no. 3, pp. 415–447.
Mahouti, P., Günes, F., Demirel, S., Uluslu, A., Belen, M.A. (2014). Efficient scattering parameter modeling of a microwave transistor using generalized regression neural network, in *IEEE Int. Conf. Microwaves, Radar, and Wireless Communication (MIKON)*, Gdansk, Poland, pp. 1–4.
Matheron, G. (1963). Principles of geostatistics, *Economic Geology*, vol. 58, no. 8, pp. 1246–1266.
Mendez Ruiz, C., Ye, C., Ye, X., Lopez Miralrio, E., Yin, M., Hsu, J., Su, T. (2013). Improve signal integrity performance by using hybrid PCB stackup, in *Proc. IEEE Int. Symp. on Electromagnetic Compatibility*, Denver, CO, pp. 317–321.
Miller, G., Timko, M., Lee, H.-S., Nestler, E., Mueck, M., Ferguson, P. (2003). Design and modeling of a 16-bit 1.5msps successive

approximation ADC with non-binary capacitor array, in *Proc. Int. Great Lakes Symp. on VLSI*, Washington, D.C, pp. 161–164.

Montgomery, C. (2014). *Design and Analysis of Experiments*, New York. NY: Wiley.

Morris, M.D., Mitchell, T.J., Ylvisaker, D. (1993). Bayesian design and analysis of computer experiments: Use of derivatives in surface prediction, *Technometrics*, vol. 35, no. 3, pp. 243–255.

Panda, B.N., Bahubalendruni, M.V.A.R., Biswal, B.B. (2014). Optimization of resistance spot welding parameters using differential evolution algorithm and GRNN, in *IEEE Int. Conf. on Intelligent Systems and Control (ISCO)*, vol. 2, no. 6, Coimbatore, India, pp. 50–55.

Patra, P. (2007). On the cusp of a validation wall, *IEEE Design and. Test of Computers*, vol. 24, no. 2, pp. 193–196.

PCI SIG Org. (2016). *Peripheral Component Interconnect Express 3.1 Specification* [Online]. Available: https://pcisig.com/specifications.

Queipo, N.V., Haftka, R.T., Shyy, W., Goel, T., Vaidyanathna, R., Tucker, P.K. (2005). Surrogate-based analysis and optimization, *Prog. Aerospace Sci.*, vol. 41, no. 1, pp. 1–28.

Rangel-Patino, F., Viveros-Wacher, A., Rayas-Sanchez, J.E., Vega-Ochoa, E.A., Duron-Rosales, I., Hakim, N. (2016). A holistic methodology for system margining and jitter tolerance optimization in post-silicon validation, in *IEEE MTT-S Latin America Microw. Conf. (LAMC)*, Puerto Vallarta, Mexico, pp. 1–3.

Rangel-Patiño, F.E., Chávez-Hurtado, J.L., Viveros-Wacher, A., Rayas-Sánchez, J.E., Hakim, N. (2017). System margining surrogate-based optimization in post-silicon validation, *IEEE Trans. Microwave Theory Techn.*, vol. 65, no. 9, pp. 3109–3115.

Rayas-Sánchez, J.E. (2001). *Neural Space Mapping Methods for Modeling and Design of Microwave Circuits*, Ph.D. Thesis, Dept. of Electrical and Comp. Eng., McMaster University, Hamilton, Canada.

Rayas-Sánchez, J.E. (2004). EM-based optimization of microwave circuits using artificial neural networks: the state of the art, *IEEE Trans. Microwave Theory Techn.*, vol. 52, no. 1, pp. 420–435.

Rayas-Sánchez, J.E., Gutiérrez-Ayala, V. (2006). EM-based Monte Carlo analysis and yield prediction of microwave circuits using linear-input neural-output space mapping, *IEEE Trans. Microwave Theory Tech.*, vol. 54, no. 12, pp. 4528–4537.

Rayas-Sánchez, J.E. (2016). Power in simplicity with ASM: tracing the aggressive space mapping algorithm over two decades of development and engineering applications, *IEEE Microw. Magaz.*, vol. 17, no. 4, pp. 64–76.

Rayas-Sánchez, J.E., Chávez-Hurtado, J.L., Brito-Brito, Z. (2017). Optimization of full-wave EM models by low-order low-dimension polynomial surrogate functionals, *Int. J. Numerical Modelling: Electron. Networks, Dev. Fields*, vol. 30, no. 3–4, e2094.

Rangel-Patino, F., Chávez-Hurtado, J.L., Viveros-Wacher, A., Rayas-Sánchez, J.E., Hakim, N. (2017). Eye diagram system margining surrogate-based optimization in a server silicon validation platform, in *European Microw. Conf. (EuMC-2017)*, Nuremberg, Germany, pp. 540–543.

Rangel-Patiño, F.E., Rayas-Sánchez, J.E., Viveros-Wacher, A., Chávez-Hurtado, J.L., Vega-Ochoa, E.A., Hakim, N. (2019). Post-silicon receiver equalization metamodeling by artificial neural networks, *IEEE Trans. Computer-Aided Design of Integrated Circuits and Systems*, vol. 38, no. 4, pp. 733–740.

Rangel-Patiño, F.E., Viveros-Wacher, A., Rayas-Sánchez, J.E., Durón-Rosales, I., Vega-Ochoa, E.A., Hakim, N., López-Miralrio, E. (2020). A holistic formulation for system margining and jitter tolerance optimization in industrial post-silicon validation, *IEEE Trans. Emerging Topics Computing*, vol. 8, no. 2, pp. 453–463.

Sacks, J., Schiller, S., Welch, W. (1989a). Designs for computer experiments, *Technometrics*, vol. 31, no. 1, pp. 41–47.

Sacks, J., Welch, W.J., Mitchell, T.J., Wynn, H.P. (1989b). Design and analysis of computer experiments, *Statistical Science*, vol. 4, pp. 409–435.

Santner, T.J., Williams, B.J., Notz, W.I. (2003). *The Design and Analysis of Computer Experiments*, New York, NY, Springer.

SATA Org. (2016). *Serial Advanced Technology Attachment 3.2 Specification* [Online]. Available: http://www.sata-io.org/.

Staum, J. (2009). Better simulation metamodeling: The why, what, and how of stochastic kriging, in *Proc. of the Winter Simulation Conf.*, Austin, TX, pp. 119–133.

Sobol, I.M. (1967). On the distribution of points in a cube and the approximate evaluation of integrals, *U.S.S.R. Comput. Mathe. Math. Phy.*, vol. 7, no. 1, pp. 86–112.

Specht, D.F. (1991). A general regression neural network, *IEEE Trans. Neural Netw.*, vol. 2, no. 6, pp. 568–576.

Thimm, G., Ra, J.B. (1997). High-order and multilayer perceptron initialization," *IEEE Trans. Neural Netw.*, vol. 8, no. 2, pp. 349–359.

Tokan, N.T. Günes, F. (2008). Analysis and synthesis of the microstrip lines based on support vector regression, in *Proc. IEEE Eur. Microw. Conf. (EuMC)*, Amsterdam, The Netherlands, pp. 1473–1476.

USB Org. (2016). *Universal Serial Bus Revision 3.1 Specification* [Online]. Available: http://www.usb.org/developers/doc.

Vicario, G., Craparotta, G., Pistone, G. (2016). Meta-models in computer experiments: Kriging versus artificial neural network, *Qual. Reliab. Engng. Int.*, vol. 32, no. 6, pp. 2055–2065.

Viveros-Wacher, A., Rayas-Sanchez, J.E. (2016). Eye diagram optimization based on design of experiments to accelerate industrial testing of high speed links, in *IEEE MTT-S Latin America Microw. Conf. (LAMC)*, Puerto Vallarta, Mexico, pp. 1–3.

Viveros-Wacher, A., Baca-Baylón, R., Rangel-Patiño, F.E., Dávalos-Santana, M.A., Vega-Ochoa, E.A., Rayas-Sánchez, J.E. (2018). Jitter tolerance acceleration using the golden section optimization technique, in *IEEE Latin American Symp. Circuits and Systems Dig. (LASCAS 2018)*, Puerto Vallarta, Mexico, pp. 1–4.

Wang, G., Shan, S. (2006). Review of metamodeling techniques in support of engineering design optimization, *J. Mech. Design*, vol. 129, no. 4, pp. 370–380.

Wang, L.T., Stroud, C.E., Touba, N.A. (2008). *System-on-Chip Test Architectures: Nanometer Design for Testability*, Burlington, MA, Elsevier.

Wang, F., Cachecho, P., Zhang, W., Sun, S., Li, X., Kanj, R., Gu, C. (2016). Bayesian model fusion: Large-scale performance modeling of analog and mixed-signal circuits by reusing early-stage data, *IEEE Trans. Computer-Aided Design Integr. Circuits Syst.*, vol. 35, no. 8, pp. 1255–1268.

Wile, B., Goss, J.C., Roesner, W. (2005). *Comprehensive Functional Verification — The Complete Industry Cycle*, Burlington, MA: Morgan Kaufmann.

Wu, C.F.J., Hamada, M. (2000). *Experiments: Planning, Analysis, and Parameter Design Optimization*, New York, NY, Wiley.

Xia, L., Meng, J., Xu, R., Yan, B., Guo, Y. (2006). Modeling of 3-D vertical interconnect using support vector machine regression, *IEEE Microw. Wireless Compon. Lett.*, vol. 16, no. 12, pp. 639–641.

Yao, W., Shi, Y., He, L., Pamarti, S. (2009). Joint design-time and post-silicon optimization for digitally tuned analog circuits, in *2009 IEEE/ACM Int. Conf. Computer-Aided Design*, San Jose, CA, pp. 725–730.

Ye, C., Ye, X., Vargas, E., Argueta, O. (2011). Full link impedance optimization for serial IOs, in *Proc. IEEE Int. Symp. Electromagnetic Compatibility*, Long Beach, CA, pp. 803–808.

Ye, X., Xiao, K., Enriquez, R. (2012). Differential far-end crosstalk cancellation — implementations and challenges, in *Proc. IEEE Int. Symp. Electromagnetic Compatibility*, Pittsburgh, PA, pp. 193–198.

Yelten, M.B., Zhu, T., Koziel, S., Franzon, P.D., Steer, M.B. (2012). Demystifying surrogate modeling for circuits and systems, *IEEE Circuits Syst. Mag.*, vol. 12, no. 1, pp. 45–63.

Zhang, Q.J., Gupta, K.C. (2000). *Neural Networks for RF and Microwave Design*, Norwood, MA, Artech House.

Chapter 6

Performance-Driven Inverse/Forward Modeling of Antennas in Variable-Thickness Domains

Slawomir Koziel and Anna Pietrenko-Dabrowska

Abstract

Design of contemporary antenna systems is a challenging endeavor. The difficulties are partially rooted in stringent specifications imposed on both electrical and field characteristics, demands concerning various functionalities, but also constraints imposed upon the physical size of the radiators. Furthermore, conducting the design process at the level of full-wave electromagnetic (EM) simulations, otherwise dictated by reliability, entails considerable computational expenses. This is particularly troublesome for the procedures involving repetitive EM analyses, e.g., parametric optimization. Utilization of fast surrogate models as a way of mitigating this issue has been fostered in the recent literature. Notwithstanding, construction of reliable surrogates is hindered by highly nonlinear antenna responses and even more by the utility requirements: design-ready models are to be valid over wide ranges of operating conditions and geometry parameters. Recently proposed performance-driven modeling, especially the nested kriging framework, addresses these difficulties by confining the surrogate model domain to a region that encapsulates the designs being optimum with respect to the relevant figures of interest. The result is a dramatic reduction of the number of training samples needed to render a usable model.

This chapter discusses an enhancement of the nested kriging methodology that involves a variable-thickness domain. This is an important advancement over the basic framework with the major benefit being a further and significant (up to 70%) reduction of the training data acquisition cost without compromising the model predictive power. It is achieved while ensuring that the model domain covers the regions containing optimum designs for various sets of performance specifications.

Our considerations are illustrated using two examples of microstrip antennas. Numerical results are supported by experimental validation of the selected designs obtained by means of the discussed models.

Keywords: Performance-drives modeling, domain confinement, nested kriging, variable-thickness domain, inverse model

6.1 Introduction

Contemporary antenna structures are designed to meet the demands pertaining to various application areas, including wireless communications (Guo *et al.*, 2016) (along with the emerging 5G technology; Zeng and Luk, 2019; Nie *et al.*, 2019), satellite communications (Mener *et al.*, 2016), medical imaging (Felicio *et al.*, 2019), or internet of things (IoT) (Jha *et al.*, 2018). In many cases, different functionalities have to be implemented including multi-band operation (Qian *et al.*, 2019), circular polarization (Ullah and Koziel, 2018), band notches (Xu *et al.*, 2018), MIMO operation (Zhao and Ren, 2019), or pattern diversity (Dong *et al.*, 2018). At the same time, reduction of the physical size of radiators is of concern for a growing number of situations, for example, for wearable (Yan *et al.*, 2015) and implantable devices (Wang *et al.*, 2018a). In pursuit of fulfilling the stringent specifications, increasingly complex antenna topologies are being developed. Their evaluation requires full-wave EM analysis because other means are either unavailable or — as equivalent network models — can only be used for yielding rough initial designs that need further tuning (Fakih *et al.*, 2019). Nowadays, EM-driven design is ubiquitous. Nevertheless, its downside is high computational cost. This becomes a practical bottleneck whenever multiple simulations are required. A representative task is parameter tuning (Hassan *et al.*, 2015; Tsukamoto and Arai, 2016). Although imperative for most antenna structures, it generates considerable CPU expenditures, particularly when reasonable initial design is not available or the global search needs to be involved (Lalbakhsh *et al.*, 2017).

Among the various ways of accelerating EM-driven design optimization procedures, expediting can be realized at the algorithmic level. Representative examples include gradient-based procedures where the speedup can be achieved by lowering the cost of antenna response Jacobian estimation using adjoint sensitivities (Wang et al., 2018b; Koziel et al., 2014), or by suppressing finite-differentiation sensitivity updates based on various criteria, e.g., monitoring design relocation (Koziel and Pietrenko-Dabrowska, 2019a) or gradient changes between the algorithm iterations (Koziel and Pietrenko-Dabrowska, 2019b). The employment of fast surrogate models provides viable alternatives. These can be realized in the context of local (Easum et al., 2018) or global optimization (Hassan et al., 2018). For the former, the surrogates are typically constructed along the optimization path, and can be data-driven (polynomial regression (Easum et al., 2017), kriging (de Villiers et al., 2017), neural networks (Dong et al., 2019), polynomial chaos expansion (Du and Roblin, 2018) or physics-based, normally obtained from underlying low-fidelity models. In the case of antennas, the surrogates most often involve coarse-mesh EM simulations (Koziel and Ogurtsov, 2014). Some of the popular physics-based optimization methods include space mapping (Rayas-Sanchez, 2016; Cervantes-González et al., 2016), response correction (Koziel and Unnsteinsson, 2018), feature-based optimization (Koziel, 2015). Global surrogate-assisted optimization often employs machine learning techniques (Alzahed et al., 2019) but also all kinds of approximation surrogates (Tak et al., 2018). In this context, sequential sampling methods are typically utilized (Torun and Swaminathan, 2019), where iterative allocation of the infill points can be aimed at improving the predictive power of the model, searching for the global optimum or the combination thereof (Queipo et al., 2005).

Utilization of the stand-alone surrogate models as replacements of CPU-heavy EM simulations has been fostered in the literature as a way of accelerating various simulation-driven design procedures, including design closure (Barmuta et al., 2015; Jacobs, 2016; Petrocchi et al., 2017). Data-driven surrogates seem to be especially attractive for these purposes due to their versatility and widespread availability (e.g., Gorissen et al., 2010; Marelli and Sudret, 2014). Well established and popular approximation techniques include polynomial regression (Queipo et al., 2005), artificial neural networks (Rawat et al., 2012), radial basis functions (RBF) (Barmuta et al., 2015), kriging interpolation (Queipo et al., 2005), Gaussian process regression (GPR) (Jacobs, 2016), or support vector regression (SVR) (Cai et al., 2018). Unfortunately, conventional methods are severely limited when applied to antenna problems. The primary challenges include nonlinearity of antenna characteristics, the curse of dimensionality affecting the modeling process due to a typically large number of geometry/material parameters that need to be handled, as well as the need for covering wide ranges of operating conditions and designable parameters. The latter is critical from the point of view of practical usefulness of the surrogate for design purposes. Alleviating these difficulties is possible in specific situations using the methods such as high-dimensional model representation (HDMR) (Yücel et al., 2015), least-angle regression (Hu et al., 2018), or incorporation of variable-fidelity models, e.g., two-stage GPR (Jacobs and Koziel, 2014), co-kriging (Kennedy and O'Hagan, 2000), Bayesian model fusion (Wang et al., 2016).

Confinement of the model domain has been recently suggested as an alternative approach to computationally efficient surrogate modeling (Koziel, 2017; Koziel and Sigurdsson, 2018; Koziel and Pietrenko-Dabrowska, 2019c). By focusing the modeling process on the region containing designs that are optimum with respect to the selected performance figures (e.g., operating frequencies of a multi-band antenna) or material parameters (e.g., relative permittivity of the dielectric substrate the antenna is implemented on), reliable surrogate can be rendered over wide ranges of operating conditions using small training data sets. Identification of such regions is realized with the help of pre-existing reference designs (Koziel and Sigurdsson, 2018), which could be available from the previous design work with the same structure or obtained as a part of the modeling procedure.

One of the most recent methods adopting the domain confinement approach is the nested kriging framework (Koziel and Pietrenko-Dabrowska, 2019c). The two kriging metamodels are utilized: the first-level (inverse) surrogate to establish the region of interest (model domain), and the second-level model being the actual surrogate. The important advantage of the technique is a straightforward arrangement of uniform training data sampling as well as model optimization, both enabled by a one-to-one mapping between the unity interval and the surrogate domain being a part of the formulation of the procedure (Koziel and Pietrenko-Dabrowska, 2019c). The critical (scalar) control parameter of nested kriging is the thickness coefficient determining the ratio between the lateral and tangential size of the domain. Its value affects the trade-off between the model predictive power and the coverage of the regions containing potentially optimum antenna designs for various sets of performance specifications.

This chapter discusses an improvement over the original nested kriging framework, which is a variable-thickness domain. According to this concept, the thickness parameter is a function of the objective space vectors. The keystones of the approach are the reference design triangulation and an appropriate mapping defined over the expansion coefficients of the objective vector with respect to the vertices of the simplexes being a result of the triangulation process. Using this mapping, the domain thickness is maintained small in the vicinity of the reference designs (which are optimal by definition) and increases towards the simplex centers (where the first-level model deviates the most from the optimum design manifold). The variable-thickness domain method permits the construction of reliable surrogates using significantly (by up to 70%) smaller data sets than required by the fixed thickness version. The computational benefits are achieved without compromising the model coverage in terms of geometry parameter and operating condition ranges. These properties are demonstrated using two antenna examples as well as benchmarking against conventional modeling methods (kriging, RBF) and the original nested kriging.

6.2 Antenna Modeling Using Nested Kriging

In order to make the chapter self-contained, we start by briefly recalling the formulation of the nested kriging modeling framework. The concept and implementation of variable-thickness domain is outlined in Section 6.3, whereas demonstration examples are provided in Section 6.4.

6.2.1 *Design space objective space: First-level (inverse) surrogate*

The nested kriging framework (Koziel and Pietrenko-Dabrowska, 2019c) takes advantage of the correlations between antenna geometry parameters corresponding to the designs that are optimum with respect to the typical figures of interest (Koziel, 2017). For example, antenna re-design form a various operating frequencies normally entails synchronized adjustment of the parameters. The parameter sets that do not adhere to these correlation patterns are of poor quality and excluding them from the modeling process may be beneficial from the point of view of the computational efficiency. This is in opposition to conventional domains determined by the lower and upper bounds for design parameters, which are indiscriminative with respect to the aforementioned correlations.

According to nested kriging, identification of the regions containing high-quality designs is realized using a set of reference designs and the first-level surrogate as elaborated on below.

We use the following notation:

- $\boldsymbol{x} = [x_1 \cdots x_n]^T$ — vector of (designable) antenna parameters;
- $\boldsymbol{l} = [l_1 \cdots l_n]^T$ — lower bounds on antenna parameters;
- $\boldsymbol{u} = [u_1 \cdots u_n]^T$ — upper bounds on antenna parameters;
- $f_k, k = 1, \ldots, N$ — figures of interest pertinent to the design task at hand, operating conditions, or material parameters (e.g., operating frequency, bandwidth, or substrate permittivity assuming that the antenna is to be re-designed for various substrates);
- X — conventional parameter space, defined by the lower and upper bounds $\boldsymbol{l}$ and $\boldsymbol{u}$, so that $l_k \leq x_k \leq u_k$ for $k = 1, \ldots, n$;
- F — objective space, defined by the ranges $f_{k.\min} \leq f_k^{(j)} \leq f_{k.\max}, k = 1, \ldots, N$; this is the region over which the surrogate model is supposed to be valid.

The design optimality is understood, for a given objective vector $\boldsymbol{f} \in F$, in the sense of minimizing the functional $U(\boldsymbol{x}, \boldsymbol{f})$, where U is the scalar merit function which quantifies the design quality. In other words, the optimum design $\boldsymbol{x}^*$ is given as

$$\boldsymbol{x}^* = U_F(\boldsymbol{f}) = \arg \min_{\boldsymbol{x}} U(\boldsymbol{x}, \boldsymbol{f}). \tag{6.1}$$

We also define

$$U_F(F) = \{U_F(\boldsymbol{f}) : \boldsymbol{f} \in F\}, \tag{6.2}$$

which is the manifold comprising the designs being optimum in the sense of (6.1) for all $\boldsymbol{f} \in F$. For the sake of clarification, let us consider an example of a multi-band antenna, where the figures of interest are the operating frequencies $f_{0.k}$, $k = 1, \ldots, N$, i.e., we have $f_k = f_{0.k}$. Assuming that the design objective is to allocate the antenna resonances at the target operating frequencies and to improve the antenna matching therein, the merit function could be defined as

$$U(\boldsymbol{x}, \boldsymbol{f}) = U\left(\boldsymbol{x}, [f_{0.1} \cdots f_{0.N}]^T\right) \quad (6.3)$$
$$= \max\{f \in \{f_{0.1}, \ldots, f_{0.N}\} : |S_{11}(\boldsymbol{x}, f)|\},$$

where $S_{11}(\boldsymbol{x}, f)$ is the antenna reflection at the design $\boldsymbol{x}$ and frequency f. If the goal is to improve the antenna gain at the operating frequencies, the merit function can be defined as

$$U(\boldsymbol{x}, \boldsymbol{f}) = U\left(\boldsymbol{x}, [f_{0.1} \cdots f_{0.N}]^T\right)$$
$$= -\min\{f \in \{f_{0.1}, \ldots, f_{0.N}\} : G(\boldsymbol{x}, f)\}$$
$$+ \beta \left[\frac{\max\{\max\{f \in \{f_{0.1}, \ldots, f_{0.N}\} : |S_{11}(\boldsymbol{x}, f)|\} - S_{\max}, 0\}}{S_{\max}} \right]^2, \quad (6.4)$$

in which $G(\boldsymbol{x}, f)$ is the antenna gain at the design $\boldsymbol{x}$ and frequency f, whereas $S_{\max}$ is the maximum acceptable reflection level at the operating frequencies (e.g., -10 dB). Minimization of the negative minimum gain over the operating frequencies amounts to maximizing the worst-case gain over $f_{0.1}$ through $f_{0.N}$. The second term in (6.4) is a penalty function that contributes to the primary objective if the condition $|S_{11}(\boldsymbol{x}, f)| \leq S_{\max}$ is violated for at least one operating frequency. The penalty coefficient β determines the constraint "hardness". The merit function oriented towards bandwidth extension or improvement of other performance figures can be defined in a similar manner.

In order to maintain computational efficiency, it is advantageous to construct the surrogate model only in the vicinity of $U_F(F)$. Within the nested kriging framework, the first approximation of this set is obtained using the reference designs $\boldsymbol{x}^{(j)} = [x_1^{(j)} \cdots x_n^{(j)}]^T$, $j = 1, \ldots, p$, optimized w.r.t. the objective vectors $\boldsymbol{f}^{(j)} = [f_1^{(j)} \cdots f_N^{(j)}]$; $\boldsymbol{f}^{(j)}$ should be allocated in a possibly uniform manner in F. As mentioned before, $\boldsymbol{x}^{(j)}$ may be available from the previous design work

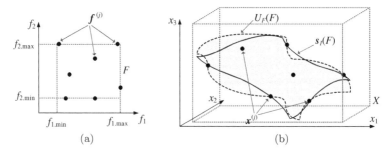

Fig. 6.1. Fundamental components of the nested kriging framework (here, illustrated for two performance figures and 3D parameter space): (a) objective space F, (b) parameter space X, the reference designs, the optimum design manifold $U_F(F)$, and the first-level model image $s_I(F)$. The manifolds $U_F(F)$ and $s_I(F)$ do not coincide for all $f \in F$ but the agreement is perfect for all the reference designs $x^{(j)}$, which are optimal in the sense of (6.1) (Koziel and Pietrenko-Dabrowska, 2019c).

on the same structure, or obtained specifically for the purpose of surrogate model construction.

The first-level surrogate $s_I(f) : F \to X$ is identified as a kriging interpolation model (Koziel and Pietrenko-Dabrowska, 2019c) using $\{f^{(j)}, x^{(j)}\}, j = 1, \ldots, p$, as the training set. As a matter of fact, it is an inverse model as its outputs are (nearly) optimum vectors of antenna geometry parameters corresponding to the objective vectors being the model input arguments. Figure 6.1 shows a graphical illustration of these concepts. It should be observed that $s_{I(F)}$ (the image of F) approximates $U_F(F)$ but it does not coincide with it because the number of the reference designs is normally limited.

6.2.2 *Surrogate model domain: Second-level surrogate*

As mentioned before, as long as the objective space F is of concern, it is sufficient to restrict the modeling process to the set $U_F(F)$. Notwithstanding, due to imperfect approximation of $U_F(F)$ by $s_{I(F)}$ it is necessary to extend the latter so that all (or most) of the designs $U_F(f)$ are included. In (Koziel and Pietrenko-Dabrowska, 2019c), the extension is implemented using the vectors normal to $s_{I(F)}$. For the sake of further considerations, the following notation is introduced:

- $\{v_n^{(k)}(\boldsymbol{f})\}, k = 1, \ldots, n - N$ — an orthonormal basis of vectors normal to $\boldsymbol{s}_{I(F)}$ at $\boldsymbol{f}$;
- $\boldsymbol{x}_{\max} = \max\{\boldsymbol{x}^{(k)}, k = 1, \ldots, p\}$;
- $\boldsymbol{x}_{\min} = \min\{\boldsymbol{x}^{(k)}, k = 1, \ldots, p\}$;
- $\boldsymbol{x}_d = \boldsymbol{x}_{\max} - \boldsymbol{x}_{\min}$ — a vector quantifying parameter variations within $\boldsymbol{s}_{I(F)}$.

Using this notation, the extension coefficients are defined as

$$\boldsymbol{\alpha}(\boldsymbol{f}) = [\alpha_1(\boldsymbol{f}) \cdots \alpha_{n-N}(\boldsymbol{f})]^T = 0.5T[|\boldsymbol{x}_d \boldsymbol{v}_n^{(1)}(\boldsymbol{f})| \cdots |\boldsymbol{x}_d \boldsymbol{v}_n^{(n-N)}(\boldsymbol{f})|]^T, \tag{6.5}$$

where T is a user-defined thickness parameter. The surrogate model domain X_S is allocated between the manifolds M_+ and M_-

$$M_{\pm} = \left\{ \boldsymbol{x} \in X : \boldsymbol{x} = \boldsymbol{s}_I(\boldsymbol{f}) \pm \sum_{k=1}^{n-N} \alpha_k(\boldsymbol{f}) \boldsymbol{v}_n^{(k)}(\boldsymbol{f}) \right\}. \tag{6.6}$$

We have

$$X_S = \left\{ \begin{array}{l} \boldsymbol{x} = \boldsymbol{s}_I(\boldsymbol{f}) + \sum_{k=1}^{n-N} \lambda_k \alpha_k(\boldsymbol{f}) \boldsymbol{v}_n^{(k)}(\boldsymbol{f}) : \boldsymbol{f} \in F, \\ -1 \leq \lambda_k \leq 1, \ k = 1, \ldots, n - N. \end{array} \right\}. \tag{6.7}$$

Figure 6.2 provides a graphical illustration of the manifolds M_+, M_-, and the domain X_S. The second-level (i.e., the actual) surrogate is constructed as a kriging interpolation model over the domain X_S. The training data is $\{\boldsymbol{x}_B^{(k)}, \boldsymbol{R}(\boldsymbol{x}_B^{(k)})\}_{k=1,\ldots,NB}$, where $\boldsymbol{x}_B^{(k)} \in X_S$ are

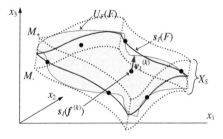

Fig. 6.2. The image $\boldsymbol{s}_{I(F)}$ of the first-level surrogate model and the normal vector $\boldsymbol{v}_1^{(k)}$ at $\boldsymbol{f}^{(k)}$; the manifolds M_- and M_+ and the surrogate model domain X_S defined as the orthogonal extension of $\boldsymbol{s}_{I(F)}$.

uniformly allocated samples, whereas R stands for the response of the EM-simulation model of the antenna. Detailed information about the design of experiments (sampling) procedure can be found in Koziel and Pietrenko-Dabrowska (2019c) (see also Section 6.3.3).

At this point, it should be mentioned that the choice of a particular modeling method (here, kriging interpolation) is not critical whatsoever. Both the first- and the second-level surrogates can be constructed using other techniques (as a matter of fact, each of these models can be rendered using a different approach). The major computational benefits result from confinement of the model domain, whereas selecting any particular approximation method is of secondary importance.

As demonstrated in Koziel and Pietrenko-Dabrowska (2019c), confining the surrogate model domain to X_S, the volume of which is significantly smaller than the volume of X, leads to considerable computational benefits. The major advantage is a possibility of constructing reliable surrogates using a small number of training samples and without formally restricting the ranges of antenna parameters and operating conditions the model is valid for. Even more importantly, these benefits are especially pronounced in higher-dimensional spaces where modeling within the conventional domain X is often computationally prohibitive.

6.3 Variable-Thickness Domain

This section discusses the importance of the surrogate model thickness parameter T in the light of the trade-offs between the model predictive power, the cost of training data acquisition, as well as the model utility. The latter is understood as the possibility of incorporating, within the domain, designs that are optimum for all objective vectors $f \in F$. This, in turn, requires that domain encapsulates the optimum design manifold $U_F(F)$. The concept and implementation of variable-thickness domain is subsequently introduced along with its incorporation into the nested kriging modeling framework.

6.3.1 *Domain thickness: Model accuracy vs. utility trade-offs*

The thickness parameter T plays an essential role in the nested kriging framework. On the one hand, it determines the volume of the

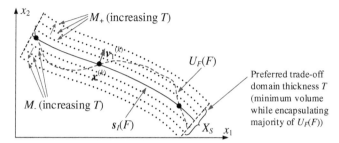

Fig. 6.3. The meaning of the domain thickness parameter T, here, explained for 2D parameter space. Increasing T enlarges the model domain which allows for encapsulating the optimum design manifold $U_F(F)$. At the same time, larger domain requires more training samples to render reliable surrogate. The desired trade-off is to find the minimum value of T enabling incorporation of the majority of $U_F(F)$ into X_S.

domain X_S. Increasing the value of T by a factor of two enlarges the domain by a factor 2^{n-N} (recall that n stands for the number of antenna parameters, whereas N is the number of figures of interest), which may be orders of magnitude for highly-dimensional space. This directly affects the number of training data samples required to yield an accurate model. In particular, keeping T at low values is beneficial for computational efficiency. On the other hand, reducing the domain thickness leads to keeping a part (or even a majority) of the optimum design manifold $U_{F(F)}$ out of X_S. This compromises the design utility of the surrogate because the true antenna optimum becomes unattainable for certain regions of the objective space as indicated in Fig. 6.3. Thus, the trade-off is to set T at the minimum value ensuring that most of $U_F(F)$ is in X_S. Clearly, this leaves a very limited room for reducing the cost of training data acquisition given a target level of the model error.

6.3.2 Variable-thickness domain: Definition and properties

The surrogate model domain defined using an appropriate value of the thickness parameter T (cf. Fig. 6.3) ensures incorporation of the majority of the optimum design manifold $U_F(F)$. At the same time, maintaining fixed T for the entire objective space is not really a necessity. In particular, as $U_F(F)$ coincides with $s_{I(F)}$ (the image of F through the first-level surrogate) at all reference designs $x^{(j)}$, at these points, the required domain thickness is zero.

Based on these observations, this section introduces a variable thickness domain as a mean to reduce the number of training data samples without compromising the model predictive power. The prerequisites are as follows:

- The thickness parameter should be a function of the objective vector f, i.e., $T = T(f)$;
- $T(f^{(j)}) = 0$ for all $j = 1, \ldots, p$, i.e., the domain thickness is zero for all reference designs $x^{(j)} = s_I(f^{(j)})$;
- $T(f) = T_{\max}$ (the T-value ensuring $U_F(F) \subset X_S$) for f corresponding to objective space locations where the expected first-level surrogate inaccuracy (w.r.t. $U_F(F)$) is the highest;
- $T(f)$ changes monotonically for intermediate locations.

The first task is to determine vectors f for which $||s_I(f) - U_F(f)||$ is maximized. One option would be to identify locations maximizing the mean square error (MSE) of the first-level model, which can be determined directly from the formulation of the kriging interpolation surrogate (Forrester and Keane, 2009). Another approach is to make a reasonable assumption that the error is the highest near the geometrical centers of the simplexes $S^{(j)}, j = 1, \ldots, N_S$, obtained by triangulating the reference designs, cf. Fig. 6.4. Here, Delaunay triangulation is employed (Borouchaki et al., 1996) to avoid degenerated simplexes. The simplex centers in the objective space will be denoted as $f_T^{(j)}$; their parameter space counterparts are $x_T^{(j)} = s_I(f_T^{(j)})$.

In order to develop the analytical form of the thickness function $T(f)$, the location of any $f \in F$ with respect to the simplexes $S^{(j)}$ has to be established. Let $x^{(j,k)}, k = 1, \ldots, N+1$, be the simplex

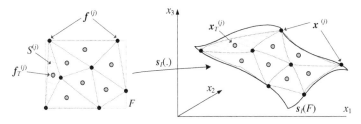

Fig. 6.4. Triangulation of the reference designs performed in the objective space. The centers $f_T^{(j)}$ of the resulting simplexes $S^{(j)}$ are mapped into the parameter space as $x_T^{(j)} = s_I(f_T^{(j)})$.

vertices, where $x^{(j.k)} \in \{x^{(1)}, \ldots, x^{(p)}\}$ (the set of reference designs) for $j = 1, \ldots, N_S$, and $k = 1, \ldots, N+1$. For any $j = 1, \ldots, N_S$, the following expansion holds:

$$f = \sum_{k=1}^{N+1} a_{j.k}(f) x^{(j.k)}, \qquad (6.8)$$

where $a_{j.1}(f) + \cdots + a_{j.N+1}(f) = 1$. Let $J(f)$ be any j for which $\min\{k : a_{j.k}(f) \geq 0\}$. It follows that $f \in H_{\text{conv}}(S^{(J(f))})$, where $H_{\text{conv}}(S)$ stands for the convex hull of S. Note that $a_{j.k}(f_T^{(i)}) = (N+1)^{-1}$ for $j = J(f_T^{(i)})$ and all $i = 1, \ldots, N_S$ (recall that $f_T^{(i)}$ were the simplex centers). At the same time, $a_{j.k}(f^{(i)}) = 1$ for a certain $k = 1, \ldots, N+1$, and zero for the remaining values of k. Furthermore, for $N > 1$, the expansion coefficients assume "intermediate" distributions for f being at the boundaries between the simplexes. For example, if $N = 2$, then, for f allocated at the center of the edge between two adjacent simplexes, we have $a_{j.k} = 1/N$ for two out of three values of k and zero for the remaining one.

The above examples indicate that it is reasonable to decide upon the analytical form of the $T(f)$ function based on the standard distribution $std([a_{j.1} \cdots a_{j.N+1}])$ of the expansion coefficients $a_{j.k}$. In particular, this distribution is zero for the simplex centers and attains its maximum equal to $std(E_{N+1})$, where the $(N+1) \times 1$ vector $E_{N+1} = [1\ 0 \cdots 0]^T$, for all reference designs.

This prompts us to the following definition

$$T(f) = T_{\max}[1 - std([a_{J(f).1} \cdots a_{J(f).N+1}])/std(E_{N+1})]. \qquad (6.9)$$

It can be observed that mapping (6.9) exhibits all properties listed at the beginning of this section. Figure 6.5 shows the exemplary $T(f)$ for a 2D objective space.

6.3.3 Nested kriging with variable-thickness domain

Incorporating the concept of variable-thickness domain into the nested kriging surrogate is straightforward. The domain definition

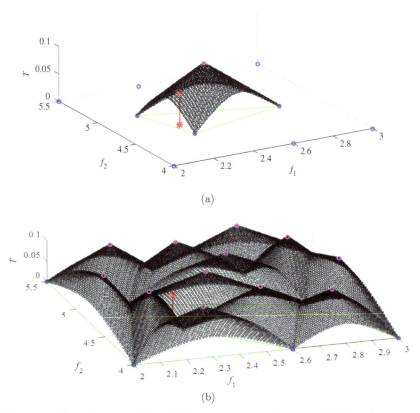

Fig. 6.5. Visualization of the thickness parameter function $T(\boldsymbol{f})$ for the exemplary objective space F: (a) a selected simplex and the corresponding thickness function. The vertical line illustrates the thickness value for a particular objective vector marked $*$, (b) $T(\boldsymbol{f})$ for the entire objective space. The reference objective vectors $\boldsymbol{f}^{(j)}$ are shown using circles on the $f_1 - f_2$ plane; the local maxima of $T(\boldsymbol{f})$ correspond to the simplex centers $\boldsymbol{f}_T^{(j)}$.

(6.7) still holds, the only difference is the definition of extension coefficients (6.5) which now takes the form of

$$\boldsymbol{\alpha}(\boldsymbol{f}) = [\alpha_1(\boldsymbol{f}) \; \cdots \; \alpha_{n-N}(\boldsymbol{f})]^T$$
$$= 0.5 T(\boldsymbol{f}) [|\boldsymbol{x}_d \boldsymbol{v}_n^{(1)}(\boldsymbol{f})| \; \cdots \; |\boldsymbol{x}_d \; \boldsymbol{v}_n^{(n-N)}(\boldsymbol{f})|]^T, \qquad (6.10)$$

with $T(\boldsymbol{f})$ given by (6.9).

A few comments should be made about the design of experiments procedure. In the original nested kriging with the fixed domain thickness parameter T, uniform sampling is straightforward owing to a one-to-one mapping between the unit interval $[0,1]^n$ onto X_S. Given the point $\boldsymbol{z} \in [0,1]^n$, the mapping h_1 (Koziel and Pietrenko-Dabrowska, 2019c)

$$\boldsymbol{y} = h_1(\boldsymbol{z}) = h_1([z_1 \cdots z_n]^T) = [f_{1.\min} + z_1(f_{1.\max} - f_{1.\min}) \\ \cdots f_{N.\min} + z_N(f_{N.\max} - f_{N.\min})] \times [-1 + 2z_{N+1} \cdots - 1 + 2z_n], \tag{6.11}$$

transforms $[0,1]^n$ onto the Cartesian product $F \times [-1,1]^{n-N}$. Subsequently, the function h_2

$$\boldsymbol{x} = h_2(\boldsymbol{y}) = h_2\left([y_1 \cdots y_n]^T\right) = \boldsymbol{s}_I\left([y_1 \cdots y_N]^T\right) \\ + \sum_{k=1}^{n-N} y_{N+k} \alpha_k\left([y_1 \cdots y_N]^T\right) \boldsymbol{v}_n^{(k)}\left([y_1 \cdots y_N]^T\right), \tag{6.12}$$

maps $F \times [-1,1]^{n-N}$ onto X_S.

Using these, uniformly distributed samples $\boldsymbol{x}_B^{(k)}$ in X_S are rendered as

$$\boldsymbol{x}_B^{(k)} = H\left(\boldsymbol{z}^{(k)}\right) = h_2\left(h_1(\boldsymbol{z}^{(k)})\right), \tag{6.13}$$

where $\{\boldsymbol{z}^{(k)}\}$, $k = 1, \ldots, N_B$, are uniformly distributed data points in $[0,1]^n$, here, obtained using Latin hypercube sampling (Beachkofski and Grandhi, 2002).

In the case of variable-thickness domain, the situation is more complex. Although the mapping (6.13) can still be used, the obtained sample allocation will not be uniform (w.r.t. to the objective space) because the sample distribution will be compressed in the vicinity of the reference designs. This would not bring any computational advantages: when using the above strategy, the sample density remains the same as for the fixed-thickness case around the simplex centers $\boldsymbol{f}_T^{(j)}$ but will be unnecessarily high around the reference points $\boldsymbol{f}^{(j)}$. In this work, a different approach is employed. More specifically, the procedure (6.11)–(6.13) is used for $T = T_{\max}$, then all data samples $\boldsymbol{x}_B^{(k)}$ such that $\boldsymbol{x}_B^{(k)} \notin X_S$ are rejected and the EM antenna model is not evaluated therein. Additionally, the training data set is supplemented by all reference designs. This allows for maintaining uniform sample arrangement throughout the domain and leads to significant computational savings as demonstrated in Section 6.4.

6.4 Demonstration Case Studies

This section discusses application of nested kriging with variable-thickness domain to modeling of antenna input characteristics. The method is compared to conventional surrogates (kriging and RBF) as well as the original nested kriging employing the fixed thickness parameter. The benchmark set includes two antenna structures, a dual-band dipole, and a broadband patch antenna with narrow ground plane.

6.4.1 Case I: Dual-band microstrip dipole antenna

Our first example is a dual-band dipole antenna shown in Fig. 6.6 (Qudrat-E-Maula and Shafai, 2014). The structure is implemented on Rogers RO4003 substrate ($\varepsilon_r = 3.38$, $h = 0.76$ mm) and described by eight independent parameters: $\boldsymbol{x} = [L_{rr}\ d\ W_s\ W_d\ S\ L_d\ L_{gr}\ W_{gr}]^T$ (all dimensions in mm except those with r-subscript which are relative). The parameters $W_r = 5, L_s = 5$, and $L_0 = 25$ are fixed. The feed line width $W_0 = 4.5$ mm is calculated to ensure 50 ohm impedance. Other parameters are $L_r = L_{rr}((W_s - W_0)/2 - W_d - d), W_g = W_{gr}W_s$, $L_g = L_{gr}(L_0 - W_g/2 + W_0/2)$, and $g = W_d$. The computational model is implemented in CST Microwave Studio (∼900,000 mesh cells, simulation time 250 s).

The surrogate model is to be constructed for the objective space parameterized by the operating frequencies f_1 and $f_2 = Kf_1$ for $2.0\,\text{GHz} \leq f_1 \leq 3.5\,\text{GHz}$, and $1.2 \leq K \leq 1.6$.

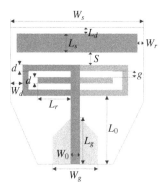

Fig. 6.6. Dual-band dipole antenna geometry (Qudrat-E-Maula and Shafai, 2014). Ground plane shown using the light-gray shade.

Nine reference designs are assigned, optimized for all combinations of $f_1 \in \{2.0, 2.75, 3.5\}$ GHz and $K \in \{1.2, 1.4, 1.6\}$. The parameter space X is determined by the lower and upper bounds for design variables, $l = [0.55\ 1.0\ 47.0\ 2.5\ 3.8\ 3.8\ 0.5\ 0.24]^T$, and $u = [1.0\ 2.5\ 77.0\ 6.5\ 6.0\ 6.0\ 1.0\ 0.5]^T$, set up using the reference points. For the sake of computational efficiency, the reference designs are obtained using the feature-based optimization framework (Koziel, 2015).

Numerical results for the method of Section 6.4 were obtained by constructing the surrogate model for several training data sets of various sizes as reported in Table 6.1. The nested kriging with variable-thickness domain has been compared to conventional surrogates (kriging and radial basis functions) as well as the (original) nested kriging with fixed domain thickness (Koziel and Pietrenko-Dabrowska, 2019c). The nested kriging models have been constructed for two values of $T_{\max}$, 0.02 and 0.05. It can be observed that both nested kriging models (with the fixed-thickness and variable-thickness domains) exhibit comparable predictive power for the training data set sizes from 200 samples up (for the original nested kriging model) and from 50 samples for the model with variable-thickness domain, which is much better than that of the conventional surrogates. For smaller data sets, the results for the model with variable-thickness domain are not representative because of very limited numbers of samples (15 and 19, respectively), which makes the estimation of the predictive power unreliable. It should be emphasized that the number of training samples required by the variable-thickness models is significantly smaller. The computational savings are as high as 75% with the average of 70% over the considered training data sets. Figure 6.7 shows the surrogate responses for the variable-thickness nested kriging model at the selected testing designs along with the corresponding EM simulation data. The visual agreement between the characteristics is satisfactory.

A supplementary validation has been conducted to determine whether variable-thickness domain affects the design utility of the surrogates. Figure 6.8 shows the initial and the optimized designs obtained for the selected objective vectors. It can be observed that the designs obtained by optimizing the fixed- and variable-thickness domain surrogates are comparable. This indicates that restricting the domain does not compromise the model ability to yield

Table 6.1. Modeling results for antenna of Fig. 6.6.

Number of training samples	Relative RMS error Conventional models (domain X)		Relative RMS error (Original) nested kriging with fixed domain thickness		Number of training samples[a]	Relative RMS error Nested kriging with variable-thickness domain	
	Kriging (%)	RBF (%)	$T_{max} = 0.02$ (%)	$T_{max} = 0.05$ (%)		$T_{max} = 0.02$ (%)	$T_{max} = 0.05$ (%)
50	55.6	58.1	14.5	18.9	15	25.7	31.2
100	46.7	46.9	11.2	13.5	19	18.30	22.5
200	41.9	44.1	6.9	7.8	55	8.9	12.1
400	36.9	41.2	6.1	6.6	108	5.9	6.9
800	35.5	37.5	4.1	4.9	208	5.0	6.7
1600	33.2	35.8	3.9	4.2	405	4.3	3.8

Note: [a]The number of training samples is determined by the relative volume of the variable-thickness and fixed-thickness domains (see the design of experiments procedure description at the end of Section 6.3).

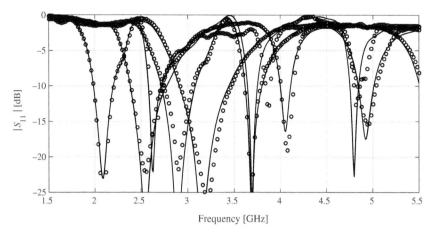

Fig. 6.7. Dipole antenna: reflection responses at the selected test designs: EM model (—), nested kriging with variable-thickness domain obtained using 400 training samples and $T_{max} = 0.02$ (o).

high-quality designs. This was to be expected because the very definition of the variable-thickness domain is founded on reducing the thickness in the vicinity of the reference designs (where the image of the first-level model agrees perfectly with the optimum design manifold) and increasing it between the reference points (where the two manifolds deviate from each other).

The design optimized for $f_1 = 2.45$ GHz and $f_2 = 3.3$ GHz has been fabricated and measured for additional validation. Figure 6.9 shows the photographs of the antenna prototype, the reflection and realized gain characteristics, as well as the H- and E-plane radiation patterns. The agreement between the simulation and measurement data is satisfactory. Small discrepancies originate mostly from assembly inaccuracies as well as the measurement setup (e.g., the 90-degree bend used to mount the antenna was not included in the computational model).

6.4.2 Case II: Broadband patch antenna

The second example is a broadband patch antenna with a narrow ground plane shown in Fig. 6.10 (Wang et al., 2018c). The structure

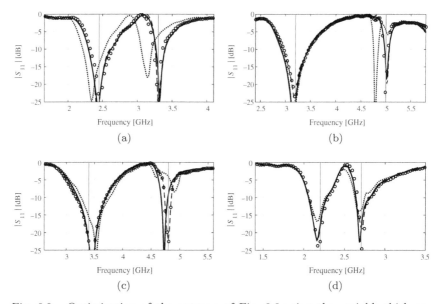

Fig. 6.8. Optimization of the antenna of Fig. 6.6 using the variable-thickness nested kriging surrogate for the four pairs of operating frequencies: (a) $f_1 = 2.45$ GHz, $f_2 = 3.3$, GHz, (b) $f_1 = 3.2$ GHz, $f_2 = 5.0$ GHz, (c) $f_1 = 3.4$ GHz, $f_2 = 4.8$ GHz, (d) $f_1 = 2.2$ GHz, $f_2 = 2.7$ GHz. Shown are: the initial design obtained from the first-level kriging model ($\cdots$), response of the optimized variable-thickness nested kriging surrogate (- - -), response of the optimized fixed-thickness surrogate (o), and EM-simulated antenna response at the variable-thickness surrogate model optimum (—).

is described by five parameters $\boldsymbol{x} = [W\ L\ dW\ W_g\ h_r]^T$. The computational model is implemented in CST Microwave Studio and evaluated using the transient solver (~400,000 cells, simulation time 94 s).

The substrate parameters, dielectric permittivity ε_r and height h are the operating conditions being a part of the objective space. The model incorporates the SMA connector. The design optimality is understood as minimization of the antenna reflection within at least 10% fractional bandwidth symmetric w.r.t. the target center frequency f_0.

We aim at constructing the surrogate model within the objective space defined by the following ranges of the center frequency and surrogate parameters: $3.0\,\text{GHz} \leq f_0 \leq 6.0\,\text{GHz}$, permittivity $2.0 \leq \varepsilon_r \leq 5.0$ and height $0.5\,\text{mm} \leq h \leq 1.0\,\text{mm}$. There are thirteen reference designs assigned, corresponding to $\{f_0, \varepsilon_r, h\} = \{3.0, 2.0, 0.5\}$,

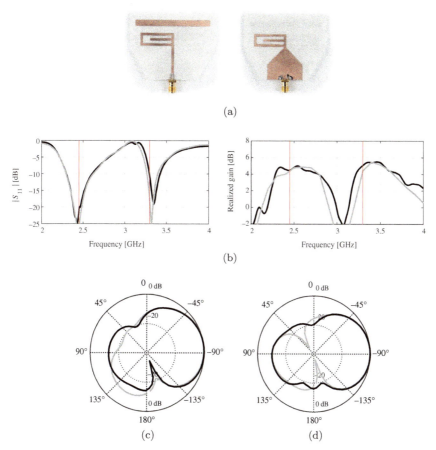

Fig. 6.9. Antenna of Fig. 6.6 optimized for $f_1 = 2.45\,\text{GHz}$, $f_2 = 3.3\,\text{GHz}$: (a) photographs of the prototype, (b) reflection $|S_{11}|$, (c) end-fire realized gain, (d) yz-plane radiation pattern at both operating frequencies (2.45 GHz and 3.3 GHz on the left- and right-hand-side, respectively). Simulated and measured characteristics shown in gray and black, respectively.

$\{3.0, 2.0, 1.0\}$, $\{3.0, 5.0, 0.5\}$, $\{3.0, 5.0, 1.0\}$, $\{4.5, 3.5, 0.75\}$, $\{4.5, 3.5, 0.5\}$, $\{4.5, 3.5, 1.0\}$, $\{4.5, 2.0, 0.75\}$, $\{4.5, 5.0, 0.75\}$, $\{6.0, 2.0, 0.5\}$, $\{6.0, 2.0, 1.0\}$, $\{6.0, 5.0, 0.5\}$, and $\{6.0, 5.0, 1.0\}$. The parameter space X is determined by the lower and upper bounds for design variables $\boldsymbol{l} = [12.5\ 10.0\ 4.0\ 8.0\ 0.02]^T$, and $\boldsymbol{u} = [40.0\ 34.0\ 17.0\ 10.0\ 0.2]^T$.

Model validation has been arranged similarly as for the first example. Table 6.2 shows the numerical data for the models constructed

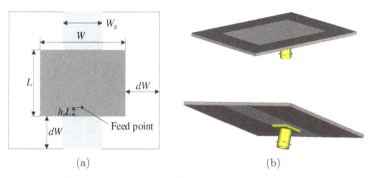

Fig. 6.10. Broadband patch antenna (Wang et al., 2018c): (a) geometry (ground plane shown using light gray shade), (b) 3D views.

using the training data sets of different sizes. The benchmark methods include conventional models established in the parameter space X (kriging interpolation and radial basis functions), as well as the nested kriging with fixed-thickness domain. The nested kriging models have been constructed for the two values of $T_{\max}$, 0.025 and 0.05. The obtained results are consistent with those reported in Section 6.4.1. In particular, the accuracy of both nested kriging models (with the fixed-thickness and variable-thickness domains) is essentially the same; however, the number of samples required by the variable-thickness domain model is considerably smaller.

The computational savings are close to 40%. The responses for the variable-thickness nested kriging model at the selected testing designs along with the corresponding EM simulation data are shown in Fig. 6.11.

The nested kriging surrogate with variable-thickness domain has been employed for antenna optimization to provide an additional validation. The objective was to assess whether variable-thickness domain affects the model capability to reach the optimum design across the objective space. The results, obtained for the selected objective vectors have been shown in Fig. 6.12. It can be noted that the designs produced by variable- and fixed-thickness domain models are of comparable quality, which corroborates the design utility of the presented technique.

One of the optimized designs, corresponding to $f_0 = 4.8\,\text{GHz}$, $\varepsilon_r = 3.38$, $h = 0.51\,\text{mm}$, has been fabricated and measured for additional validation. The antenna structure has been implemented on

Table 6.2. Modeling results for antenna of Fig. 6.9.

Number of training samples	Relative RMS error					Number of training samples[a]	Relative RMS error	
	Conventional models (domain X)		(Original) nested kriging with fixed domain thickness				Nested kriging with variable-thickness domain	
	Kriging (%)	RBF (%)	$T_{max} = 0.025$ (%)	$T_{max} = 0.05$ (%)			$T_{max} = 0.025$ (%)	$T_{max} = 0.05$ (%)
50	43.4	52.0	10.1	17.8		30	13.0	14.8
100	21.2	25.4	7.9	9.0		63	7.8	9.1
200	16.0	18.2	7.2	6.5		125	7.3	6.4
400	12.8	14.7	4.7	6.1		247	4.9	5.3
800	9.8	11.3	3.3	4.8		505	3.8	5.1

Note: [a]The number of training samples is determined by the relative volume of the variable-thickness and fixed-thickness domains (see the design of experiments procedure description at the end of Section 6.3).

236   *Surrogate Modeling for High-Frequency Design*

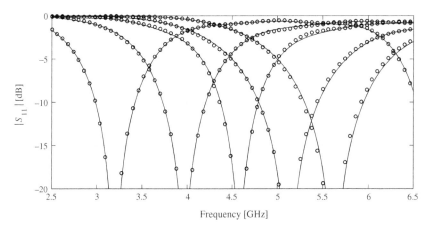

Fig. 6.11. Patch antenna: reflection responses at the selected test designs: EM model (—), nested kriging with variable-thickness domain obtained using 800 training samples and $T_{\max} = 0.05$ (o).

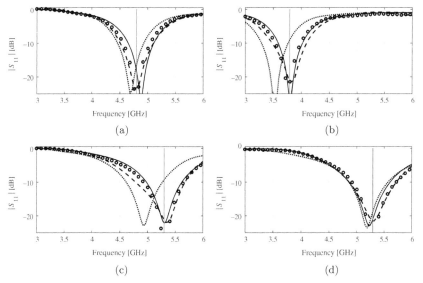

Fig. 6.12. Optimization of the antenna of Fig. 6.10 using the variable-thickness nested kriging surrogate set up using 800 samples ($T_{\max} = 0.05$): initial design obtained from the first-level model ($\cdots$), surrogate model response at the optimized design obtained using variable thickness model (- - -), response of the optimized fixed-thickness surrogate (o), EM simulated response at the variable-thickness surrogate model optimum (—): (a) $f_0 = 4.8$ GHz, $\varepsilon_r = 3.38$, $h = 0.51$ mm, (b) $f_0 = 3.8$ GHz, $\varepsilon_r = 2.5$, $h = 0.76$ mm, (c) $f_0 = 5.3$ GHz, $\varepsilon_r = 3.38$, $h = 0.81$ mm, (d) $f_0 = 5.3$ GHz, $\varepsilon_r = 4.4$, $h = 1.0$ mm.

Performance-Driven Inverse/Forward Modeling of Antennas   237

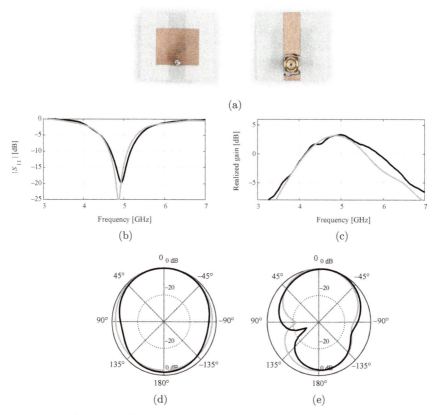

Fig. 6.13. Antenna of Fig. 6.10 optimized for $f_0 = 4.8\,\text{GHz}$, $\varepsilon_r = 3.38$ and $h = 0.51\,\text{mm}$, implemented on RO4003C substrate: (a) antenna prototype, (b) reflection, (c) realized gain, (d) H-plane pattern at 4.8 GHz, (e) E-plane pattern at 4.8 GHz; simulations (gray) and measurements (black).

Rogers RO4003C substrate. Figure 6.13 shows the photographs of the antenna prototype, the reflection and realized gain characteristics, as well as the H- and E-plane radiation patterns. The agreement between the simulation and measurement data is satisfactory.

6.5 Summary and Discussion

This chapter discussed a variation of the nested kriging framework, the recent performance-driven modeling method (Koziel and

Pietrenko-Dabrowska, 2019c). The keystone of the approach was to introduce a functional dependence of the lateral size (thickness) of the surrogate model domain on the objective space location. This was motivated by the fact that the first-level model uncertainty (in terms of representing the optimum design manifold) ranges from zero at the reference designs used to set up the surrogate, to its maxima between these designs. From this perspective, maintaining fixed domain thickness, as in the original nested kriging, is not imperative.

The variable-thickness domain enables a significant reduction of the training data set size without compromising the model accuracy. The presented concept has been explained, formalized, and its implementation has been validated in a comprehensive manner using two examples of microstrip antennas. The numerical results conclusively demonstrate that the computational savings pertinent to training data acquisition can be as high as 70% while maintaining the predictive power of the surrogates essentially intact as compared to the fixed-thickness version. The same has been observed concerning the design utility of the models, i.e., their ability to encapsulate the optimum designs across the entire objective space.

Acknowledgments

This work was supported in part by the Icelandic Centre for Research (RANNIS) Grant 206606 and by National Science Centre of Poland Grant 2018/31/B/ST7/02369.

References

Alzahed, A.M., Mikki, S.M., Antar, Y.M.M. (2019). Nonlinear mutual coupling compensation operator design using a novel electromagnetic machine learning paradigm, *IEEE Ant. Wireless Prop. Lett.*, vol. 18, no. 5, pp. 861–865.

Barmuta, P., Ferranti, F., Gibiino, G.P., Lewandowski, A., Schreurs, D.M.M.P. (2015). Compact behavioral models of nonlinear active devices using response surface methodology, *IEEE Trans. Microw. Theory Tech.*, vol. 63, no. 1, pp. 56–64.

Beachkofski, B., Grandhi, R. (2002). Improved distributed hypercube sampling, *American Institute of Aeronautics and Astronautics*, paper AIAA 2002-1274.

Borouchaki, H., George, P.L., Lo, S.H. (1996). Optimal Delaunay point insertion, *Int. J. Numer. Meth. Eng.*, vol. 39, no. 20, pp. 3407–3437.

Cai, J., King, J., Yu, C., Liu, J., Sun, L. (2018). Support vector regression-based behavioral modeling technique for RF power transistors, *IEEE Microw. and Wireless Comp. Lett.*, vol. 28, no. 5, pp. 428–430.

Cervantes-González, J.C., Rayas-Sánchez, J.E., López, C.A., Camacho-Pérez, J.R., Brito-Brito, Z., Chávez-Hurtado, J.L. (2016). Space mapping optimization of handset antennas considering EM effects of mobile phone components and human body, *Int. J. RF Microw. CAE*, vol. 26, no. 2, pp. 121–128.

de Villiers, D.I.L., Couckuyt, I., Dhaene, T. (2017). Multi-objective optimization of reflector antennas using kriging and probability of improvement, *Int. Symp. Ant. Prop.*, pp. 985–986, San Diego, USA.

Dong, Y., Choi, J., Itoh, T., (2018). Vivaldi antenna with pattern diversity for 0.7 to 2.7 GHz cellular band applications, *IEEE Ant. Wireless Prop. Lett.*, vol. 17, no. 2, pp. 247–250.

Dong, J., Qin, W., Wang, M. (2019). Fast multi-objective optimization of multi-parameter antenna structures based on improved BPNN surrogate model, *IEEE Access*, vol. 7, pp. 77692–77701.

Du, J., Roblin, C. (2018). Stochastic surrogate models of deformable antennas based on vector spherical harmonics and polynomial chaos expensions: application to textile antennas, *IEEE Trans. Ant. Prop.*, vol. 66, no. 7, pp. 3610–3622.

Easum, J.A., Nagar, J., Werner, D.H. (2017). Multi-objective surrogate-assisted optimization applied to patch antenna design, *Int. Symp. Ant. Prop.*, pp. 339–340, San Diego, USA.

Easum, J.A., Nagar, J., Werner, P.L., Werner, D.H. (2018). Efficient multi-objective antenna optimization with tolerance analysis through the use of surrogate models, *IEEE Trans. Ant. Prop.*, vol. 66, no. 12, pp. 6706–6715.

Fakih, M.A., Diallo, A., le Thuc, P., Staraj, R., Mourad, O., Rachid, E.A. (2019). Optimization of efficient dual band PIFA system for MIMO half-duplex 4G/LTE and full-duplex 5G communications, *IEEE Access*, vol. 7, pp. 128881–128895.

Felicio, J.M., Bioucas-Dias, J.M., Costa, J.R., Fernandes, C.A. (2019). Antenna design and near-field characterization for medical microwave imaging applications, *IEEE Trans. Ant. Prop.*, vol. 67, no. 7, pp. 4811–4824.

Forrester, A.I.J., Keane, A.J. (2009), Recent advances in surrogate-based optimization, *Prog. Aerospace Sci.*, vol. 45, pp. 50–79.

Gorissen, D., Crombecq, K., Couckuyt, I., Dhaene, T., Demeester, P. (2010). A surrogate modeling and adaptive sampling toolbox for computer based design, *J. Mach. Learn. Res.*, vol. 11, pp. 2051–2055.

Guo, D., He, K., Zhang, Y., Song, M. (2016). A multiband dual-polarized omnidirectional antenna for indoor wireless communication systems, *IEEE Ant. Wireless Prop. Lett.*, vol. 16, pp. 290–293.

Hassan, E., Noreland, D., Augustine, R., Wadbro, E., Berggren, M. (2015). Topology optimization of planar antennas for wideband near-field coupling, *IEEE Trans. Ant. Prop.*, vol. 63, no. 9, pp. 4208–4213.

Hassan, A.K.S.O., Etman, A.S., Soliman, E.A. (2018). Optimization of a novel nano antenna with two radiation modes using kriging surrogate models, *IEEE Photonic J.*, vol. 10, no. 4, art. no. 4800807.

Hu, R., Monebhurrun, V., Himeno, R., Yokota, H., Costen, F. (2018). An adaptive least angle regression method for uncertainty quantification in FDTD computation, *IEEE Trans. Ant. Prop.*, vol. 66, no. 12, pp. 7188–7197.

Jacobs, J.P. (2016). Characterization by Gaussian processes of finite substrate size effects on gain patterns of microstrip antennas, *IET Microwaves Ant. Prop.*, vol. 10, no. 11, pp. 1189–1195.

Jacobs, J.P., Koziel, S. (2014). Two-stage framework for efficient Gaussian process modeling of antenna input characteristics, *IEEE Trans. Antennas Prop.*, vol. 62, no. 2, pp. 706–713.

Jha, K.R., Bukhari, B., Singh, C., Mishra, G., Sharma, S.K. (2018). Compact planar multistandard MIMO antenna for IoT applications, *IEEE Trans. Ant. Prop.*, vol. 66, no. 7, pp. 3327–3336.

Kennedy, M.C., O'Hagan, A. (2000). Predicting the output from complex computer code when fast approximations are available, *Biometrika*, vol. 87, pp. 1–13.

Koziel, S. (2015). Fast simulation-driven antenna design using response-feature surrogates, *Int. J. RF & Micr. CAE*, vol. 25, no. 5, pp. 394–402.

Koziel, S. (2017). Low-cost data-driven surrogate modeling of antenna structures by constrained sampling, *IEEE Antennas Wireless Prop. Lett.*, vol. 16, pp. 461–464.

Koziel, S., Ogurtsov, S. (2014). *Antenna Design by Simulation-Driven Optimization. Surrogate-Based Approach*, New York, Springer.

Koziel, S., Ogurtsov, S., Cheng, Q.S., Bandler, J.W. (2014). Rapid EM-based microwave design optimization exploiting shape-preserving response prediction and adjoint sensitivities, *IET Microwaves, Ant. Prop.*, vol., 8, no. 10, pp. 775–781.

Koziel, S., Sigurdsson, A.T. (2018). Triangulation-based constrained surrogate modeling of antennas, *IEEE Trans. Ant. Prop.*, vol. 66, no. 8, pp. 4170–4179.

Koziel, S., Unnsteinsson, S.D. (2018). Expedited design closure of antennas by means of trust-region-based adaptive response scaling, *IEEE Ant. Wireless Propag. Lett.*, vol. 17, no. 6, pp. 1099–1103.

Koziel, S., Pietrenko-Dabrowska, A. (2019a). Reduced-cost electro-magnetic-driven optimization of antenna structures by means of trust-region gradient-search with sparse Jacobian updates, *IET Microwaves Ant. Prop.*, vol. 13, no. 10, pp. 1646–1652.

Koziel, S., Pietrenko-Dabrowska, A. (2019b). Variable-fidelity simulation 9 models and sparse gradient updates for cost-efficient optimization of 10 compact antenna input characteristics, *Sensors*, vol. 19, no. 8, Article. no. 1806.

Koziel, S., Pietrenko-Dabrowska, A. (2019c). Performance-based nested surrogate modeling of antenna input characteristics, *IEEE Trans. Ant. Prop.*, vol. 67, no. 5, pp. 2904–2912.

Lalbakhsh, A., Afzal, M.U., Esselle, K.P. (2017). Multiobjective particle swarm optimization to design a time-delay equalizer metasurface for an electromagnetic band-gap resonator antenna, *IEEE Ant. Wireless Prop. Lett.*, vol. 16, pp. 915–915.

Marelli, S., Sudret, B. (2014). UQLab: a framework for uncertainty quantification in Matlab, *The 2nd Int. Conf. on Vulnerability and Risk Analysis and Management (ICVRAM 2014)*, University of London, UK, July 13–15, pp. 2554–2563.

Mener, S., Gillard, R., Roy, L. (2016). A dual-band dual-circular-polarization antenna for Ka-band satellite communications, *IEEE Ant. Wireless Prop. Lett.*, vol. 16, pp. 274–277.

Nie, Z., Zhai, H., Liu, L., Li, J., Hu, D., Shi, J. (2019). A dual-polarized frequency-reconfigurable low-profile antenna with harmonic suppression for 5G application, *IEEE Ant. Wireless Prop. Lett.*, vol. 18, no. 6, pp. 1228–1232.

Qian, J.F., Chen, F.C., Xiang, K.R., Chu, Q.X. (2019). Resonator-loaded multi-band microstrip slot antennas with bidirectional radiation patterns, *IEEE Trans. Ant. Prop.*, vol. 67, no. 10, pp. 6661–6666.

Qudrat-E-Maula, M., Shafai, L. (2014). A dual band microstrip dipole antenna, *Int. Symp. Antenna Tech. Applied Electrom. (ANTEM)*, Victoria, BC, Canada, July 13–16.

Queipo, N.V., Haftka, R.T., Shyy, W., Goel, T., Vaidynathan, R., Tucker, P.K. (2005). Surrogatebased analysis and optimization, *Progress in Aerospace Sciences*, vol. 41, no. 1, p. 128.

Petrocchi, A., Kaintura, A., Avolio, G., Spina, D., Dhaene, T., Raffo, A., Schreurs, D.M.P.-P. (2017). Measurement uncertainty propagation in transistor model parameters via polynomial chaos expansion, *IEEE Microwave Wireless Comp. Lett.*, vol. 27, no. 6, pp. 572–574.

Rawat, A., Yadav, R.N., Shrivastava, S.C. (2012). Neural network applications in smart antenna arrays: A review, *AEU – Int. J. Elec. Comm.*, vol. 66, no. 11, pp. 903–912.

Rayas-Sanchez, J.E. (2016). Power in simplicity with ASM: tracing the aggressive space mapping algorithm over two decades of development and engineering applications, *IEEE Microw. Mag.*, vol. 17, no. 4, pp. 64–76.

Tak, J., Kantemur, A., Sharma, Y., Xin, H. (2018). A 3-D-printed W-band slotted waveguide array antenna optimized using machine learning, *IEEE Ant. Wireless Prop. Lett.*, vol. 17, no. 11, pp. 2008–2012.

Torun, H.M., Swaminathan, M. (2019). High-dimensional global optimization method for high-frequency electronic design, *IEEE Trans. Microwave Theory Techn.*, vol. 67, no. 6, pp. 2128–2142.

Tsukamoto, K., Arai, H. (2016). Optimization of smooth walled horn antenna using multilevel fast multipole method, *Int. Symp. Ant. Prop. (ISAP)*, Okinawa, Japan, Oct. 24–28.

Ullah, U., Koziel, S. (2018). A broadband circularly polarized wide-slot antenna with a miniaturized footprint, *IEEE Ant. Wireless Prop. Lett.*, vol. 17, no. 12, pp. 2454–2458.

Wang, F., Cachecho, P., Zhang, W., Sun, S., Li, X., Kanj, R., Gu, C. (2016). Bayesian model fusion: large-scale performance modeling of analog and mixed-signal circuits by reusing early-stage data, *IEEE Trans. on Computer-Aided Design of Integrated Circuits and Systems (TCAD)*, vol. 35, no. 8, pp. 1255–1268.

Wang *et al.* (2018a). Efficient gradient-based optimization of pixel antenna with large-scale connections, *IET Microwaves Ant. Prop.*, vol. 12, no. 3, pp. 385–389.

Wang *et al.* (2018b). An implantable and conformal antenna for wireless capsule endoscopy, *IEEE Ant. Wireless Prop. Lett.*, vol. 17, no. 7, pp. 1153–1157.

Wang *et al.* (2018c). Broadband patch antenna with narrow width ground plane, *IEEE Int. Symp. Ant. Prop.*, Boston, MA, USA, July 8–13.

Xu, Y., Wang, J., Ge, L., Wang, X., Wu, W. (2018). Design of a notched-band Vivaldi antenna with high selectivity, *IEEE Ant. Wireless Prop. Lett.*, vol. 17, no. 1, pp. 62–65.

Yan, S., Soh, P.J., Vandenbosch, G.A.E. (2015). Wearable dual-band magneto-electric dipole antenna for WBAN/WLAN applications, *IEEE Trans. Ant. Prop.*, vol. 63, no. 9, pp. 4165–4169.

Yücel, A.C., Bağcı, H., Michielssen, E. (2015). An ME-PC enhanced HDMR method for efficient statistical analysis of multiconductor transmission line networks, *IEEE Trans. Comp. Packaging and Manufacturing Techn.*, vol .5, no. 5, pp. 685–696.

Zeng, J., Luk, K.M. (2019). Single-layered broadband magnetoelectric dipole antenna for new 5G application, *IEEE Ant. Wireless Prop. Lett.*, vol. 18, no. 5, pp. 911–915.

Zhao, A., Ren, Z. (2019). Size reduction of self-isolated MIMO antenna system for 5G mobile phone applications, *IEEE Ant. Wireless Prop. Lett.*, vol. 18, no. 1, pp. 152–156.

© 2022 World Scientific Publishing Europe Ltd.
https://doi.org/10.1142/9781800610750_0007

Chapter 7

Sampling Methods for Surrogate Modeling and Optimization

Qingsha S. Cheng and Zhen Zhang

Abstract

In recent years, surrogate-assisted optimization algorithm is commonly used in antenna and microwave component design optimization. Sampling method is the first step in data-driven surrogate model establishment. The purpose of the chapter is to explain and clarify various sampling methods and to inspire new implementations and applications. Four widely used conventional sampling methods are reviewed including the full factorial design, Monte Carlo sampling, Latin hypercube sampling, and space-infill sampling method. Two adaptive sampling methods are reviewed and explained. A hybrid sampling method considers both superiority of objective function values and uniformity of sample distribution, therefore, improving the performance of surrogate modeling and optimization at the same time. An adaptive sampling region updating strategy establishes model and performs optimization iteratively in local region(s). The size and the direction of movement of each local region are adaptively changed. As a result, the sampling efficiency is improved. These methods are valuable for EM-simulation-based data-driven surrogate modeling and optimization.

Keywords: EM simulation, data-driven surrogate, Latin hypercube sampling, Monte Carlo sampling, adaptive sampling, adaptive region updating

7.1 Introduction

Full-wave electromagnetic (EM) simulations are required for high-frequency component design. This makes the design procedure likely time-consuming. Various surrogate models such as polynomial regression(Chavez-Hurtado and Rayas-Sanchez, 2016), support vector machine (Jacobs *et al.*, 2013), artificial neural networks (ANN) (Washington, 2002; Kim *et al.*, 2007; Zhang *et al.*, 2018), and Gaussian process regression (Jacobs, 2015; Wang *et al.*, 2018; Jacobs and Koziel, 2014; Liu *et al.*, 2014, 2019) are established to reduce the computational cost of microwave component modeling and surrogate-assisted optimization. For any data-driven surrogate modeling method, the quality of samples affects the efficiency and accuracy of surrogate modeling and surrogate-assisted optimization processes.

Conventional (static) sampling methods widely used for surrogate-assisted microwave component modeling and optimization include Monte Carlo sampling (MCS) (Metropolis and Ulam, 1949), and Latin hypercube sampling (LHS) (Mckay *et al.*, 1979), etc. These sampling methods focus on improving the uniformity of distributed samples in the design space. For microwave components, in certain regions of the design space, the performance with respect to design parameters may be very sensitive. Therefore, "good" samples satisfying the design specifications may be located in small regions in the design space. If uniform sampling methods are used, the good samples (satisfying design specifications) and the poor samples (not satisfying design specifications) are imbalanced. Surrogate models based on these imbalanced samples may not allow for accurate optimization. High-quality adaptive sampling methods are desired for surrogate modeling and optimization to improve the efficiency of sampling (Zhang *et al.*, 2020; Na and Zhang, 2013). In addition to sampling methods, the selection of sampling space is also the key to achieve the sampling efficiency. An adaptive sampling region updating strategy (Song *et al.*, 2019) changes sampling space during an optimization process for higher efficiency.

In this chapter, the available sampling methods for surrogate modeling and optimization are categorized in Fig. 7.1. Conventional sampling methods are reviewed in Section 7.2. A hybrid sampling

Sampling Methods for Surrogate Modeling and Optimization

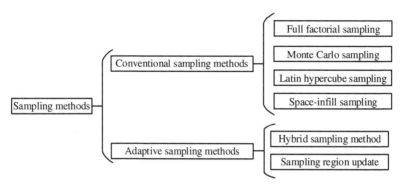

Fig. 7.1. Category of sampling methods for surrogate modeling and optimization.

method is introduced in Section 7.3. An adaptive sampling region updating strategy is described in Section 7.4. Section 7.5 concludes the chapter.

7.2 Conventional Sampling Methods

Sampling methods are responsible for filling the n-dimensional hypercube $[lb; ub]^n$ for surrogate modeling and surrogate-assisted optimization. Usually, the samples are assigned in an n-dimensional unit hypercube $[0; 1]^n$, then they are denormalized to the n-dimensional hypercube $[lb; ub]^n$ (Sola and Sevilla, 1997). There are many denormalization methods, and the most commonly used method is linear denormalization, as defined by

$$x_i^{(j)} = s_i^{(j)} \cdot (x_{ub}^{(j)} - x_{lb}^{(j)}) + x_{lb}^{(j)}, \qquad (7.1)$$

where $x_i^{(j)}$ is the jth dimension element of the ith sample in $[lb; ub]^n$. $s_i^{(j)} x$ is the jth dimension element of the ith sample in $[0; 1]^n$. $x_{ub}^{(j)}$ and $x_{lb}^{(j)}$ are the upper bound and the lower bound of the jth dimension element of the sample.

In this section, four conventional sampling methods are reviewed including the full factorial design, Monte Carlo sampling, Latin hypercube sampling, and space-infill sampling method.

7.2.1 Full factorial sampling method

Full factorial sampling is a popular sampling method (Bechhofer et al., 1995). It is a deterministic sampling method. A full factorial sampling is simply defined as the nodes of an L-level grid ($L > 1$). A 3-level grid and a 10-level grid are shown in Figs. 7.2(a) and 7.2(b) respectively. For n-dimensional input variable $\boldsymbol{x} = (x_1, \ldots, x_n)^T$, a full factorial sampling in $[0; 1]^n$ is obtained as follows:

$$\hbar_{FF} = \left\{ \frac{i_1 - 1}{L - 1}, \frac{i_2 - 1}{L - 1}, \ldots, \frac{i_n - 1}{L - 1} \right\}, \quad \forall i_1, i_2, \ldots, \quad i_n = 1, \ldots, L. \tag{7.2}$$

The practical use of full factorial designs is usually limited to no more than 2 or 3 qualitative levels because sample set size ($m = L^n$) drastically increases with both the dimension n and the number of levels L.

7.2.2 Monte Carlo sampling method

Monte Carlo sampling is another technique that is straightforward to implement (Gerstner and Heinz, 2012). The classical Monte Carlo sampling (MCS) was proposed by Metropolis and Ulam in 1949. MCS is a random sampling method. An MCS is simply defined as:

$$\hbar_{\text{MCS}} = \left\{ x_j^{(i)}, \quad i = 1, \ldots, m, \quad j = 1, \ldots, n \right\}^n, \tag{7.3}$$

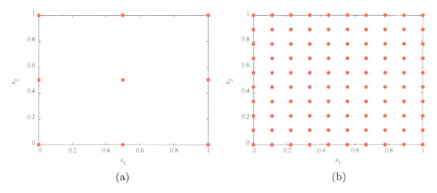

Fig. 7.2. Illustration of full factorial sampling in two-dimensional space: (a) with nine samples, and (b) with 100 samples.

Sampling Methods for Surrogate Modeling and Optimization   249

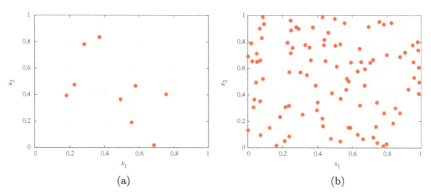

Fig. 7.3. Illustration of MCS in two-dimensional space: (a) with 9 samples, and (b) with 100 samples.

where $x_j^{(i)}$ is the jth dimension of ith samples and it is a uniform random number in [0; 1]. In order to guarantee the samples truly fill in the hypercube, a large number of samples m is required. Illustration of MCS in two-dimensional space is shown in Fig. 7.3.

7.2.3 Latin hypercube sampling method

Latin hypercube sampling (LHS) is one of the most popular sampling techniques with better uniformity than the MCS method. LHS (Mckay et al., 1979) ensures the margin uniformity of the samples for input variables.

LHS has a rather simple definition that makes it a very popular technique to obtain space-filling samples. The samples are uniformly distributed in [0;1] for each dimension. A uniform LHS is defined as

$$\hbar_{\text{LHS}} = \left\{ x_j^{(i)} = \frac{\pi_j^{(i)} - 1 + u_j^{(i)}}{m}, i = 1, \ldots, m, \ j = 1, \ldots, n \right\}^n, \quad (7.4)$$

where $x_j^{(i)}$ is the jth dimension of ith samples, $\pi_j^{(i)}$ is a random integer in the sequence $\{1, \ldots, m\}$. $u_j^{(i)}$ is a random variable uniformly distributed over [0; 1]. Illustration of LHS in two-dimensional space is shown in Fig. 7.4.

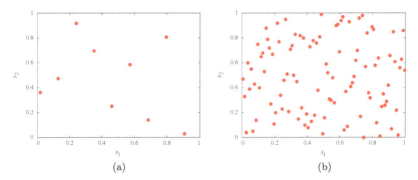

Fig. 7.4. Illustration of Latin hypercube sampling in two-dimensional space: (a) with nine samples, and (b) with 100 samples.

7.2.4 Space-infill sampling method

The above sampling methods are all based on Euclidean distance between samples (Villagran et al., 2015). In addition to the Euclidean distance-based sampling methods, the space-infill sampling methods such as Delaunay triangulation (Delaunay, 1934) and Voronoi tessellations/diagram (Voronoi, 1908) are also popular. They are used to establish surrogate for antennas (Koziel et al., 2017). Voronoi diagram method aims at filling the space uniformly by placing samples in Voronoi cells. The larger the size of a Voronoi cell, the more under-sampled the region is. Hence, the samples can be placed in the large Voronoi cells to enhance the homogeneity of sample numbers in the design space. A simple Voronoi diagram construction based on 10 randomly generated sample points in the two-dimensional domain is illustrated in Fig. 7.5(a). Each sample is surrounded by a cell boundary.

The Delaunay triangulation method minimizes the maximum angle for all the triangles in the triangulation. For example, a set of K points is a triangulation such that no point is inside the boundary of any triangle formed by the points in two-dimension space. The Delaunay triangulation method can be generalized to an N-dimensional case where a triangle becomes an N-dimensional simplex. Further details can be found in Garud et al. (2017) and Klein and Lee (2013). The illustration of the Delaunay triangulation method in two-dimensional space with 10 samples is shown in Fig. 7.5(b).

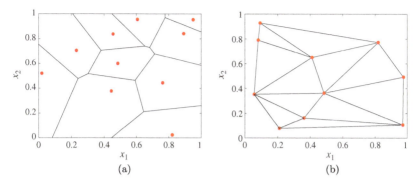

Fig. 7.5. Illustration of the space-filling sampling method: (a) Voronoi diagram method, and (b) Delaunay triangulation method.

7.3 A Hybrid Sampling Method for Surrogate Modeling and Optimization

Recently, in order to improve the modeling and optimization performance of microwave components, a hybrid sampling method has been proposed in Zhang et al. (2020). The set of samples includes the initial samples and the infilling samples. The initial samples are obtained using a conventional sampling method, such as MCS or LHS. The infilling samples are added through an iterative infilling process using a local sampling method and a global sampling method. The flowchart of the hybrid sampling method is shown in Fig. 7.6.

7.3.1 Local sampling

Since microwave components are sensitive to parameter variations in certain regions of design space, in many cases, design samples satisfying the design specifications fall in a small region of design space. For example, in the landscape of a multiplexer, the design objective function value is lower than −10 dB (Liu et al., 2018) in a small area in the design space. More high-quality local samples in that region may be required. The local sampling method determines the number and locations of the good samples using the following steps.

Step 1: Formulate the design objective function (F_{obj}) of the microwave component problem.

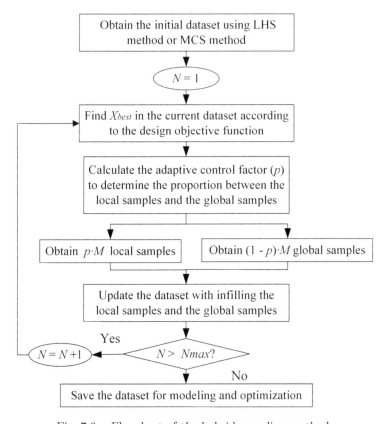

Fig. 7.6. Flowchart of the hybrid sampling method.

Step 2: Calculate the adaptive control factor p to determine the number of infilling local samples in each iteration. p increases with the number of iteration. p is obtained as

$$p = 0.1 + 0.8 \frac{N}{N_{\max}} \zeta, \qquad (7.5)$$

where N is the current iteration. $N_{\max}$ is the maximum number of iteration. $\zeta \in [0; 1]$ is a random number.

Step 3: Find the best sample in the current dataset. The best sample corresponds to the smallest value of F_{obj}.

$$\boldsymbol{X}_{\text{best}} = \arg\min(F_{\text{obj}}). \qquad (7.6)$$

Step 4: Generate $p \cdot M$ local samples $\boldsymbol{S}_{l \cdot k}(k = 1 \cdots p \cdot M)$ using the greedy strategy as in Equation (7.3), M is the number of infilling samples of the current iteration.

$$S_{l.k} = X_{\text{best}} + F \cdot (X_{r1} - X_{r2}), \tag{7.7}$$

where X_{r1} and X_{r2} are two random samples in the current dataset. F is a weighting factor, which determines the weight of the $X_{r1} - X_{r2}$, and F is set to 0.7.

In each iteration, through the above four steps, local samples are added to the dataset.

7.3.2 Global sampling

In each iteration, $(1 - p) \cdot M$ global (infilling) samples are obtained using the MCS with a minimum distance rejection principle. If the minimum of Euclidean distances between an infilling sample and all samples of the current dataset is smaller than the parameter μ (e.g., 0.001), the sample is rejected. The minimum distance $D_{\min}$ is defined as

$$D_{\min} = \min_{1 \leq j \leq L} (\|X_{rs} - X_j\|), \tag{7.8}$$

where L is the number of samples in the current dataset, X_{rs} is the infilling samples obtained using the MCS method, and X_j is the jth sample of the current dataset.

The samples generated by the adaptive hybrid sampling method for a two-dimensional Ackley function (Liu et al., 2017) are shown in Fig. 7.7.

7.3.3 Microwave verification example

A bandstop microstrip filter with quarter-wavelength resonant open stubs (Na et al., 2017; Feng et al., 2014) is used as a test example. The filter's geometry is shown in Fig. 7.8. A substrate with a thickness of $H = 0.635$ mm and dielectric constant $\varepsilon_r = 9.4$ is used and the width of a 50Ω feed line is $W_0 = 0.635$ mm. The design variables are $X = [W_1, W_2, L_0, L_1, L_2]^T$ mm. The response of the filter is the S_{21} characteristic; the frequency range of interest is 5–15 GHz. The number of initial samples is 30, and the number of infilling samples at each iteration is $M = 20$. The performance of the hybrid sampling method is tested in the context of surrogate modeling and

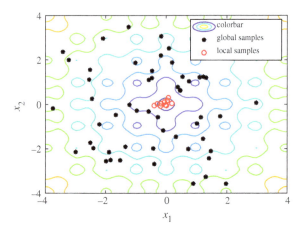

Fig. 7.7. Illustration of the hybrid adaptive sampling method.

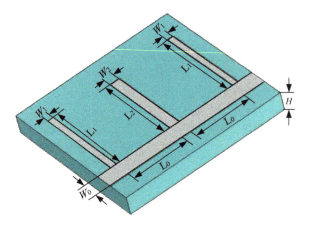

Fig. 7.8. Geometry of a bandstop microstrip filter.

optimization. The EM simulation samples are obtained using CST MWS (Studio, 2018) (27,000 meshes, 2 min).

The range of the original design space is defined as $\boldsymbol{LB} \leq \boldsymbol{X} \leq \boldsymbol{UB}$ ($\boldsymbol{LB}$ and $\boldsymbol{UB}$ are the lower bound vector and upper bound vector of design space, respectively) as shown in Table 7.1, which is the same as Metropolis and Ulam (1949). An enlarged design space $\boldsymbol{LB}' \leq \boldsymbol{X} \leq \boldsymbol{UB}'$ ($\boldsymbol{LB}' = 0.95\,\boldsymbol{LB}$ and $\boldsymbol{UB} = 1.05\,\boldsymbol{UB}'$) is also defined to show the effectiveness of the method.

$$F_{\text{obj}}(\boldsymbol{X}) = \max_{f}\{|S_{21}(\boldsymbol{X},f)|\}, \qquad (7.9)$$

where f is 9.6–10.4 GHz the stopping band.

7.3.3.1 *Modeling performance*

For verifying the modeling performance, multilayer perceptron (MLP) neural network models with three hidden layers (10 nodes in each layer) are established using various numbers of samples based on MCS, LHS, and the adaptive hybrid sampling methods. The number of testing EM samples is 120. The testing error is evaluated using root mean square (RMS) at all testing samples and frequencies. The testing errors of the models are shown in Table 7.2.

Table 7.1. Range of design space/unit (mm).

	Bounds	W_1	W_2	L_0	L_1	L_2
Original design space	**LB**	0.15	0.23	2.54	2.54	2.54
	UB	0.19	0.27	3.35	3.35	3.35
Enlarged design space	**LB'**	0.14	0.22	2.41	2.41	2.41
	UB'	0.2	0.28	3.52	3.52	3.52

Table 7.2. Testing error comparison using various sampling methods.

Design space	Number of training samples	MCS (%)	LHS (%)	Hybrid Sampling Method (%)
[LB UB]	50	3.84	5.99	**3.26**
	100	2.07	2.34	**1.81**
	200	1.53	1.59	**1.39**
	400	1.39	1.47	**1.31**
	800	1.37	1.45	**1.31**
[LB' UB']	50	8.78	6.51	**4.74**
	100	6.93	6.86	**3.67**
	200	4.76	5.12	**2.91**
	400	3.98	4.01	**2.86**
	800	3.01	3.27	**2.82**

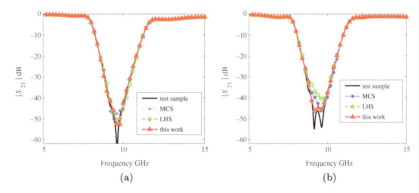

Fig. 7.9. Modeling results with 200 different samples: (a) original design space $[\boldsymbol{LB}\ \boldsymbol{UB}]$, and (b) enlarged design space $[\boldsymbol{LB'}\ \boldsymbol{UB'}]$. This work refers to hyper sampling method (Zhang et al., 2020).

It can be seen from Table 7.2 that the MLP model based on the hybrid sampling method achieves a better overall modeling performance than the MLP models based on MCS and LHS sampling methods. For the same number of samples, the testing error of the hybrid sampling method is much smaller than that of the MCS and the LHS. In the enlarged design space $\boldsymbol{LB'} \leq \boldsymbol{X} \leq \boldsymbol{UB'}$, the hybrid method achieves a more significant performance enhancement than in the original design space. When the number of samples is larger than 400, the modeling performance does not improve much for all three methods. This is because there are enough samples in the dataset. The square error of test samples with 50 training samples are shown in Fig. 7.9. It illustrates the modeling results of two designs in the original design space and the enlarged design space, both with 200 samples.

7.3.3.2 *Optimization performance*

The established MLP models are optimized using differential evolution (DE) algorithm (Storn and Price, 1997) and the objective function (7.9). The settings of the DE algorithm are as follows: the maximum iteration number 1000, the population size of 50, the scaling factor of 0.6, and the crossover probability of 0.8. The final design is verified using CST. The optimization results are shown in Table 7.3.

From Table 7.3, it is observed that the optimal designs obtained using the hybrid sampling method achieve the best objective value

Table 7.3. Optimal objective value comparison using various sampling methods.

Design space	Number of training samples	MCS (dB)	LHS (dB)	Hybrid Sampling Method (dB)
[**LB UB**]	50	−30.98	−32.76	**−39.91**
	100	−36.56	−35.92	**−43.87**
	200	−39.66	−40.05	**−46.34**
	400	−40.93	−40.87	**−46.41**
	800	−41.75	−43.64	**−46.44**
[**LB′ UB′**]	50	−28.98	−28.13	**−37.81**
	100	−29.57	−31.68	**−39.09**
	200	−32.90	−35.74	**−39.81**
	400	−36.92	−39.87	**−40.63**
	800	−39.04	−39.90	**−42.65**

compared with the MCS and LHS methods in both the original design space [**LB UB**] and the enlarged space [**LB′ UB′**]. This indicates that the hybrid sampling method has exploited a larger number of high-quality samples in the vicinity of the near-optimal design. Figure 7.10 shows the EM responses of the designs obtained in the two design spaces with 200 samples. To produce the same or better design, fewer samples are required using the hybrid sampling method, therefore, higher computational efficiency is achieved.

The hybrid sampling method combines the advantage of both local sampling and global sampling. The numerical results show that the hybrid sampling method renders more accurate MLP neural network models compared with MCS and LHS using the same number of samples. Furthermore, the MLP neural network models based on the hybrid sampling method lead to better objective function values when applied to design optimization.

7.4 Adaptive Sampling Region Updating for Surrogate-Assisted Optimization

In the surrogate-assisted optimization process, surrogate accuracy is improved through the infilling sample criterion method (Liu et al., 2017, 2019) performed on a high-fidelity EM model (fine

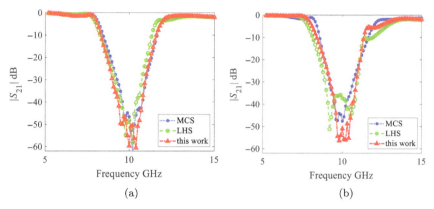

Fig. 7.10. EM responses of the optimal design with 200 different samples: (a) original design space $[\boldsymbol{LB}\ \boldsymbol{UB}]$, and (b) enlarged design space $\boldsymbol{LB'}\ \boldsymbol{UB'}$. This work refers to hyper sampling method (Zhang et al., 2020).

model). However, the efficiency of optimization is low due to a large number of fine model samples to be acquired in the sampling region. An adaptive sampling region updating method is proposed to allow better performance of surrogate-assisted optimization (Song et al., 2019).

7.4.1 Adaptive sampling region updating strategy

The adaptive sampling region updating algorithm is an iterative process of local (region) modeling and optimization. Therefore, the size and the direction of movement of each local region are two important factors. The adaptive sampling region updating strategy takes advantage of multi-fidelity surrogates (Koziel et al., 2012) building on a series of coarse models (coarse-discretization EM models $\{\boldsymbol{R}_{c.j}\}$, $j = 1, \ldots, K$, where a larger value of j means higher fidelity). $\boldsymbol{R}_{c,i}$ is evaluated by the same EM solver as a high-fidelity model but faster.

First, a local surrogate model $s^{(0,1)}$ is established using first multi-fidelity $\boldsymbol{R}_{c.1}$ and optimized in the region containing the initial design similar to Koziel et al. (2012), until the new local optimal design lies within the interior of the current region. At this point, we assume as the optimal design of the surrogate model at the lowest accuracy has been found.

Then, this optimal design is used as the regional center of the next higher accuracy local surrogate model (related to second multi-fidelity $R_{c.2}$) and the new surrogate model is established and optimized using a similar procedure. The process of establishing and optimizing the local surrogate model at the same accuracy is defined as one iteration of our algorithm.

The fidelity of the local surrogate increases with the number of iterations. In addition, the current optimal design serves as a starting point for the next iterative process. The size of the local region at the beginning of each iteration is adaptively updated according to a judgment factor. The judgment factor of the $(i+1)$th surrogate can be expressed as $\delta_r^{(i+1)}$, defined as

$$\delta_r^{(i+1)} = \frac{1}{m} \sum_{l=1}^{m} \beta_l r_l^{(i+1)}, \tag{7.10}$$

where β_l can adjust the weight of the judgment factor at different frequencies if needed (here, $\beta_l = 1$, $l = 1, \ldots, m$). $r_l^{(i+1)}$ representing the similarity between the surrogate model and coarse model at the lth frequency is defined for,

$$r_l^{(i+1)} = \frac{R_{c.j}^l\left(x^{(i+1,j)}\right) - R_{c.j}^l\left(x^{(i,j)}\right)}{s_l^{(i+1,j)}\left(x^{(i+1,j)}\right) - s_l^{(i,j)}\left(x^{(i,j)}\right)}, \tag{7.11}$$

where $R_{c.j}^l$ is the lth component of the coarse model response vector $R_{c.j}$.

The judgment factor $\delta_r^{(i+1)}$ represents the overall degree of similarity (for all frequency points) between the current local surrogate model and its corresponding coarse model. If $\delta_r^{(i+1)} \geq \delta_1$ (here, $\delta_1 = 0.7$), it means that the current surrogate model is accurate, and the next local region size can be expanded ($d^{(i+1,j)} = 2d^{(i,j)}$). If $\delta_r^{(i+1)} \leq \delta_2$ (here, $\delta_2 = 0.3$), it indicates the current surrogate may not be sufficiently accurate to reflect the true response of the current coarse model. Therefore, the next local region size is reduced ($d^{(i+1,j)} = d^{(i,j)}/2$). When $\delta_2 \leq \delta_r^{(i+1)} \leq \delta_1$ the next region size is kept the same as the current one and the starting point of the $(i+1)$th iteration is taken from the local optimal design found in the ith iteration.

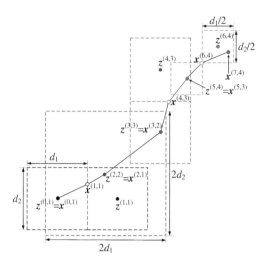

Fig. 7.11. Multi-fidelity local surrogate model optimization (conceptual illustration for design parameter dimension $n = 2$). A local surrogate model $s^{(0,1)}$ (3) is established using $\mathbf{R}_{c,1}$ in the initial region of size $[d_1\ d_2]$ (corresponding to $\mathbf{d} = [d_1/2\ d_2/2]^T$). It is optimized to yield a new design $\mathbf{x}^{(1,1)}$, where the next surrogate model $\mathbf{s}^{(1,1)}$ is established and centered at $\mathbf{z}^{(1,1)}$. After $\mathbf{z}^{(2,2)} = \mathbf{x}^{(2,1)}$ is found, the algorithm enters the second iteration and the next surrogate model is built on $\mathbf{R}_{c,2}$ with an enlarged local region size assuming a judgment factor $\sigma_r > 0.7$. The process continues and terminates after 4 iterations $\mathbf{x}_s^* = \mathbf{x}^{(7,4)}$.

The multi-fidelity surrogate models are built using coarse models of various fidelities. Each iteration of the algorithm corresponds to a coarse model $\mathbf{R}_{c.j}$ at a specific fidelity. The algorithm converges in K iterations and K coarse models of different fidelity are used. Figure 7.11 shows the conceptual illustration of the process. The process exploits four coarse models: $\mathbf{R}_{c.1}$, $\mathbf{R}_{c.2}$, $\mathbf{R}_{c.3}$, and $\mathbf{R}_{c.4}$. For each iteration, a particular color is used to mark local regions, their centers, and local optimal designs. The dashed line is used to mark the local regions in which the surrogate model is established. The solid dot $\mathbf{z}^{(i,j)}$ is the center of the local region and the hollow circles are the current local optimal design $\mathbf{x}^{(i,j)}$. The red dot represents the optimal design of the surrogate model obtains after $K = 4$ iterations.

7.4.2 Surrogate refinement using fine model

The local surrogate modeling is established using polynomial regression. The optimized design $\mathbf{x}_s$ of the last local surrogate model based on $R_{c.K}$ (closest to the fine model accuracy) is a good estimation of the optimal design of the fine (high-fidelity) model. To improve the accuracy of the optimal design for the fine model, the last surrogate model is further refined. The refinement process is performed in the local region centered at x_s. The surrogate model $s^*(\bullet)$ is established using a coarse model $R_{c.K}$ within this region.

In order to improve the surrogate model $s^*(\bullet)$, an input space mapping method (Cheng et al., 2012) is applied. A vector c^* is obtained to improve the matching between s^* and fine model R_f in the parameter extraction process

$$c^* = \arg\min_{c} \|R_f(x_s^*) - s^*(x_s^* + c)\|. \quad (7.12)$$

Then an output space mapping (Song et al., 2019) correction term $R_f(x_s^*) - s^*(x_s^* + c^*)$ is added to the surrogate model to ensure the zero-order consistency between the surrogate and the fine model at x_s^* (Sola and Sevilla, 1997). The refined optimal design is then found as

$$x^* = \arg\min_{x_s^* - d \leq x \leq x_s^* + d} U(s^*(x + c^*) + [R_f(x_s^*) - s^*(x_s^* + c^*)]), \quad (7.13)$$

where d is the same as the last local region. The refinement for local surrogate requires only one evaluation of R_f.

The flowchart of the surrogate-assisted optimization based on an adaptive sampling region updating is shown in Fig. 7.12.

7.4.3 Antenna verification examples

The monopole antenna structure, shown in Fig. 7.13, is optimized. The structure consists of a rectangular radiating patch fed through a microstrip line, and a ground plane with a small rectangular slot and a long ground stub. According to Liu et al. (2013), the small rectangular slot with the size of $s_1 \times s_2$ is to improve the impedance matching at high frequencies, and the long stub for isolation enhancement is folded to reduce the overall antenna footprint area. The structure is designed on a Rogers substrate, RO4350B ($\varepsilon_r = 3.55, \tan\delta = 0.0037$, and $h = 0.8\,\text{mm}$).

In this case, 11 design variables $x = [lg, g, a_1, a_2, l_1, l_2, w_1, s_1, s_2, o_1, o_3]^T$ mm are considered. The feedline width is fixed at $w_0 = 1.8\,\text{mm}$. The initial design is $x^{(0,1)} = [7, 1, 10, 10, 11, 3, 1, 3, 1, 6, 1]^T$ mm. The size of the initial local region is

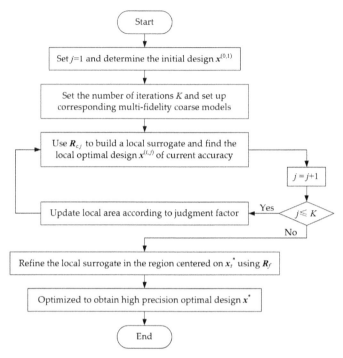

Fig. 7.12. Surrogate-assisted optimization based on adaptive sampling region updating.

defined as $\boldsymbol{d}^{(0,1)} = [1, 1, 0.1, 0.1, 1, 1, 0.1, 1, 0.1, 1, 0.1]^T$ mm. The fine model $\boldsymbol{R}_f$ is evaluated using the CST (1168614 mesh cells, evaluation time is 6.5 min). We optimized this example using three low-fidelity models: $\boldsymbol{R}_{c.1}$ (134560 meshes, 1 min), $\boldsymbol{R}_{c.2}$ (187110 meshes, 1.5 min), and $\boldsymbol{R}_{c.3}$ (338680 meshes, 2.8 min). As a comparison, this example is also optimized using the original algorithm, in which the low-fidelity model $\boldsymbol{R}_c$ is the same as $\boldsymbol{R}_{c.2}$ with 187110 meshes and an evaluation time of 1.5 min.

The optimization goal is to meet the following specifications: reflection coefficients $|S_{11}| \leq -12$dB for 3.1–10.6 GHz. In addition, the antenna footprint $S(\boldsymbol{x}) = (o_1 + 0.5 \cdot a_2 + o_3) \times (l_g + l_1 + w_1)$ is also considered while maintaining the above reflection coefficient value. The size reduction is selected as the main optimization problem, and a good reflection coefficient value is achieved using the penalty function approach. Thus, the local region optimization is

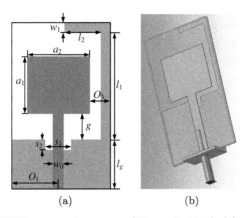

Fig. 7.13. The UWB monopole antenna (Liu et al., 2013): (a) geometrical details of the radiator and the ground plane; and (b) three-dimensional visualization of the antenna with the SMA connector in CST (The design is electrically small. In order to ensure the reliability of the simulation results, the EM model is simulated with an SMA connector attached).

defined as

$$x^{(i,j)} = \arg\min(S(x) + \beta \cdot c(s^{(i,j)}(x))), \qquad (7.14)$$

where β is a penalty factor (here, $\beta = 1000$) and $s^{(i,j)}(\bullet)$ is the ith multi-fidelity surrogate model built using $R_{c.j}$, $(j = 1, 2, 3)$. And c is a penalty function defined as

$$c(s^{(i,j)}(x)) = \max\{\max\{|S_{11}|_{3.1\,\text{GHz to}\,10.6\,\text{GHz}}\} + 12, 0\}. \qquad (7.15)$$

After the optimal solution of the final local surrogate model is found, the surrogate model is refined. The optimal design $x^* = [11.2, 0.7, 8.85, 8.77, 13.8, 5.06, 1.01, 2.3, 1.9, 6.96, 1.71]^T$ obtained by the refined local surrogate model (8) is very close to that of the fine model and they both satisfy the specification, as shown in Fig. 7.14.

Figure 7.15 shows the response of fine model R_f at the initial design $x^{(0,1)}$ and the optimal design obtained x^* by the proposed algorithm. The corresponding antenna size is 13 mm × 26 mm (338 mm^2). In addition, the figure also shows the results obtained by the original algorithm and the direct optimization algorithm (genetic algorithm). It can be seen that the

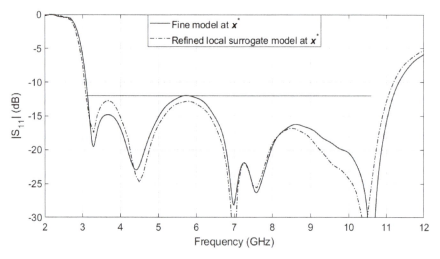

Fig. 7.14. A UWB monopole antenna: response of the optimal design in the refined local surrogate model and the fine model.

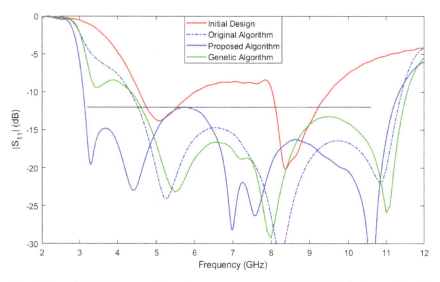

Fig. 7.15. A UWB monopole antenna: responses of the initial design and optimal designs obtained by various algorithms. The proposed algorithm refer to the method proposed by Song *et al.* (2019).

$x_1^* = [8.96, 0.64, 7.78, 8.46, 13.16, 3.7, 0.97, 2.18, 2.07, 6.7, 1.8]^T$ optimal design of the original algorithm does not satisfy the design specification. One of the optimization results of the genetic algorithm

Table 7.4. A UWB monopole antenna: Optimization cost.

Method	Required the number of evaluations		Optimization cost Absolute(h)	Relative to R_f
Direct optimization	Fine model R_f	497	53.8	497
Original algorithm	Coarse model R_c	161	4	38
	Fine model R_f	2	0.2	2
	Total optimization time	N/A	4.2	40
Proposed algorithm (Song et al., 2019)	Coarse model $R_{c.1}$	69	1.1	10
	Coarse model $R_{c.2}$	46	1.1	10
	Coarse model $R_{c.3}$	69	3.2	30
	Fine model R_f	2	0.2	2
	Total optimization time	N/A	5.6	52

$x_2^* = [10.84, 0.64, 9.13, 8.68, 14.04, 4.82, 1.14, 2.13, 1.84, 7.12, 1.59]^T$ is also shown in Fig. 7.13 (three runs of the genetic algorithm are performed and none of the results satisfies the design specification).

The optimization costs of the UWB monopole antenna using the three algorithms are compared in Table 7.4. The original algorithm terminated after 4.2 h without finding the design satisfying the specifications. The direct optimization (genetic algorithm) is not able to obtain a design satisfying the specification after an average of 53.8 h (CST genetic algorithm terminates after 497 fine model evaluations by default). Our proposed algorithm obtains a design satisfying the specification and the total optimization cost is 5.6 h.

7.5 Conclusions

In this chapter, a variety of sampling methods for surrogate modeling and surrogate-assisted optimization are reviewed, including conventional sampling, hybrid sampling, and the adaptive sampling region strategy. The conventional sampling methods are reviewed. These methods focus on the uniformity of sample distribution in the design space. The hybrid sampling method is reviewed and explained. It is found robust since both uniformity and superiority of objective function values are considered, which can improve the effectiveness of surrogate modeling and optimization at the same time. The adaptive sampling method is presented. It is combined with multi-fidelity local surrogates to improve the sampling efficiency and to reduce the cost of EM simulation-based optimization.

Acknowledgments

This work was supported in part by the National Natural Science Foundation of China Grant 62071211 and the University Key Research Project of Guangdong Province under Grant 2018KZDXM063.

References

Bechhofer, R.E., Santner, T.J., Goldsman, D.M. (1995). Design and analysis of experiments for statistical selection, screening and multiple comparisons, *Technometrics*, vol. 38, no. 3, pp. 289–290.

Chavez-Hurtado, J.L., Rayas-Sanchez, J.E. (2016). Polynomial-based surrogate modeling of RF and microwave circuits in frequency domain exploiting the multinomial theorem, *IEEE Trans. Microw. Theory Tech.*, vol. 64, no. 12, pp. 4371–4381.

Cheng, Q.S., Bandler, J.W., Koziel, S. (2012). Space mapping: The state of the art. *Int. J. RF Microw. Comput. Aided Eng.*, vol. 22, no. 6, pp. 639–651.

Delaunay, B. (1934). Sur la sphere vide. a la memoire de georges voronoi. Bulletin de l'Académie des Sciences de l'URSS, *Classe des sciences mathématiques et na*, pp. 793–800.

Feng, F., Zhang, C., Gongal-Reddy, V., Zhang, Q., Ma, J. (2014). Parallel space-mapping approach to EM optimization, *IEEE Trans. Microw. Theory Tech.*, vol. 62, no. 5, pp. 1135–1148.

Garud, S.S., Karimi, I.A., Kraft, M. (2017). Design of computer experiments: A review, *Comput. Chem. Eng.*, vol. 106, pp. 71–95.

Gerstner, T., Heinz, S. (2012). Dimension- and time-adaptive multilevel Monte Carlo methods, *Lecture Notes in Computational ence and Engineering*, vol. 88, pp. 107–120.

Jacobs, J.P., Koziel, S., Ogurtsov, S. (2013). Computationally efficient multi-fidelity Bayesian support vector regression modeling of planar antenna input characteristics, *IEEE Trans. Antennas Propag.*, vol. 61, no. 2, pp. 980–984.

Jacobs, J.P., Koziel, S. (2014). Two-stage framework for efficient Gaussian process modeling of antenna input characteristics, *IEEE Trans. Antennas Propag.*, vol. 62, no. 2, pp. 706–713.

Jacobs, J.P. (2015). Efficient resonant frequency modeling for dual-band microstrip antennas by Gaussian process regression, *IEEE Antennas and Wireless Propagation Lett.*, vol. 14, pp. 337–341.

Klein, R., Lee, D. (2013). *Voronoi Diagrams and Delaunay Triangulations* (Singapore: World Scientific).

Kim, Y., Keely, S., Ghosh, J., Ling, H. (2007). Application of artificial neural networks to broadband antenna design based on a parametric frequency model, *IEEE Trans. Antennas Propag.*, vol. 55, no. 3, pp. 669–674.

Koziel, S., Ogurtsov, S., Szczepanski, S. (2012). Local response surface approximations and variable-fidelity electromagnetic simulations for computationally efficient microwave design optimisation, *IET Microw. Antennas Propag.*, vol. 6, no. 9, pp. 1056–1062.

Koziel, S., Sigurdsson, A.T., Szczepanski, S. (2017). Uniform sampling in constrained domains for low-cost surrogate modeling of antenna input characteristics, *IEEE Antennas and Wireless Propagation Letters*, vol. 17, no. 1, pp. 164–167.

Liu, L., Cheung, S.W., Yuk, T.I. (2013). Compact MIMO antenna for portable devices in UWB applications, *IEEE Trans. Antennas Propag.*, vol. 61, no. 8, pp. 4257–4264.

Liu, B., Aliakbarian, H., Ma, Z.K., Vandenbosch, G.A.E., Gielen, G., Excell, P. (2014). An efficient method for antenna design optimization based on evolutionary computation and machine learning techniques, *IEEE Trans. Antennas Propag.*, vol. 62, no. 1, pp. 7–18.

Liu, B., Yang, H., Lancaster, M.J. (2017). Global optimization of microwave filters based on a surrogate model-assisted evolutionary algorithm, *IEEE Trans. Microw. Theory Tech.*, vol. 65, no. 6, pp. 1976–1985.

Liu, B., Yang, H., Lancaster, M.J. (2018). Synthesis of coupling matrix for diplexers based on a self-adaptive differential evolution algorithm, *IEEE Trans. Microw. Theory Tech.*, vol. 66, no. 2, pp. 813–821.

Liu, B., Akinsolu, M.O., Ali, N., Abd-Alhameed, R. (2019). Efficient global optimisation of microwave antennas based on a parallel surrogate model-assisted evolutionary algorithm, *IET Microw. Antennas Propag.*, vol. 13, no. 2, pp. 149–155.

Mckay, M.D., Beckman, R.J., Conover, W.J. (1979). A comparison of three methods for selecting values of input variables in the analysis of output from a computer code, *Technometrics*, vol. 21, no. 2, pp. 239–245.

Metropolis, N., Ulam, S. (1949). The Monte Carlo method, *J. Am. Statal Assoc.*, vol. 44, no. 247, pp. 335–341.

Na, W., Zhang, Q. (2013). Automated parametric modeling of microwave components using combined neural network and interpolation techniques, *IEEE MTT-S International Microwave Symposium Digest (MTT)*, pp. 1–3, in Seattle, WA.

Na, W., Feng, F., Zhang, C., Zhang, Q. (2017). A unified automated parametric modeling algorithm using knowledge-based neural network and L_1 optimization, *IEEE Trans. Microw. Theory Tech.*, vol. 65, no. 3, pp. 729–745.

Sola, J., Sevilla, J. (1997). Importance of input data normalization for the application of neural networks to complex industrial problems, *IEEE Trans. Nuclear Sci.*, vol, 44, no. 3, pp. 1464–1468.

Song, Y., Cheng, Q.S., Koziel, S. (2019). Multi-fidelity local surrogate model for computationally efficient microwave component design optimization, *Sensors*, vol. 19, no. 13.

Storn, R., Price, K. (1997). Differential Evolution – A Simple and Efficient Heuristic for global Optimization over Continuous Spaces, *J. Global Optim.*, vol. 11, no. 4, pp. 341–359.

Studio, C.M. (2018). *CST Studio Suite*, Dearborn, MI, USA.

Villagran, A., Huerta, G., Vannucci, M., Jackson, C.S., Nosedal, A. (2015). Non-parametric sampling approximation via voronoi tessellations, *Commun. Statist. — Simul. Comput.*, vol. 1–2, pp. 717–736.

Voronoi, G. (1908). Nouvelles applications des paramètres continus à la théorie des formes quadratiques. Premier mémoire. Sur quelques propriétés des formes quadratiques positives parfaites, *Journal Für Die Rne Und Angewandte Mathematik*, vol. 1908, no. 133, pp. 97–102.

Wang, M., Lv, W., Yang, F., Yan, C., Cai, W., Zhou, D., Zeng, X. (2018). Efficient Yield optimization for analog and SRAM circuits via Gaussian process regression and adaptive yield estimation, *IEEE Trans. Computer-Aided Design Integr. Circ. Syst.*, vol. 37, no. 10, pp. 1929–1942.

Washington, G. (2002). Aperture antenna shape prediction by feedforward neural networks, *IEEE Trans. Antennas Propag.*, vol. 45, no. 4, pp. 683–688.

Zhang, C., Jin, J., Na, W., Zhang, Q., Yu, M. (2018). Multivalued neural network inverse modeling and applications to microwave filters, *IEEE Trans. Microw. Theory Tech.*, vol. 66, no. 8, pp. 3781–3797.

Zhang, Z., Cheng, Q.S., Chen, H., Jiang, F. (2020). An efficient hybrid sampling method for neural network-based microwave component modeling and optimization, *IEEE Microw. Wireless Comp. Lett.*, vol. 30, no. 7, pp. 625–628.

© 2022 World Scientific Publishing Europe Ltd.
https://doi.org/10.1142/9781800610750_0008

Chapter 8

Statistical Design Centering of Microwave Systems via Space Mapping Technology and Modified Trust Region Algorithm

Abdel-Karim S.O. Hassan and Ahmed E. Hammad H. Elqenawy

Abstract

System design centering process searches for nominal values of system designable parameters which maximize the probability of satisfying the design specifications (yield function). Statistical design centering implements a statistical analysis method such as Latin hypercube sampling (LHS) for yield function estimation, and explicitly optimizes it. In this chapter, we introduce a new statistical design centering technique for microwave system design. The technique combines a modified surrogate-based derivative-free trust region (TR) optimization algorithm, and the generalized space mapping (GSM) technique. The modified TR algorithm is a derivative-free optimization algorithm that employs quadratic surrogate models to replace the computationally expensive yield function in the optimization process. The modified TR algorithm allows the shape of the TR to be dynamically adapted to a hyper-elliptic shape accompanied with the quadratic model. This improves the accuracy and convergence properties of the algorithm. Generally, TR algorithms exhibit global convergence features irrespective of the starting point setting. The new design centering approach utilizes the GSM technique to approximate the feasible region in the design parameter space with a sequence of iteratively updated space mapping (SM) surrogates. At each SM iteration, the modified TR algorithm optimizes the yield function for the current SM region approximation to get a better center. Two microwave circuit examples are used to show the effectiveness of the new design centering technique to obtain an optimal design in few SM iterations. In the design process, we employ Sonnet *em* for the bandstop microstrip filter design and CST Studio Suite for the ultra-wideband (UWB) multiple-input–multiple–output (MIMO) antenna.

Keywords: Design centering, microwave systems design, space mapping, trust region methods, yield optimization

8.1 Introduction

8.1.1 *Design centering problem*

In general, the system parameters are subject to known but unavoidable statistical fluctuations. This phenomenon may be a result of manufacturing process variations, model uncertainties or environmental variations during operations. Therefore, the values of the system parameters may deviate from their assigned nominal values. As a result, some of the underlined systems may violate the design specifications. Design centering is one of the common problems concerning optimal system design. The task is to seek for the values of system designable parameters which maximize the probability of satisfying the design specifications (yield function). This process makes the design more robust against system parameter variations. Due to its importance, design centering has been discussed and investigated by many researchers in Hassan and Mohamed (2013), Hassan *et al.* (2006, 2015b, 2015c, 2016, 2017), Antreich *et al.* (1994), Abdel-Malek and Bandler (1980), Abdel-Malek *et al.* (1999, 2006), Hassan (2003), Zhang *et al.* (2019), Hassan and Abdel-Naby (2011).

Systems are usually characterized by certain design parameters and given performance measures over predefined specifications. The required performance of these systems is decided by the designer and usually described by specifying bounds on the performance measures (for example, upper and lower specification on S-parameters). Generally, the performance measures are functions of the designable system parameters. Hence, the design specifications define what is called the feasible region (*FR*) over the design parameter space, and can be defined as

$$FR = \{x \in \mathbb{R}^n | \boldsymbol{F}(\boldsymbol{R}(\boldsymbol{x})) \leq 0\}, \tag{8.1}$$

where $\boldsymbol{x} \in \mathbb{R}^n$ is the vector of design parameters, n is the number of design parameters, $\boldsymbol{R} : \mathbb{R}^n \to \mathbb{R}^m$, $\boldsymbol{R}(\boldsymbol{x})$ is the response vector, and $\boldsymbol{F} : \mathbb{R}^m \to \mathbb{R}^l$ is the vector function that contains all the constraints where l is the number of constraints.

In practice, the system parameters are often subject to known but unavoidable statistical fluctuations (Hassan and Mohamed, 2013). The yield function Y can be defined as

$$Y = \text{Prob}\{x \in FR\}. \tag{8.2}$$

It represents the probability of satisfying the specifications. To simulate the statistical fluctuations that affect the system, the parameters are treated as random variables with a given probability density function (PDF), say $\varphi(\boldsymbol{x}, \boldsymbol{\theta})$ where $\boldsymbol{\theta}$ is a vector of the distribution parameters, e.g., means, variances and correlation coefficients. Accordingly, the yield function and can be given as

$$Y(\boldsymbol{\theta}) = \int_{FR} \varphi(\boldsymbol{x}, \boldsymbol{\theta}) \, d\boldsymbol{x}. \tag{8.3}$$

For design centering, it is assumed that, the PDF $\varphi(\boldsymbol{x}, \boldsymbol{\theta})$ is dependent only on the nominal parameter vector $x_0 \in \mathbb{R}^n$. Hence, The design centering problem can be formulated as

$$\max_{x_0} \left[Y(x_0) = \int_{FR} \varphi(\boldsymbol{x}, x_0) \, d\boldsymbol{x} \right]. \tag{8.4}$$

It is clear that the yield integral in Equation (8.4) cannot be evaluated analytically, since it requires the evaluation of n-dimensional integral over a "non-explicitly defined" region. Instead, it can be estimated. One of the popular methods used for estimating the yield integral is the Monte Carlo method (Hassan and Mohamed, 2013). To verify this, we define the following acceptance index function

$I_a : \mathbb{R}^n \to \mathbb{R}$, as

$$I_a(x^j) = \begin{cases} 1 & \text{if } x^j \in FR \\ 0 & \text{if } x^j \notin FR, \end{cases} \quad (8.5)$$

where FR is the feasible region defined by (8.1). Then, the yield integral in (8.4) can be rewritten as

$$Y(x_0) = \int_{\mathbb{R}^n} I_a(x)\varphi(x, x_0) dx = E\{I_a(x)\}, \quad (8.6)$$

where, $E\{\cdot\}$ denotes the expectation.

Hence, the yield function value at a nominal parameter vector x_0 can be estimated by generating a set of sample points x^j, $j = 1, 2, \ldots, K$ in the designable parameter space using the PDF of the parameters. The system is simulated for each sample point x^j, and the acceptance index function (8.5) is evaluated. Hence, the yield function at the nominal parameter vector x_0 can be estimated as

$$Y(x_0) \approx \frac{1}{K} \sum_{j=1}^{K} I_a(x^j) = \frac{m_a}{K}, \quad (8.7)$$

where m_a is number of sample points satisfying the design specifications, i.e., the percentage of acceptable circuits gives an estimate of the yield value at x_0. The error in estimating a yield value (estimation variance) is given by the following formula (Hassan and Mohamed, 2013):

$$V(Y(x_0)) \approx \frac{Y(x_0)(1 - Y(x_0))}{K - 1}. \quad (8.8)$$

It is clear that the estimation variance is inversely proportional to the number of samples considered. Hence, the most straightforward way for improving the accuracy and obtaining a low variance estimator is to increase the number of system simulations. This can result in prohibitively large expenses in terms of the computing time. Furthermore, there is no assurance that all regions of the design parameter space will be explored equally well. Several techniques have been proposed for improving the accuracy of a Monte Carlo yield estimate without increasing the number of samples. These methods are called variance reduction techniques. Examples include Latin hypercube sampling, stratified Monte Carlo method and importance sampling (Hassan and Mohamed, 2013). The main objective of variance reduction techniques is to spread the sample data points as evenly as possible around the interior design space. This has the benefit of needing a fewer number of runs to achieve the same level of confidence than the number required for the Monte Carlo approach because we have guaranteed that the entire probability range will be explored.

Generally, design centering approaches can be categorized into two main classes, namely geometrical and statistical approaches. For geometrical approaches, the yield function is implicitly optimized by fetching the feasible region center and using it to approximate the design center. In this regard, the feasible region could be approximated by a convex body, e.g., hyper-sphere, a hyper-cube or a hyper-ellipsoid, then use the center of this body as the design center (Hassan and Mohamed, 2013; Antreich et al., 1994; Abdel-Malek and Bandler, 1980; Abdel-Malek et al., 1999, 2006; Hassan, 2003; Hassan et al., 2015c; Hassan and Abdel-Naby, 2011). On the other hand, statistical approaches utilize statistical analysis approaches to explicitly optimize the yield function with no restrictions on the problem size (Hassan and Mohamed, 2013; Hassan et al., 2006, 2015b, 2017; Zhang et al., 2019). As previously stated, a yield function value at a given nominal parameter values could be estimated by generating a set of sample points in the design parameter space using a predefined probability distribution of the system parameters. The system is simulated for each sample point where the acceptable systems percentage provides a suitable estimate of the yield function value. To have an acceptable accuracy, a sufficiently large number of the system simulations should be performed, which makes this approach, in general, CPU-intensive.

Thus, in practice, any statistical design centering approach exhibits some technical difficulties. Firstly, the computational effort of evaluating the computationally expensive yield function values several times through the optimization process may represent an obstacle against any optimization method. Secondly, the gradient information of the yield function cannot be obtained analytically. Moreover, the required simulation cost in estimating the gradient, instead, may be prohibitive. For computationally expensive objective functions such as the yield function, the derivative free optimization (DFO) algorithms could be the most practical and applicable scenarios.

In this respect, derivative free trust region (TR) methods always have been one of the most efficient DFO techniques. Generally, the TR framework guarantees global convergence to a solution starting from any arbitrary initial point (Nocedal and Wright, 1999). In addition, the derivative free TR methods use computationally cheap surrogate-based models which are constructed using only function evaluations at selected sample points. There are several types of surrogate models in the literature such as response surfaces, neural networks, SM-surrogate, radial basis functions, kriging models, etc. Recently, the algorithmic acceleration of TR routines through sparse Jacobian matrix updates is presented (Koziel and Pietrenko-Dabrowska, 2019a, 2019b).

8.1.2 *Microwave design centering*

Regarding microwave systems design centering, there is another big challenge besides the aforementioned design centering difficulties. In any EM-based system, to get only one value of the yield function at a given design point; a computationally expensive full wave electromagnetic simulator is to be invoked many times, at the generated sample points. So, the high expense of system simulations required, may obstruct the design centering process. This makes the design centering of the microwave systems practically prohibitive (Hassan and Mohamed, 2013). One of the intelligent solutions that always have been used to overcome this problem is to implement computationally cheap surrogate-based models for approximating the response functions. The surrogate-based models may be for example, space mapping (SM), response surfaces, Kriging models and Neural Networks. The surrogate-based model are initially constructed and iteratively updated during the design centering process. For microwave systems, the computationally expensive true model response, would be replaced by a computationally cheaper and efficient surrogate model whose response will be denoted by $\boldsymbol{R}_s : \mathbb{R}^n \to \mathbb{R}^m$. Therefore, an approximate feasible region FR_s is defined to represent the design constraints imposed on the cheap surrogate model response as

$$FR_s = \{\boldsymbol{x} \epsilon \mathbb{R}^n | \boldsymbol{F}(\boldsymbol{R}_s(\boldsymbol{x})) \leq 0\}. \tag{8.9}$$

Accordingly, the yield function $Y(\theta)$ can be defined as

$$Y(\boldsymbol{\theta}) = \int_{FR_S} \varphi(\boldsymbol{x}, \boldsymbol{\theta}) \, d\boldsymbol{x}. \tag{8.10}$$

Assuming that, the PDF $\varphi(\boldsymbol{x}, \boldsymbol{\theta})$ is dependent only on the nominal parameter vector $x_0 \epsilon \mathbb{R}^n$. The microwave design centering problem can be given as

$$\max_{x_0} \left[Y(x_0) = \int_{FR_s} \varphi(\boldsymbol{x}, x_0) \, d\boldsymbol{x} \right]. \tag{8.11}$$

In this respect, the exact calculations using the simulator (high fidelity accurate model) could be performed, instead, using a cheaper surrogate model $\boldsymbol{R}_s$ (Hassan and Mohamed, 2013; Hassan et al., 2015c, 2017; Abdel-Malek et al., 2006). Space mapping (SM) technology (Koziel et al., 2006; Bandler et al., 2004), introduces SM surrogate models that can be used to reduce the optimization cost. SM technology exploits fast, yet inaccurate, cheap coarse models to construct SM surrogate models that replace the time-consuming, computationally-CPU intensive full-wave fine models.

8.2 New Statistical Design Centering Technique for Microwave Systems

This chapter is mainly devoted to introducing a new microwave design centering technique. The technique integrates the modified surrogate-based TR optimization algorithm (Hameed et al., 2017) with the generalized space mapping (GSM) methodology (Koziel et al., 2006). The technique implements a modified surrogate-based derivative-free TR optimization algorithm that locally fits the yield function with quadratic surrogate models in TRs around the current iterates (Hameed et al., 2017; Hassan et al., 2015a). Basically, the modified algorithm constructs and successively updates quadratically-modeled surrogates to be optimized instead of the CPU time-intensive objective function over hyper-elliptic TRs (Hameed et al., 2017). The shape of the TR is dynamically adapted to a hyper-elliptic shape accompanied with the quadratic model. This improves the accuracy and the convergence properties of the algorithm (Hameed et al., 2017). The TR optimization techniques are generally characterized by their global convergence to an optimal point irrespective the initial point setting (Nocedal and Wright, 1999). Moreover, only the objective function values are utilized to construct and, hence, efficiently update the quadratically-modeled surrogates replacing the objective function over the TRs during the optimization process. The advantages of TR algorithms include high reliability, and fast convergence.

The modified surrogate-based TR optimization algorithm (Hameed *et al.*, 2017) is integrated with the GSM methodology (Koziel *et al.*, 2006). In our approach, computationally cheap SM surrogate models are constructed and iteratively updated to approximate the response functions. For each SM surrogate model, an approximation to the feasible region is determined. Consequently, a sequence of iteratively updated SM feasible region approximations is generated during the design centering process. For each SM iteration, the modified TR algorithm is implemented to optimize the yield function over the current SM feasible region approximation, hence, a better design center point is obtained. The procedure is continued until convergence and until a satisfactory design center is reached.

The modified TR algorithm including the mathematical basis needed to formulate the algorithm is introduced in Section 8.2.1. Also, steps for the modified TR algorithm are presented. The GSM technique is presented in Section 8.2.2. An overall algorithm of the presented statistical design centering technique of microwave systems is outlined in Section 8.2.3.

8.2.1 *Modified trust region algorithm*

The optimization algorithm proposed in this work to optimize the yield function, belongs to the derivative free TR optimization class. In this class, the computationally expensive objective function $f(x)$ is optimized by locally approximating it around a current iterate $\boldsymbol{x}_k$ using a computationally cheaper quadratic model ($\boldsymbol{s}$). The model is then optimized, instead of the objective function, over a current TR. The quadratic model $M(\boldsymbol{s})$ can be put at the kth iteration in the form (Hassan *et al.*, 2015a; Powell, 2002, 2006):

$$M^{(k)}(\boldsymbol{s}) = c^{(k)} + \boldsymbol{b}^{(k)^T}\boldsymbol{s} + \frac{1}{2}\boldsymbol{s}^T\boldsymbol{H}^{(k)}\boldsymbol{s}, \qquad (8.12)$$

where $s = x - x_k \in \mathbb{R}^n$; $c^{(k)} \in \mathbb{R}$, the vector $b^{(k)} \in \mathbb{R}^n$, and the symmetric matrix $H^{(k)} \in \mathbb{R}^{n \times n}$ are the unknown parameters of $M^{(k)}(s)$.

Initially, the matrix H is assumed to be diagonal, so the number of the initial model parameters will be $m = 2n + 1$. By generating m points x_i, $i = 1, 2, \ldots, m$, and evaluating the objective function $f(x)$ at these points the initial model parameters can be obtained by satisfying the following matching conditions (Hassan et al., 2015a):

$$M^{(1)}(x_i) = f(x_i), \quad i = 1, 2, \ldots, m. \tag{8.13}$$

Conn et al. (2000), and Gould and Nocedal (1998) confirmed that the shape of an ideal TR should reflect the geometry of the model to improve the accuracy and convergence properties. Consequently, a modification of the algorithm in Hassan et al. (2015a), is introduced in Hameed et al. (2017), which allows the shape of the TR to be dynamically adapted to a hyper-elliptic shape accompanied with the quadratic model. In this modification, it is assumed that x_k is the current solution point, at the kth iteration. The quadratic model $M^{(k)}(s)$ is then optimized, instead of the objective function, over the current hyper-elliptic TR and a new point is then obtained by solving the following TR sub-problem (Liu et al., 2013):

$$\min_S M^{(k)}(s), \text{ subject to } \|s\|_{|H^{(k)}|} \leq \Delta_k, \tag{8.14}$$

where $\|s\|_{|H^{(k)}|}^2 = s^T |H^{(k)}| s$, Δ_k is a parameter that controls the TR size at each iteration, and $|H^{(k)}|$ is a matrix with the same eigenvectors of $H^{(k)}$ and with the absolute values of the corresponding eigenvalues, as $H^{(k)}$ is not necessarily positive definite. The TR sub-problem (8.14) can be solved using the method of Gould and Nocedal (1998).

Let s^* denote the solution of the TR sub-problem (8.14). Then, a new solution point is obtained as $x_n = x_k + s^*$. According to the traditional TR optimization work, the actual reduction in the objective function $f(x)$ is compared to the reduction obtained by using the quadratic model $M(s)$ by evaluating the following reduction ratio:

$$r_k = \frac{f(x_k) - f(x_n)}{M^{(k)}(x_k) - M^{(k)}(x_n)}. \tag{8.15}$$

This ratio reflects how much the quadratic model agrees with the objective function within the TR. Then, Δ_k and x_k are updated using the following formulas (Hameed et al., 2017):

$$r_k \begin{cases} r_k < 0.1: & \Delta_{k+1} = \frac{1}{2}\Delta_k \\ 0.1 \leq r_k \leq 0.7: & \Delta_{k+1} = \Delta_k \\ r_k \geq 0.7 \begin{cases} \|s^*\| < \Delta_k: & \Delta_{k+1} = \Delta_k \\ \|s^*\| \geq \Delta_k: & \Delta_{k+1} = 1.5 * \Delta_k, \end{cases} \end{cases}$$
(8.16)

$$x_{k+1} = \begin{cases} x_k + s^*, & \text{if } r_k > 0 \\ x_k & \text{otherwise} \end{cases}.$$
(8.17)

To complete the iteration of the TR optimization process, the parameters of the current model $M^{(k)}(x)$ will be updated. Namely, the parameters $c^{(k)}$, $b^{(k)}$ and $H^{(k)}$ will be updated. The constant $c^{(k)}$ is assigned the value of $f(x_k)$. Then the model will be updated in two steps (Hassan et al., 2015a). First, the vector $b^{(k)}$ is evaluated so that the model fits the last $2n$ points obtained, x_i, $i = 1, 2, \ldots, 2n$. More precisely, the vector $b^{(k)}$ is obtained by using the previous value of the model Hessian matrix, $H^{(k-1)}$ and solving the following system of linear equations using the least-squares approximation (Hassan et al., 2015a):

$$\begin{bmatrix} s_1^T \\ s_2^T \\ \vdots \\ s_{2n}^T \end{bmatrix} b^{(k)} = \begin{bmatrix} f(s_1) - c^{(k)} - \frac{1}{2}s_1^T H^{(k-1)} s_1 \\ f(s_2) - c^{(k)} - \frac{1}{2}s_2^T H^{(k-1)} s_2 \\ \vdots \\ f(s_{2n}) - c^{(k)} - \frac{1}{2}s_{2n}^T H^{(k-1)} s_{2n} \end{bmatrix}.$$
(8.18)

The model Hessian matrix is then updated using the following formula (Hassan et al., 2015a):

$$H^{(k)} = aH^{(k-1)} + uv^T,$$
(8.19)

where a is a constant satisfying, $0.5 < a < 1$, and the vector $v \in \mathbb{R}^n$ and

$$u = [\text{sign}(\text{diag}(H^{(k-1)}))] * \sqrt{(1-a) * |\text{diag}(H^{(k-1)})|}.$$
(8.20)

The vector v is evaluated such that the model $M^{(k)}(x)$ attempts to fit all the available q points obtained so far, x_i, $i = 1, 2, \ldots, q$ i.e., the vector v is obtained by solving the weighted system of linear equations using the least-squares approximation (Powell, 2006):

$$\begin{bmatrix} \frac{1}{2}s_1^T u s_1^T w_1 \\ \frac{1}{2}s_2^T u s_2^T w_2 \\ \vdots \\ \frac{1}{2}s_m^T u s_m^T w_m \end{bmatrix} v = \begin{bmatrix} w_1 * \left(f(s_1) - c^{(k)} - s_1^T b^{(k)} - \frac{1}{2}s_1^T a H^{(k-1)} s_1 \right) \\ w_2 * \left(f(s_2) - c^{(k)} - s_2^T b^{(k)} - \frac{1}{2}s_2^T a H^{(k-1)} s_2 \right) \\ \vdots \\ w_m * \left(f(s_m) - c^{(k)} - s_m^T b^{(k)} - \frac{1}{2}s_m^T a H^{(k-1)} s_m \right) \end{bmatrix},$$
(8.21)

where w_i, $i = 1, 2, \ldots, q$ are weights assigned to the different available points according to their distances from the TR center (Hassan et al., 2015a). In the proposed algorithm, the weight w_i, associated with each equation takes the form:

$$w_i = \begin{cases} 1 & \text{if } \|s_i\| \leq c_1 \Delta \\ \dfrac{c_1 \Delta}{\|s_i\|} & \text{if } \|s_i\| > c_1 \Delta \end{cases}, \quad i = 1, 2, \ldots, q, \qquad (8.22)$$

where c_1 is a positive constant and $c_1 \geq 1$.

After calculating the term uv^T, the matrix is made symmetric by resetting the off-diagonal elements to their average values, and the new Hessian matrix $H^{(k)}$ is obtained using equation (8.19).

8.2.1.1 Steps for the modified TR algorithm

The steps of the modified TR algorithm are given below:

1. Given $x_0 \in \mathbb{R}^n$ (initial design parameter values); $\Delta_1 > 0$ (initial TR size); $0.5 < a < 1$; $c_1 \geq 1$; $Nmax$ (the maximum number of stored points); δ (a termination criterion). Set $N = 0$ (the number of objective function evaluations) (Hameed et al., 2017).
2. Set $k = 1$, construct the initial quadratic model using Equations (8.12) and (8.13).
3. Solve the hyper-elliptic TR sub-problem (8.14) using the method of Gould and Nocedal (1998), to obtain s^*, then find a new solution point $x_n = x_k + s^*$ of the model $M^{(k)}(x)$ over the TR.
4. Evaluate $f(x_n)$ and compute the reduction ratio using Equation (8.15).

5. Update the TR size to obtain Δ_{k+1} using Equation (8.16).
6. Determine the TR center of the next iteration x_{k+1} using Equation (8.17).
   If $\|f(x_{k+1}) - f(x_k)\| \leq \delta$, the algorithm will be terminated with $x_{\text{opt}} = x_{k+1}$ and $f_{\text{opt}} = f(x_{k+1})$. For two successive iterations, if the reduction ratio is negative, go to Step 9, else continue.
7. Add the point x_n to the set of available points, if the number of available points exceeds $Nmax$, exclude the farthest point from x_{k+1}.
8. Update the quadratic model using the updating formulas (8.18)–(8.22), then set $k = k + 1$ and go to Step 3.
9. Generate a new point s_{new} (Hameed et al., 2017), add it to the set of available points, and then go to Step 8.

8.2.2 Generalized space mapping technique

The new approach for statistical microwave system design centering combines the modified surrogate-based TR optimization algorithm presented in Section 8.2.1, with the GSM methodology (Koziel et al., 2006). In the GSM approach, computationally cheap space mapping surrogate models are constructed and iteratively updated for approximating the computationally expensive response functions (fine model). The computationally expensive "fine" model can be replaced by a simpler and fast yet inaccurate "coarse" model to accelerate the design process. The "coarse" model accuracy could be improved sequentially by creating the GSM surrogate suggested by Koziel et al. (2006). The GSM-based surrogate employed a computationally fast coarse model along with input and output mappings in the form:

$$R_s^i(x) = A^i \cdot R_c(B^i \cdot x + c^i) + d^i + E^i \cdot (x - x_0^i), \quad (8.23)$$

where x_0^i is the current nominal parameter vector, $A^i \in M_{m \times m}$ is a diagonal matrix, $R_c : X_c \to \mathbb{R}^m$ is the coarse model response vector, $B^i \in M_{n \times n}$, $c^i \in M_{n \times 1}$, and $d^i \in M_{m \times 1}$ is given by

$$d^i = R_f(x_0^i) - A^i \cdot R_c(B^i \cdot x_0^i + c^i), \quad (8.24)$$

where $R_f : X_f \to \mathbb{R}^m$ is the fine model response vector, and $E^i \in M_{m \times n}$ is given by

$$E^i = J_f(x_0^i) - A^i \cdot J_c(B^i \cdot x_0^i + c^i) \cdot B^i, \quad (8.25)$$

where $\boldsymbol{J}_f : \boldsymbol{X}_f \to \mathbb{R}^{m \times n}$ and $\boldsymbol{J}_c : \boldsymbol{X}_c \to \mathbb{R}^{m \times n}$ are the fine and coarse model responses' Jacobian matrices with respect to the corresponding points, respectively. The parameter extraction (PE) optimization process is defined by

$$(\boldsymbol{A}^i, \boldsymbol{B}^i, \boldsymbol{c}^i) = \arg \min_{A,B,C} \boldsymbol{e}^i(\boldsymbol{A}, \boldsymbol{B}, \boldsymbol{c}). \tag{8.26}$$

The PE process of (8.26) is employed to obtain the mapping parameters $\boldsymbol{A}^i$, $\boldsymbol{B}^i$ and $\boldsymbol{c}^i$, where $\boldsymbol{e}^i$ represents the response difference between the fine model and its surrogate and it is given by

$$\boldsymbol{e}^i(\boldsymbol{A}, \boldsymbol{B}, c) = \sum_{k=0}^{i} w_k \| \boldsymbol{R}_f(\boldsymbol{x}^k) - \boldsymbol{A} \cdot \boldsymbol{R}_c(\boldsymbol{B} \cdot \boldsymbol{x}^k + c) \|$$

$$+ \sum_{k=0}^{i} v_k \| \boldsymbol{J}_f(x^k) - \boldsymbol{A} \cdot \boldsymbol{J}_c(\boldsymbol{B} \cdot \boldsymbol{x}^k + c) \cdot \boldsymbol{B} \|, \tag{8.27}$$

where the weight given by the coefficients w_k and v_k are user-defined weights $\in \{0, 1\}$.

8.2.3 The statistical design centering algorithm

Step 1: Let $k = 0$, Δ_1, $\boldsymbol{x}_{\text{start}}$, $\boldsymbol{H}$ (a user-defined covariance matrix) and δ (a stopping parameter).

Step 2: Find the estimated yield value Y of the fine model at $\boldsymbol{x}_{\text{start}}$ exploiting the LHS technique assuming normally distributed parameters with covariance matrix $\boldsymbol{H}$.

Step 3: Using the same starting point $\boldsymbol{x}_{\text{start}}$, apply parameter extraction (PE) (Koziel et al., 2006; Bandler et al., 2004) to obtain the SM parameters and construct the initial SM surrogate model $\boldsymbol{R}_s^1(x)$, put $k = 1$.

Step 4: Apply the modified TR algorithm to the SM surrogate in Step 3 until a better yield value $Y(\boldsymbol{x}_0^k)$ at point $\boldsymbol{x}_0^k$ is obtained.

Step 5: Reconstruct the SM surrogate $\boldsymbol{R}_s^{k+1}(x)$ at the point $\boldsymbol{x}_0^k$, utilizing PE.

Step 6: Restart the modified algorithm with $\boldsymbol{x}_0^k$ to obtain $\boldsymbol{x}_0^{k+1}$ and $Y(\boldsymbol{x}_0^{k+1})$.

Step 7: If $\|Y(\boldsymbol{x}_0^{k+1}) Y(\boldsymbol{x}_0^k)\| \leq 1\%$ or $\|\boldsymbol{x}_0^{k+1} - \boldsymbol{x}_0^k\| \leq \delta$, go to step 8, else set $k = k + 1$ and go to step 5.

Step 8: Evaluate the fine model yield at the final point x_0^k using LHS, then stop.

8.3 Practical Examples

The proposed microwave design centering technique is applied to two microwave examples. The first system is a symmetrical bandstop microstrip filter with open stubs, evaluated using Sonnet *em*. The second one is the UWB multiple–input-multiple-output (MIMO) antenna and designed by employing the full-wave CST EM solver.

8.3.1 *Bandstop microstrip filter with open stubs*

The presented technique is applied to a symmetrical bandstop microstrip filter with three open stubs (Bakr *et al.*, 2000) shown in Fig. 8.1. The open stub lengths are L_1, L_2, L_1 and the corresponding stub widths are W_1, W_2, W_1. An alumina substrate, with width W_o, thickness $H = 25$ mil, loss tangent $= 0.001$, dielectric constant $\varepsilon_r = 9.4$, is used for a $50\,\Omega$ feeding line. A full-wave EM solver Sonnet *em* is employed to evaluate both the coarse and fine models with different discretization densities. The design parameters are $x = [W_1, W_2, L_o, L_1, L_2]$. The given design specifications on $|S_{21}|$ at 21 frequency points are formulated as design constraints on feasible region where it is given by

$$|S_{21}| \leq 0.05 \text{ for } 9.3\,\text{GHz} \leq \omega \leq 10.7\,\text{GHz},$$
$$|S_{21}| \geq 0.9 \text{ for } 12\,\text{GHz} \leq \omega \text{ and } \omega \leq 8\,\text{GHz}.$$

The initial design point at $10\,\text{GHz}$ (center frequency) is taken as $[6.00, 9.01, 106.45, 110.15, 108.81]^T$ (in mils). The fine model is simulated with high resolution grid $1\,\text{mil} \times 1\,\text{mil}$ cell size while the coarse model is simulated with $5\,\text{mils} \times 5\,\text{mils}$ cell size. 100 LHS sample points are used to obtain the yield values assuming normally and independent distributed parameters. The yield values assuming independent parameters with parameter spreads $\sigma = [0.02, 0.2, 2.0, 1.0, 1.0]$ (in mils). The final design center is $[5.5958, 13.0635, 122.0072, 119.3062, 108.7415]$ (in mils). The yield values at the initial and final design centers are 0% and 72%, respectively, as shown in Figs. 8.2(a) and 8.2(b), respectively. It is worth

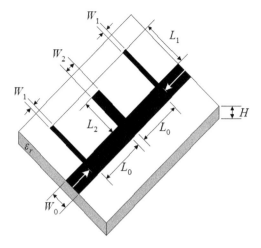

Fig. 8.1. Structure of bandstop microstrip filter (Bakr et al., 2000).

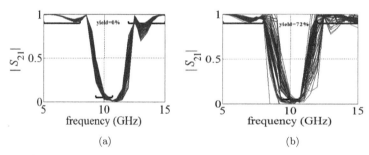

Fig. 8.2. (a) The initial yield, and (b) the final yield of the bandstop microstrip filter.

mentioning that the final yield has been improved to 72% yield using only 4 SM iterations when implementing the proposed new approach (Hameed et al., 2017), while the original approach (Hassan et al., 2015a) reached 51% yield after 5 SM iterations (Hassan et al., 2017). These results demonstrate the efficacy of the modified algorithm.

8.3.2 Ultra-wideband multiple-input–multiple-output antenna

The new technique for statistical microwave circuit design centering is used to obtain the optimal design of the UWB MIMO antenna shown

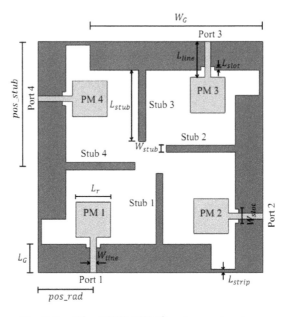

Fig. 8.3. The UWB MIMO antenna structure.

in Fig. 8.3. The antenna has a compact size of 65×65 mm^2 on a Rogers substrate, RO4350B with a bandwidth from 3.1 GHz to 10.6 GHz. The thickness of the substrate is taken as 0.8 mm with a dielectric constant ε_r of 3.5 and a loss tangent of 0.004 (Liu et al., 2013). Four identical planar monopole (PM) antenna elements are printed on the top layer of the substrate to form the antenna and act as radiators. Each one has a square shape with side length L_r and is fed by a $50-\Omega$ microstrip line with length L_{line} and width W_{line} as shown in Fig. 8.3. On the bottom layer of the substrate, four identical ground planes are printed. Each ground has a length L_G and a width W_G with a rectangular slot on the upper edge to enhance the matching (Liu et al., 2013). The slot has a length L_{slot} and a width W_{slot}. The distance from the center of each radiator to the beginning of its ground is taken as pos_rad as illustrated in Fig. 8.3. The four ground planes are connected together using a strip with thickness L_{strip} between any two adjacent grounds. Further, a stub with a length L_{stub} and a width W_{stub} is added to each ground plane to achieve better isolation and large impedance bandwidth (Liu et al., 2013). The distance from the center of each stub to the beginning of its ground is taken as *pos_stub* as shown in Fig. 8.3. The coarse and fine models employ CST EM solver with different discretization mesh size. The design parameters $\boldsymbol{x}$ are $[L_r, pos\_rad, L_\text{line}, W_\text{line}, L_\text{slot}, W_\text{slot}, L_G, W_G, L_\text{stub}, W_\text{stub}, pos\_stub, L_\text{strip}]$. The feasible region is characterized by the following constraints:

$$|S_{11}| < -10\,\text{dB}, \quad \text{for} \quad 3.1\,\text{GHz} \leq \omega \leq 10.6\,\text{GHz},$$
$$|S_{21}| < -15\,\text{dB}, \quad \text{for} \quad 3.1\,\text{GHz} \leq \omega \leq 10.6\,\text{GHz},$$
$$|S_{31}| < -15\,\text{dB}, \quad \text{for} \quad 3.1\,\text{GHz} \leq \omega \leq 10.6\,\text{GHz},$$
$$pos\_rad - L_r/2 > L_G,$$
$$pos\_rad + L_r/2 < pos\_stub - W_\text{stub}/2,$$
$$L_\text{line} + L_r < 65 - pos\_stub - W_\text{stub}/2,$$
$$L_\text{line} > L_G,$$
$$pos\_stub + W_\text{stub}/2 < W_G,$$
$$L_G + L_\text{stub} < 32.5,$$
$$L_G + L_\text{stub} < pos\_stub - W_\text{stub}/2,$$
$$L_G + W_G < 65.$$

The initial design point is $[10, 15, 10, 2, 1, 4, 8, 50, 10, 2, 1]^T$ (in mm). The fine model is simulated with a fine grid mesh while the coarse model is simulated with a coarse grid mesh. 100 LHS sample points are used to obtain the yield values assuming normally and independent distributed parameters. Good results are achieved in 5 SM iterations using the new design centering technique. The yield values are computed assuming independent parameters with parameter spreads $\boldsymbol{\sigma} = [0.2, 0.2, 0.2, 0.2, 0.2, 0.02, 0.02, 0.02, 0.2, 0.2, 0.02, 0.02]$ (in mm). The final design centre is [9.0145, 17.185, 9.6040, 1.6172, 0.9562, 3.1834, 8.6060, 42.9350, 19.7780, 1.9521, 33.1800, 1.14607] (in mm). The yield values at the initial and final design centres are 0% and 88%, respectively. Further, Fig. 8.4 shows the simulated S-parameters $|S_{11}|$, $|S_{21}|$ and $|S_{31}|$ at the initial and final centers. It can be observed that the new design centering technique allows for a significant improvement of the matching and the isolation properties of the designed antenna.

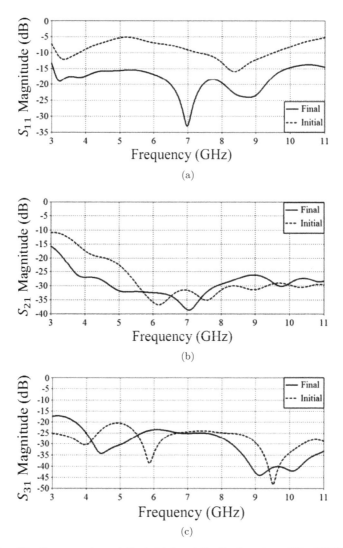

Fig. 8.4. The S-parameters at the initial and the final centers of the UWB MIMO antenna: (a) $|S_{11}|$, (b) $|S_{21}|$, (c) $|S_{31}|$.

8.4 Conclusion

In this chapter, a novel statistical design centering technique for microwave systems is presented and verified. The proposed approach employs a TR framework for yield optimization and exploits the space mapping (GSM) technique to reduce the computational effort required during the design process. The proposed technique could be considered as a stochastic optimization technique and it is well suited for problems whose objective functions exhibit some uncertainties in their values, in particular, the yield maximization problem. The new design centering technique is employed to obtain the optimal design center of two microwave systems. The first one is a symmetrical bandstop microstrip filter with open stubs employing Sonnet *em* solver with five design parameters, where design specifications of $|S_{21}|$ at 21 frequency points are used to constrain the feasible region. The second system is an UWB MIMO antenna where the CST Studio Suite is used for simulation process of the UWB MIMO antenna with 12 design parameters and more than 14 design specifications. A significant improvement in the yield values in both examples has been be obtained using a few SM iterations.

Acknowledgments

Authors would like to thank Prof. Slawomir Koziel, School of Science and Engineering, Reykjavik University, for his invitation to contribute to this book. The authors also thank Prof. Tamer M. Abuelfadl, Faculty of Engineering, Cairo University, for contributions to the original work reviewed in this chapter.

References

Abdel-Malek, H.L., Bandler, J.W. (1980). Yield optimization for arbitrary statistical distributions: Part I-Theory, *IEEE Transactions Circuits and Systems*, vol. 27, pp. 245–253.

Abdel-Malek, H.L., Hassan, A.S.O., Bakr, M.H. (1999). A boundary gradient search technique and its applications in design centering, *IEEE Transaction Computer-Aided Design Integrated Circuits System*, vol. 18, no. 11, pp. 1654–1660.

Abdel-Malek, H.L., Hassan, A.S.O., Soliman, E.A., Dakroury, S.A. (2006). The ellipsoidal technique for design centering of microwave circuits exploiting space-mapping interpolating surrogates, *IEEE Transactions Microwave Theory Techniques*, vol. 54, no. 10, pp. 3731–3738.

Antreich, K.J., Graeb, H.E., Wieser, C.U. (1994). Circuit analysis and optimization driven by worst-case distances, *IEEE Transactions*

Computer-Aided Design Integrated Circuits System, vol. 13, no. 1, pp. 57–71.

Bakr, M.H., Bandler, J.W., Ismail, M.A., Rayas-Sánchez, J.E., Zhang, Q.J. (2000). Neural space mapping optimization for EM-based design, *IEEE Transactions on Microwave Theory and Techniques*, vol. 48, no. 12, pp. 2307–2315.

Bandler, J.W., Cheng, Q.S., Dakroury, S.A., Mohamed, A.S., Bakr, M.H., Madsen, K., Søndergaard, J. (2004). Space mapping: The state of the art, *IEEE Transactions Microwave Theory Techniques*, vol. 52, no. 1, pp. 337–361.

Conn, A.R., Gould, N.I.M., Orban, D., Toint, P.L. (2000). A Primal-Dual trust region algorithm for minimizing a non-convex function subject to bound and linear equality constraints, *Mathematical Programming*, vol. 87, pp. 215–249.

Gould, N.I.M., Nocedal, J. (1998). On the modified absolute-value factorization norm for trust region minimization, in Renato, L.De, Murli, A., Pardalos, P.M., Torlado, G., (eds.), *High Performance Algorithms and Software in Nonlinear Optimization*, Kluwer, pp. 225–241.

Hameed, M.F.O., Hassan, A.S.O., Elqenawy, A.E., Obayya, S.S.A. (2017). Modified trust region algorithm for dispersion optimization of photonic crystal fibers, *Journal of Lightwave Technology*, vol. 35, no. 17, pp. 3810–3818.

Hassan, A.S.O. (2003). Normed distances and their applications in optimal circuit design, *Journal of Optimization and Engineering*, vol. 4, no. 3, pp. 197–213.

Hassan, A.S.O., Abdel-Naby, A. (2011). A new hybrid method for optimal circuit design using semi-definite programming, *Engineering Optimization*, vol. 44, no. 6, pp. 1–16.

Hassan, A.S.O., Mohamed A.S.A. (2013). Surrogate-based circuit design centering, in Koziel, S., Leifsson, L. (eds.), *Surrogate-Based Modeling and Optimization*, New York, Springer, pp. 27–49.

Hassan, A.S.O., Abdel-Malek, H.L., Rabie, A.A. (2006). Non-derivative design centering algorithm using trust region optimization and variance reduction, *Engineering Optimization*, vol. 38, no. 1, pp. 37–51.

Hassan, A.S.O., Abdel-Malek, H.K., Mohamed, A.S.A., Abuelfadl, T.M., Elqenawy, A.E. (2015a). RF cavity design exploiting a new derivative-free trust region optimization approach, *Journal of Advanced Research*, vol. 6, no. 6, pp. 915–924.

Hassan, A.S.O., Mohamed, A.S.A., El-Sharabasy, A.Y. (2015b). EM-based yield optimization exploiting trust-region optimization and space mapping technology, *International Journal of RF and Microwave Computer-Aided Engineering*, vol. 25, no. 6, pp. 474–484.

Hassan, A.S.O., Mohamed, A.S.A., Rabie, A.A., Etman, A.S. (2015c). A novel surrogate-based approach for optimal design of electromagnetic-based circuits, *Engineering Optimization*, vol. 48, no. 2, pp. 185–198.

Hassan, A.S.O., Rafat, N.H., Mohamed, A.S.A. (2016). Optimal design of photonic crystal nanostructures, in Koziel, S., Leifsson, L., Yang, X.-S. (eds.), *Simulation-driven Modeling and Optimization*, Springer, pp. 233–260.

Hassan, A.S.O., Abdel-Malek, H.L., Mohamed, A.S.A., Elqenawy, A.E. (2017). Space mapping surrogate-based microwave circuit design centering using a new statistical technique, *International Journal Numerical Modelling*, vol. 30, no. 3–4.

Koziel, S., Bandler, J.W., Madsen, K. (2006). Space-mapping-based interpolation for engineering optimization, *IEEE Transactions Microwave Theory and Techniques*, vol. 54, no. 6, pp. 2410–2420.

Koziel, S., Pietrenko-Dabrowska, A. (2019a). Reduced-cost electromagnetic-driven optimisation of antenna structures by means of trust-region gradient-search with sparse Jacobian updates, *IET Microwaves Ant. Prop.*, vol. 13, no. 10, pp. 1646–1652.

Koziel, S., Pietrenko-Dabrowska, A. (2019b). Variable-fidelity simulation models and sparse gradient updates for cost-efficient optimization of compact antenna input characteristics, *Sensors*, vol. 19, no. 8, pp. 1806.

Liu, L., Cheung, S.W., Yuk, T.I. (2013). Compact MIMO antenna for portable devices in UWB applications, *IEEE Transactions Antennas and Propagation*, vol. 61, no. 8, pp. 4257–4264.

Nocedal, J., Wright, S.J. (1999). *Numerical Optimization*, Verlag, Springer.

Powell, M.J.D. (2002). UOBYQA: Unconstrained optimization by quadratic approximation, *Mathematical Programming*, vol. 92, pp. 555–582.

Powell, M.J.D. (2006). The NEWUOA Software for unconstrained optimization without derivatives, *Large-Scale Nonlinear Optimization*, Springer, pp. 225–297.

Zhang, J., Feng, F., Na, W., Yan, S., Zhang, Q.J. (2019). Parallel space-mapping based yield-driven EM optimization incorporating trust region algorithm and Polynomial Chaos expansion, *IEEE Access*, doi: 10.1109/ACCESS.2019.2944415.

© 2022 World Scientific Publishing Europe Ltd.
https://doi.org/10.1142/9781800610750_0009

Chapter 9

Expedited Yield-Driven Design of High-Frequency Structures by Kriging Surrogates in Confined Domains

Anna Pietrenko-Dabrowska and Slawomir Koziel

Abstract

Uncertainty quantification is an important aspect of engineering design, also pertaining to the development and performance evaluation of high-frequency structures systems. Manufacturing tolerances as well as other types of uncertainties, related to material parameters (e.g., substrate permittivity) or operating conditions (e.g., bending) may affect the characteristics of antennas or microwave devices. For example, in the case of narrow- or multi-band antennas, this usually leads to frequency shifts of the operating bands. Quantifying these effects is imperative to adequately assess the design quality, either in terms of the statistical moments of the performance parameters or the yield. Reducing the system sensitivity to parameter deviations is even more essential when increasing the probability of the system satisfying the prescribed requirements is of concern. The prerequisite of such procedures is statistical analysis, normally carried out at the level of full-wave electromagnetic (EM) analysis. Although necessary to ensure reliability, it entails considerable computational expenses, often prohibitive. Following the recently fostered concept of constrained modeling, this chapter discusses a simple technique for rapid surrogate-assisted yield optimization of high-frequency structures. The keystone of the approach is an appropriate definition of the optimization domain. This is realized by considering a few pre-optimized designs that represent the directions featuring maximum variability of the circuit responses (particularly the parts thereof that affect the yield value in the most significant way) with respect to its geometry parameters. Due to a small volume of such a domain, an accurate replacement model can be established therein using a small number of training samples, and employed to improve the yield. The implementation details are tailored to a particular type of device. Verification results obtained for several antenna structures and a miniaturized rat-race coupler indicate that the optimization process can be accomplished at low cost of a few dozen of EM simulations. The result reliability is validated through comparisons with EM-based Monte Carlo simulations.

Keywords: Uncertainty quantification, statistical analysis, yield optimization, surrogate-assisted optimization, domain confinement

9.1 Introduction

The vast majority of high-frequency design procedures aim at finding the nominal designs, i.e., obtained under the assumption that the fabricated prototype retains the values of geometry and/or material parameters equal to those rendered in the course of antenna development (e.g., through parametric optimization). In practice, uncertainties of various types may affect the system operation in an undesirable manner. Their appropriate quantification may be therefore crucial to verify whether the system is likely to satisfy the prescribed performance specifications. There are generally two types of uncertainties pertaining to high-frequency structures. The most common ones are deviations of dimensions originating from imperfect fabrication processes (e.g., chemical etching in the case of microstrip or antenna components, or mechanical milling in the case of waveguide structures) as well as the material parameters (e.g., substrate permittivity) from their nominal values. These are the inherent (or aleatory) uncertainties, most often related to manufacturing tolerances (Biernacki *et al.*, 2012) and typically characterized by probability distributions. Due to their stochastic nature, quantification requires performing statistical analysis (Hassan *et al.*, 2015; Koziel *et al.*, 2005; Syrytsin *et al.*, 2017). Reducing the effects of tolerances normally entails stochastic design that aims at improving statistical performance measures, e.g., the yield (Leifsson *et al.*, 2020). The second type of uncertainties are systematic (or epistemic) ones, related to the lack of knowledge of the operating conditions (temperature, bending radius of a wearable antenna, etc.). These may often be handled by ensuring that the acceptable system performance is secured for the specified ranges of the conditions.

The effects of tolerances may lead to the circuit failing to meet the prescribed performance requirements. Consequently, quantification of such effects and, eventually, reducing them already at the design stage is essential to ensure the structure robustness (Hassan et al., 2015; Prasad et al., 2016). The latter normally means diminishing statistical moments of the system outputs, especially their variance (Kim et al., 2015). However, for microwave or antenna components, design specifications are often expressed in a minimax form, i.e., using upper/lower bounds for the selected figures of interest (e.g., maximum acceptable level of reflection, minimum acceptable bandwidth, maximum acceptable power split error, etc.) (Budimir and Goussetis, 2003); therefore, a popular statistical performance metric is a yield (Koziel et al., 2005).

Regardless of the assumed metric, uncertainty quantification requires carrying out statistical analysis (Vidal et al., 2005; Li et al., 2015). The latter, when executed at the level of full-wave electromagnetic (EM) simulation models, is an expensive procedure. In particular, conventional routines, such as Monte Carlo analysis, require a large number of circuit evaluations, the cost of which may be prohibitive. At the same time, the use of EM analysis is imperative to ensure reliability. It is especially critical for miniaturized components, where considerable cross-coupling effects cannot be adequately accounted for by simpler methods, e.g., network equivalents (Jin et al., 2017; Wu and Chiu, 2016). The literature offers several major approaches aiming at mitigation of the aforementioned difficulties. One of the simplest is the worst-case analysis (Sengupta et al., 2005), which assumes the most disadvantageous scenarios (e.g., combinations of maximum parameter deviations from their nominal values). Although being computationally tractable, the performance estimations produced by this method are overly pessimistic. Nowadays, the most popular methods of choice are those involving fast replacement models, where the operations normally performed through EM analysis are executed on the surrogate (Ma et al., 2018; Ren et al., 2016). The specific techniques include response surface approximation (Matoglu et al., 2004), artificial neural networks (Zhang et al., 2008) as well as polynomial chaos expansion (PCE) (Du and Roblin, 2017; Rossi et al., 2014; Petrocchi et al., 2017). PCE is particularly attractive when statistical moments of the system output (e.g., the variance) are of interest, because they can be calculated directly from the expansion coefficients without the necessity of running Monte Carlo simulation (Petrocchi et al., 2017). A practical problem of surrogate-assisted statistical analysis is a possibly high cost of model construction. This becomes apparent for structures described by larger numbers of parameters, or when the model domain needs to cover a larger portion of the design space. The recent techniques such as PC-kriging (Leifsson et al., 2020) promise partial mitigation of this issue by using a more complex trend function (i.e., PCE instead of low-order polynomials). Other possible methods rely on reducing the design space dimensionality, e.g., principal component analysis (Ochoa and Cangellaris, 2013), the employment of variable-fidelity simulations (space mapping; Rayas-Sanchez and Gutierrez-Ayala, 2006a; co-kriging; Kennedy and O'Hagan, 2000), but also hybridization (e.g., surrogate modeling methods combined with model order reduction techniques Spina et al., 2014).

Although estimating the effects of uncertainties is an important design consideration, reducing them is even more essential. Robust design (also referred to as tolerance-aware design or yield-driven design) aims at reducing the sensitivity of performance parameters (operating frequency, power split ratio, bandwidth) to, e.g., manufacturing tolerances (Kouassi et al., 2016; Biernacki et al., 2012; Scotti et al., 2005). In practice, this is realized through parameter tuning with the maximization of appropriately defined statistical performance metrics. As mentioned before, in the case of minimax specifications, a suitable metric is a yield, i.e., the probability of satisfying the prescribed specifications under the assumed deviations of geometry and/or material parameters. Robust design typically entails significant computational expenditures due to the necessity of multiple yield estimations. In particular, direct EM-based stochastic optimization is not doable in most practical situations. Instead, the widely used approaches rely on fast surrogate models (Matoglu et al., 2004; Zhang et al., 2008; Du and Roblin, 2017; Rossi et al., 2014; Petrocchi et al., 2017; Leifsson et al., 2020; Ochoa and Cangellaris, 2013). Similarly, as in the case of statistical analysis, the popular techniques include response surface approximations (Matoglu et al., 2004), space mapping (Abdel-Malek et al., 2006), neural networks (Rayas-Sanchez and Gutierrez-Ayala, 2006b), and polynomial chaos expansion (PCE) (Zhang et al., 2018).

Notwithstanding, due to a high cost of setting up the surrogate valid within broader ranges of the system parameters, otherwise necessary to conduct the optimization process, iterative methods seem to be more economical. Sequential approximate optimization (SAO) (Koziel and Bekasiewicz, 2018a) is a generic concept, where the surrogate is constructed in the domain being a small vicinity of the current design, locally optimized, and relocated into the new domains defined along the optimization path. Here, the necessity of reconstructing the model is compensated for by a significantly lower cost of training data acquisition within each region considered during the process. An alternative option is the utilization of response features (Koziel and Bandler, 2015), where the surrogate modeling process is focused on selected characteristic points of the circuit responses. The latter is sufficient for evaluating the system performance, yet their dependence on geometry parameters is much less nonlinear than the dependence of the original outputs. Modeling at the level of features has been demonstrated to require considerably smaller training data sets (Koziel and Bekasiewicz, 2018b).

This chapter discusses an alternative approach to low-cost yield optimization of high-frequency structures. It is based on recently fostered performance-driven modeling (Koziel, 2017), where the surrogate model domain is focused on carefully selected regions of the parameter space, specifically those corresponding to designs that are of high-quality with respect to the relevant figures of interest. As demonstrated in (Koziel and Pietrenko-Dabrowska, 2019), this allows for constructing reliable surrogates using small training data sets and without formally restricting the ranges of the system parameters. Here, this concept is adopted to define the domain of the surrogate model established for yield optimization purposes. The domain is spanned by the directions corresponding to the maximum variability of the circuit responses (particularly the parts thereof that affect the yield value in the most significant way) with respect to its geometry parameters. The small volume of the domain determined this way allows for setting up an accurate model using a fraction of data samples required by conventional methods. For the sake of verification, the technique is applied to several microstrip antenna structures as well as a 1-GHz equal-power split microstrip rat-race coupler. The yield-optimized design is obtained at the cost of a few dozen EM analyses of the circuit. EM-based Monte Carlo simulations corroborate the reliability of the approach. Computational benefits are demonstrated by comparisons with benchmark surrogate-assisted methods.

9.2 Yield Optimization Problem and Benchmark Algorithms

This section recalls the yield optimization problem statement, illustrated using two specific cases, multi-band antennas, and equal power split coupler. Subsequently, two basic surrogate-assisted yield optimization algorithms are described. These will be used as benchmark methods in Section 9.4.

9.2.1 *Yield optimization problem*

The choice of the statistical figures of merit determines the formulation of the tolerance-aware design task. One option would be to manipulate the statistical moments of the system outputs, e.g., to reduce their variance. However, in the case of antennas or microwave components, design specifications are often expressed in a minimax form, through appropriately chosen upper/lower acceptance levels (e.g., maximum in-band reflection, minimum in-band transmission, maximum power split error at the operating frequency or band, etc. (Budimir and Goussetis, 2003)). Consequently, a typically applied figure of merit to be estimated and then maximized in the course of the optimization process is the yield (Koziel *et al.*, 2005), i.e., the percentage of designs satisfying the prescribed performance specifications under the assumed deviations of parameters.

Let us consider two examples, later discussed in more detail in Section 9.4. The first example concerns multi-band antennas. The minimax specifications are assumed for the input characteristics, specifically, given the target operating frequencies f_{0k}, $k = 1, \ldots, N$, and the target fractional bandwidth B. The antenna at the design $\boldsymbol{x}$ ($\boldsymbol{x}$ stands for a vector of adjustable parameters) is said to satisfy the requirements if the following condition holds:

$$\max\left\{ f \in \bigcup_{k=1}^{N}\left[\left(1-\frac{B}{2}\right)f_{0k},\left(1+\frac{B}{2}\right)f_{0k}\right] : |S_{11}(\boldsymbol{x},f)| \right\} \\ \leq S_{\max}, \tag{9.1}$$

where f is the frequency, whereas $S_{\max}$ is typically $-10\,\text{dB}$. The condition (9.1) means that the antenna matching is no worse than $S_{\max}$ within all target fractional bandwidths.

Typically, the initial design for yield optimization is the nominal design $\boldsymbol{x}^{(0)}$, which may be obtained by solving a standard minimax problem of the form

$$\boldsymbol{x}^{(0)} = \arg\min_{\boldsymbol{x}} \left\{ \max\left\{ f \in \bigcup_{k=1}^{N} \left[\left(1 - \frac{B}{2}\right) f_{0k}, \right.\right.\right.$$
$$\left.\left.\left. \left(1 + \frac{B}{2}\right) f_{0k} \right] : |S_{11}(\boldsymbol{x}, f)| \right\} \right\}. \tag{9.2}$$

The solution to (9.2) determines the design that exhibits the best possible antenna matching within the operating bands of interest.

The second case are microwave couplers. The considered coupler outputs are the scattering parameters, $S_{k1}(\boldsymbol{x}, f)$, $k = 1, \ldots, 4$, being functions of the design parameter (geometry) vector $\boldsymbol{x}$ and frequency f. For the sake of example, we consider the following performance figures:

- Power split error $dS(\boldsymbol{x}) = ||S_{21}(\boldsymbol{x}, f)| - |S_{31}(\boldsymbol{x}, f)||$;
- $-20\,\text{dB}$ bandwidth $B(\boldsymbol{x})$, defined as

$$B(\boldsymbol{x}) = 2\min\{f_0 - \max\{f_{11.L}(\boldsymbol{x}), f_{41.L}(\boldsymbol{x})\},$$
$$\min\{f_{11.H}(\boldsymbol{x}), f_{41.H}(\boldsymbol{x})\} - f_0\}, \tag{9.3}$$

where f_0 is the operating frequency of the coupler, whereas $f_{11.L}$ and $f_{11.H}$ are the frequencies corresponding $|S_{11}(\boldsymbol{x}, f)| = -20\,\text{dB}$ (lower and upper ends of the S_{11} bandwidth, respectively); similar definitions hold for $f_{41.L}$ and $f_{41.H}$.

In other words, $B(\boldsymbol{x})$ is the symmetric part (with respect to f_0) of the frequency range for which both the coupler matching and isolation characteristics are at or below $-20\,\text{dB}$.

We assume the following minimax specifications for the coupler:

$$dS(\boldsymbol{x}) \leq dS_{\max}, \tag{9.4}$$

and

$$B(\boldsymbol{x}) \geq B_{\min}. \tag{9.5}$$

For the purpose of subsequent considerations, the nominal design $\boldsymbol{x}^{(0)}$ will be the one optimized to obtain equal power split and to minimize $|S_{11}|$ and $|S_{41}|$ for the frequencies corresponding to $B_{\min}$. An exemplary formulation of the optimization task is

$$x^{(0)} = \arg\min_{x} \left\{ \max \left\{ \begin{array}{c} f_0 - \frac{B_{\min}}{2} \leq f \leq f_0 + \frac{B_{\min}}{2} : \\ \max\{|S_{11}(x,f)|, |S_{41}(x,f)|\} \end{array} \right\} \right.$$
$$\left. + \beta.[dS(x)]^2 \right\}, \qquad (9.6)$$

where the primary objective is a minimization of the maximum of matching and isolation characteristic within the frequency range from $f_0 - B_{\min}/2$ and $f_0 + B_{\min}/2$, whereas the second component is a penalty factor introduced to enforce $dS(x) = 0$. The value of the penalty factor determines the trade-off between the accuracy of reaching the required power split and bandwidth improvement.

Let dx denote a vector of parameter deviations, e.g., manufacturing tolerances. The deviations are described by the assumed probability distributions, e.g., joint Gaussian with zero mean and variance σ, or uniform with maximum deviation $d_{\max}$. In general, the distributions of the various parameters may be correlated (Koziel and Bekasiewicz, 2016) with the correlations described, e.g., by an appropriate covariance matrix. In this chapter, for simplicity, it is assumed that the parameter deviations are statistically independent.

Given the performance specifications and parameter deviations, the yield $Y(x)$ at the design x (in particular, at the nominal design $x = x_{\text{nom}}$) can be estimated using the Monte Carlo analysis as

$$Y(x) = \frac{1}{p} \sum_{k=1}^{p} H(x^{(k)}), \qquad (9.7)$$

where $x^{(k)} = x + dx^{(k)}$, $k = 1, \ldots, p$, are observables with $dx^{(k)}$ being the random deviations as described earlier. The function $H(x)$ is defined as

$$H(x) = \begin{cases} 1 & \text{if design specifications are satisfied} \\ 0 & \text{otherwise.} \end{cases} \qquad (9.8)$$

The yield optimization problem is formulated as

$$x^* = \arg\min_{x}\{-Y(x)\}. \qquad (9.9)$$

Typically, the initial design for (9.9) is the nominal design $x^{(0)}$, which may be obtained by solving the minimax problem (9.2) (for the example of multi-band antennas) or (9.6) (for the example of microwave couplers).

9.2.2 Surrogate-based yield optimization — benchmark Algorithm 1: One-shot optimization

Nowadays, the most common approach to realize EM-based statistical analysis and yield optimization is through the employment of fast surrogate models. A brief characterization of available methods has been provided in Section 9.1. Here, we consider two techniques utilized in Section 9.4 as benchmark methods to be compared with the methodology discussed in this chapter (Section 9.3). In both cases, the surrogates are constructed using kriging interpolation (Simpson et al., 2001), but, in general, any other method can be used as well (RBF; Queipo et al., 2005; PCE; Petrocchi et al., 2017, etc.).

The overall idea is to construct a single surrogate model that covers a sufficiently large neighborhood of the nominal design so that reliable yield estimation can be carried out within this entire region. Let $d = [d_1 \cdots d_n]^T$ be the size vector. Upon setting up the surrogate in the interval $[x^{(0)} - d, x^{(0)} + d]$, the yield is optimized by solving the problem (9) therein. To create a sufficient room for yield improvement, the size vector should be relatively large. Here, we assume that $d_k = 10 d_{\max}$, $k = 1, \ldots, n$, where $d_{\max}$ is the maximum deviation in the case of uniform distribution, or 3σ for Gaussian distribution of variance σ. The advantage of this method is simple implementation; however, the cost of training data acquisition may be considerable due to a domain size.

9.2.3 Surrogate-based yield optimization — benchmark Algorithm 2: Sequential approximate optimization

An alternative strategy is to employ sequential approximate optimization (SAO). In particular, we replace solving of (9.9) by an iterative process

$$x^{(i+1)} = \arg\min_{x}\{-Y_s^{(i)}(x)\}, \qquad (9.10)$$

where $x^{(i)}$, $i = 0, 1, \ldots$, are the approximations of x^*, whereas $Y_s^{(i)}$ is the yield estimated using the ith surrogate model constructed in the vicinity $x^{(i)} - d_l \leq x \leq x^{(i)} - d_l$ of the current design $x^{(i)} = [x_1^{(i)} \cdots x_n^{(i)}]^T$. The main idea is to construct the surrogate within a smaller domain, which is then relocated between the algorithm iterations. For our numerical experiments, we use the size of $3d_{\max}$. Solving the problem (9.10) is subject to constraints $x_k^{(i)} - d_{l.k} + d_{\max} \leq x_k \leq x_k^{(i)} + d_{l.k} - d_{\max}$, $k = 1, \ldots, n$, to ensure that the point x is at least at the distance $d_{\max}$ from the surrogate model domain boundary (in all directions). A clear advantage of the method is the lower computational cost of constructing a reliable surrogate (as compared to the algorithm of Section 9.2.2); however, a few iterations are necessary to conclude the optimization process. The algorithm is terminated if the current iteration does not improve the yield, i.e., if $Y_s^{(i+1)}(x^{(i+1)}) \leq Y_s^{(i)}(x^{(i)})$.

9.3 Surrogate-Based Yield Optimization with Domain Confinement

This section formulates an alternative yield optimization technique considered in this chapter. We discuss the main components of the optimization framework, including a definition of the surrogate model domain, as well as the optimization algorithm flow. As explained below, the appropriate confinement of the parameter space region for surrogate construction is critical to ensure good predictive power of the model without using the excessive numbers of training data samples, while retaining sufficient flexibility (in terms of the parameter ranges covered by the surrogate). The latter is essential for reliable identification of the yield-optimum design.

The reference algorithms outlined in Section 9.2 represent the two extreme strategies for yield optimization, i.e., single surrogate model constructed over a larger portion of the parameter space versus an iterative process with the surrogates constructed over smaller regions relocated along the optimization path. One of the objectives is to develop a technique that maintains the simplicity of the one-shot approach while constructing the surrogate at a reasonable cost. This is realized here using the concept of performance-driven modeling (Koziel, 2017; Koziel and Pietrenko-Dabrowska, 2019). The idea of domain confinement was presented in (Koziel and Pietrenko-Dabrowska, 2019) in a generic setup, i.e., with the purpose of constructing fast design-ready surrogates over broad ranges of geometry and operating parameters. Here, it is adopted for rendering more localized models, the domain of which is spanned by the parameter space directions that affect the system characteristics in the most significant manner (from the perspective of the yield values). At the same time, the domain size remains small in all orthogonal directions. In the following, the procedure for finding the domain spanning directions is introduced in the context of multi-band antenna design (Section 9.3.1) and microwave coupler design (Section 9.3.2), along with the domain definitions. These two methods share the same main idea, yet they are tailored to particular sets of performance specifications as considered in Section 9.2.1.

9.3.1 Yield optimization of multi-band antennas

We aim at narrowing down the surrogate model domain so that it is oriented along the path corresponding to the maximum changes of the relevant antenna responses (here, reflection characteristics at and around the target operating frequencies). The directions of essential changes of antenna responses are identified by executing two optimization runs, one aiming at maximization of the antenna fractional bandwidths (symmetric with respect to the operating frequencies), the second aiming at minimizing the antenna reflection at the operating frequencies. Thus, two additional designs are obtained (apart from the nominal design $x^{(0)}$) as

$$x^{(1)} = \arg\min_{x}\{-\min\{B_1(x), \ldots, B_N(x)\}\}, \quad (9.11)$$

$$x^{(2)} = \arg\min_{x}\{\max\{|S_{11}(x, f_{01})|, \ldots, |S_{11}(x, f_{0N})|\}\}. \quad (9.12)$$

In (9.11), $B_k(x)$ is a symmetric part of the kth bandwidth, i.e., $B_k(x) = 2\min\{f_{0k}-f_{1k}(x), f_{2k}(x)-f_{0k}\}$, $k = 1, \ldots, N$, with f_{1k} and f_{2k} being the frequencies corresponding to -10 dB level of $|S_{11}|$ (left- and right-hand-side ends of the kth resonance). Note that both (9.11) and (9.12) are formulated in a minimax sense, i.e., the improvement of the worst case, the bandwidth in (9.11) and the reflection levels at the operating frequencies in (9.12).

The problems (9.11), (9.12) are solved using trust-region gradient search (Conn et al., 2000) with the Jacobian matrix updated using the rank-one Broyden (1965) formula. The latter is sufficient because the expected design relocations $\|x^{(0)} - x^{(1)}\|$ and $\|x^{(0)} - x^{(2)}\|$ are limited. Consequently, the optimization process can be realized at a very low cost of around $1.5n$ EM analyses, where n is the parameter space dimensionality.

Let $s(t) = [s_1(t) \cdots s_n(t)]^T$ be a t-parameterized curve such that

$$s_j(t) = a_{j0} + a_{j1}t + a_{j2}t^2, \qquad (9.13)$$

for $0 \leq t \leq 1$, so that $s(0) = x^{(1)}$, $s(0.5) = x^{(0)}$, and $s(1) = x^{(2)}$. The model coefficients can be found as

$$\begin{bmatrix} a_{10} & \cdots & a_{n0} \\ a_{11} & \cdots & a_{n1} \\ a_{12} & \cdots & a_{n2} \end{bmatrix} = \begin{bmatrix} 1 & 0 & 0 \\ 1 & 0.5 & 0.25 \\ 1 & 1 & 1 \end{bmatrix}^{-1} \begin{bmatrix} (x^{(1)})^T \\ (x^{(0)})^T \\ (x^{(2)})^T \end{bmatrix}. \qquad (9.14)$$

Let $S(t)$ be the interval with the center at $s(t)$ and the size $d_c = [d_{c1} \cdots d_{cn}]^T$, where d_{cj} is a small multiplicity of the maximum design deviation $d_{\max}$, e.g., $2d_{\max}$. We define the surrogate model domain X_S as the set-theory union of the intervals $S(t)$ for $0 \leq t \leq 1$, i.e.,

$$X_S = \bigcup_{0 \leq t \leq 1} S(t). \qquad (9.15)$$

This domain contains the designs $x^{(0)}$, $x^{(1)}$, $x^{(2)}$, and a vicinity of the entire curve $s(t)$ of the size d_c. The set is of small size yet it covers the directions of significant changes of the antenna responses, which are the most important from the point of view of manipulating the shape of the resonances, and, consequently, the performance figures such as yield. Due to a limited volume, a reliable surrogate model can be established in X_S using a small number of training data samples. Furthermore, the entire optimization process can be performed within X_S without the necessity of iterating the process, upon domain relocation, as in the algorithm Section 9.2.2. Figure 9.1 provides a graphical illustration of the reference designs and the domain X_S.

Expedited Yield-Driven Design of High-Frequency Structures 305

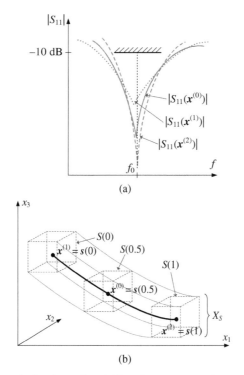

Fig. 9.1. Yield optimization of multi-band antennas using performance-driven surrogates: (a) reflection responses of an exemplary narrow-band antenna at the nominal design $x^{(0)}$, maximum bandwidth design $x^{(1)}$, and best matching (at f_0) design $x^{(2)}$. These designs determine the directions of the most significant response changes (from the point of view of the target operating bandwidth), (b) the reference designs $x^{(0)}$ through $x^{(2)}$ form a path (a parameterized curve $s(t)$). The union of intervals $S(t)$ (cf. (9.15) form the surrogate model domain X_S.

Having defined the domain, the surrogate is constructed using kriging interpolation. The yield optimization is then carried out by directly solving the problem (9.9) at the level of surrogate, similarly as in the algorithm of Section 9.2.1.

9.3.2 Yield optimization of microwave couplers

Given the design specifications (9.4) and (9.5) concerning the power split error and the coupler bandwidth, consider the bandwidth-defining frequencies $f_{11.L}(x)$, $f_{11.H}(x)$, $f_{41.L}(x)$, $f_{41.H}(x)$, and the transmission levels at the operating frequency $l_1(x) = |S_{21}(x, f_0)|$

and $l_2(\boldsymbol{x}) = |S_{31}(\boldsymbol{x}, f_0)|$. Let us define the vector $\boldsymbol{F}(\boldsymbol{x})$ as

$$\boldsymbol{F}(\boldsymbol{x}) = [f_{11.L}(\boldsymbol{x}) f_{11.H}(\boldsymbol{x}) f_{41.L}(\boldsymbol{x}) f_{41.H}(\boldsymbol{x}) l_1(\boldsymbol{x}) l_2(\boldsymbol{x})]^T. \quad (9.16)$$

The Jacobian $\boldsymbol{J}_F$ of $\boldsymbol{F}(\boldsymbol{x})$ at $\boldsymbol{x}$ can be obtained from the Jacobian $\boldsymbol{J}_S$ of the coupler frequency characteristics (the latter estimated using finite differentiation).

Let

$$\boldsymbol{L}_F(\boldsymbol{x}) = [L_1(\boldsymbol{x}) L_2(\boldsymbol{x}) \cdots L_6(\boldsymbol{x})]^T$$
$$= \boldsymbol{F}(\boldsymbol{x}^{(0)}) + \boldsymbol{J}_F(\boldsymbol{x}^{(0)}) \cdot (\boldsymbol{x} - \boldsymbol{x}^{(0)}), \quad (9.17)$$

be a linear expansion model of $\boldsymbol{F}$ at the nominal design $\boldsymbol{x}^{(0)}$. Further, let $\boldsymbol{x}^{(1)}$ and $\boldsymbol{x}^{(2)}$ be the two designs obtained by solving the following minimization problems:

$$\boldsymbol{x}^{(1)} = \arg\min_{\boldsymbol{x}} \left\{ L_5(\boldsymbol{x}) - L_6(\boldsymbol{x}) + \beta_2 \left\| \begin{bmatrix} L_1(\boldsymbol{x}) \\ \vdots \\ L_4(\boldsymbol{x}) \end{bmatrix} \right. \right.$$
$$\left. \left. - \begin{bmatrix} L_1(\boldsymbol{x}^{(0)}) \\ \vdots \\ L_4(\boldsymbol{x}^{(0)}) \end{bmatrix} \right\|^2 \right\}, \quad (9.18)$$

$$\boldsymbol{x}^{(2)} = \arg\min_{\boldsymbol{x}} \left\{ \begin{array}{l} -2\min\{f_0 - \max\{L_1(\boldsymbol{x}), L_3(\boldsymbol{x})\}, \\ \min\{L_2(\boldsymbol{x}), L_4(\boldsymbol{x})\} - f_0\} \\ + \beta_1 \cdot [L_5(\boldsymbol{x}) - L_6(\boldsymbol{x})]^2 \end{array} \right\}. \quad (9.19)$$

Both (9.18) and (9.19) are subject to a constraint $\|\boldsymbol{x} - \boldsymbol{x}^{(0)}\| \leq D$, where D is a user-defined parameter. In the numerical experiments of Section 9.4, we use $D = 0.5\,\text{mm}$. Note that the overall cost of solving (9.18) and (9.19) is only n EM analyses of the coupler, which corresponds to the cost of estimating the Jacobian $\boldsymbol{J}_F$ in (9.17).

The vectors $\boldsymbol{x}^{(1)}$ and $\boldsymbol{x}^{(2)}$ determine the directions of the maximum change of the coupler power split (while retaining the bandwidth) and its bandwidth (while retaining the power split), respectively. In particular, the vectors $\boldsymbol{v}_1 = \boldsymbol{x}^{(1)} - \boldsymbol{x}^{(0)}$ and $\boldsymbol{v}_2 = \boldsymbol{x}^{(2)} - \boldsymbol{x}^{(0)}$ will be used to span the domain of the surrogate model.

Let $S(t) = [S_1(t) \cdots S_n(t)]^T$ be a t-parameterized surface such that

$$s(t) = S\left([t_1\ t_2]^T\right) = x^{(0)} + t_1 v_1 + t_2 v_2, \quad (9.20)$$

for $-1 \leq t_1, t_2 \leq 1$. We have $S([0\ 0]^T) = x^{(0)}$, $S([1\ 0]^T) = x^{(1)}$, and $S([0\ 1]^T) = x^{(2)}$.

Let $S_I(t)$ be the interval with the centre at $S(t)$ and the size $d_c = [d_{c1} \cdots d_{cn}]^T$, where d_{cj} is a multiplicity of the maximum design deviation $d_{\max}$, e.g., $2d_{\max}$.

The surrogate model domain X_S is then defined as the union of all intervals $S_I(t)$ for $-1 \leq t_1, t_2 \leq 1$, i.e.,

$$X_S = \bigcup_{-1 \leq t_1, t_2 \leq 1 S_I} \left([t_1 t_2]^T\right). \quad (9.21)$$

By definition, the domain contains the nominal design $x^{(0)}$, vectors $x^{(1)}$ and $x^{(2)}$, the points $x^{(0)} - v_k$, $k = 1, 2$, as well as the neighborhood of the entire plane $S(t)$ of the size d_c. It should be noted that the size of X_S is small, yet it covers the directions of essential changes of the coupler responses, i.e., those that are important to manipulate the yield value. Because of the small volume, the cost of setting up the surrogate model within X_S is low. On the other hand, due to the mentioned coverage of important directions, there is no need to iterate the process of constructing and optimizing the surrogate as in the algorithm of Section 9.2.2. A graphical illustration of the domain X_S has been provided in Fig. 9.2.

Given X_S, the surrogate model is constructed using kriging interpolation. The yield optimization process is conducted by solving (9.9), i.e., similarly as in the algorithm of Section 9.2.1.

9.4 Demonstration Case Studies

This section discusses numerical validation of the yield optimization procedure discussed in Section 9.3. It involves three antenna structures, a ring-slot antenna, a dual-band uniplanar dipole, and a triple-band dipole, as well as a miniaturized rat-race coupler.

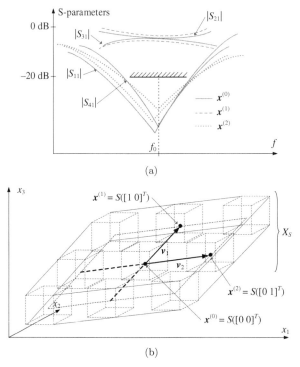

Fig. 9.2. Yield optimization of microwave couplers by means of performance-driven surrogates: (a) scattering parameters of an exemplary coupler at the nominal design $x^{(0)}$, design $x^{(1)}$ (spoiled power split), and design $x^{(2)}$ (improved -20 dB bandwidth); for clarity, only the selected S-parameters are shown for $x^{(1)}$ ($|S_{21}|, |S_{31}|$) and $x^{(2)}$ ($|S_{11}|, |S_{41}|$). These designs determine the directions of the most significant response changes (from the point of view of yield manipulation), (b) the designs $x^{(0)}$ through $x^{(2)}$ form a parameterized surface $S(t)$. The union of intervals $S_I(t)$ (cf. (9.21)) forms the surrogate model domain X_S.

The methodology of Section 9.3 is compared to the surrogate-assisted algorithms of Section 9.2. At the same time, the reliability of surrogate-based statistical analysis is validated using EM-driven Monte Carlo analysis run at the initial and optimized designs.

9.4.1 Case I: Ring-slot antenna

The first verification example is a ring slot antenna shown in Fig. 9.3 (Koziel and Bekasiewicz, 2017), implemented on 0.76-mm-thick substrate of relative permittivity $\varepsilon_r = 2.0$. The structure is excited

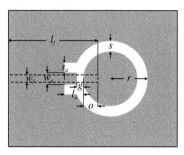

Fig. 9.3. Geometry of the ring slot antenna with a microstrip feed (dashed line) (Koziel and Bekasiewicz, 2017).

through a microstrip line feeding a circular ground plane slot with defected ground structure (DGS). The latter is employed to suppress the antenna harmonic frequencies (Koziel and Bekasiewicz, 2017). There are eight geometry parameters $x = [l_f\ l_d\ w_d\ r\ s\ s_d\ o\ g]^T$. The computational model of the antenna is implemented in CST Microwave Studio (∼300,000 cells, simulation time 90 s).

The design objective is minimization of the in-band reflection for the frequency range 4.15–4.85 GHz (center frequency $f_0 = 4.5$ GHz). The nominal design is $x^{(0)} =$[20.28 6.54 0.24 11.83 2.95 6.77 7.85 2.23]^T. The other reference designs $x^{(1)} = $ [20.03 6.30 0.20 11.84 2.94 6.74 7.89 2.43]^T, and $x^{(2)} = $ [20.26 6.51 0.20 11.68 2.92 6.47 7.49 2.24]^T, were obtained by solving (9.11) (maximum bandwidth design) and (9.12) (best reflection at $f_0 = 4.5$ GHz design), respectively. The cost of obtaining the reference designs is only 13 and 14 EM simulations, respectively.

Geometry parameter deviations are described by independent uniform probability distributions with the maximum deviation $d_{\max} = 0.05$ mm. Table 9.1 shows the results of yield estimation at the initial design, yield optimization results, yield estimation at the final design for the algorithm of Section 9.3, as well as the reference algorithms (Sections 9.2.1 and 9.2.2). The size parameters $d_{c.k}$ defining the surrogate model domain were set to $2d_{\max}$, and the model itself has been set up using 35 training samples (relative RMS error 0.5%).

The surrogate model for the algorithm of Section 9.2.1 has been set using 400 samples within the domain of size $10d_{\max}$ (relative RMS error 0.7%). The surrogate models for the algorithm of Section 9.2.2 have been set with 50 samples within the domain of size

Table 9.1. Yield optimization of the ring slot antenna of Fig. 9.3.

Optimization algorithm	Initial yield — Estimated by surrogate model (%)	Initial yield — EM-based (%)	Optimized yield — Estimated by surrogate model (%)	Optimized yield — EM-based (%)	CPU cost[a]
Algorithm of Section 9.2.1	81	81	92	93	400
Algorithm of Section 9.2.2	81	81	91	91	150[b]
Algorithm of Section 9.3.1	81	81	91	91	62[c]

Notes: [a] Optimization cost in number of EM analyses of the antenna structure.
[b] The algorithm convergence after three iterations (surrogate setup cost 50 training samples per iteration).
[c] The cost includes training data acquisition (35 EM analyses) and generation of the reference designs $x^{(1)}$ and $x^{(2)}$ (27 EM simulations in total).

$3d_{\max}$ (relative RMS error 0.4% at the first domain centered at $x^{(0)}$). The optimum design produced by the considered approach is $x^* = [20.186.430.2111.852.956.787.902.31]^T$. Figure 9.4 shows a visualization of the Monte Carlo analysis at the nominal design and at x^*. In each case, Monte Carlo analysis was performed using 500 samples generated according to the respective probability distribution.

The results of Table 9.1 indicate that appropriate constraining of the surrogate model domain as in the considered approach allows for a significant reduction of the computational cost of the yield optimization process. On the one hand, the cost of setting a reliable model itself is low due to the small volume of the domain. On the other hand, because the domain covers the relevant directions within the parameter space (i.e., those corresponding to the essential changes of the antenna response), there is no need to iterate the procedure: a one-shot approach is normally sufficient.

This is corroborated by comparisons with algorithms of Sections 9.2.1 and 9.2.2. Both render the results of similar quality in terms of the final value of the yield, although the quality of results produced by the algorithm of Section 9.2.1 is slightly degraded due to a larger domain of the surrogate. The latter affects the

Expedited Yield-Driven Design of High-Frequency Structures 311

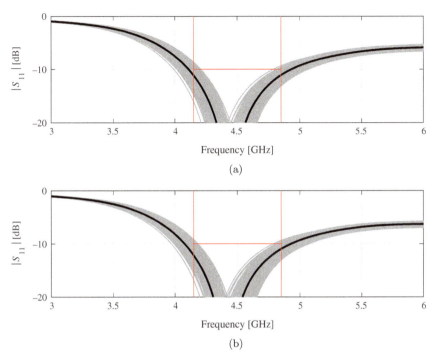

Fig. 9.4. Monte Carlo analysis of antenna of Fig. 9.3 using EM simulations (gray plots): (a) nominal design, (b) yield-optimized design obtained using the algorithm of Section 9.3.1. Black plots show the antenna response at the nominal and optimized designs, respectively.

model predictive power despite using a considerably larger number of training samples. Furthermore, EM-based Monte Carlo analysis executed for validation confirms reliability of yield estimation obtained from the surrogates.

9.4.2 Case II: Dual-band uniplanar dipole antenna

The second verification example is a dual-band uniplanar dipole antenna shown in Fig. 9.5 (Chen *et al.*, 2006). The antenna is implemented on a Rogers RO4350 substrate ($\varepsilon_r = 3.5$, $h = 0.76$ mm) and fed by a 50 Ohm coplanar waveguide (CPW). The adjustable variables are $\boldsymbol{x} = [l_1\ l_2\ l_3\ w_1\ w_2\ w_3]^T$. Other parameters are fixed: $l_0 = 30$, $w_0 = 3$, $s_0 = 0.15$ and $o = 5$ (all dimensions in mm). The computational model of the antenna is implemented in

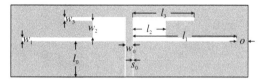

Fig. 9.5. Geometry of a dual-band uniplanar dipole antenna (Chen et al., 2006).

CST Microwave Studio and evaluated using its time-domain solver (~100,000 cells; simulation time 60 s).

The design objective is to minimize the in-band reflection within 8% symmetric bandwidths centered at the frequencies $f_{01} = 3.0$ GHz and $f_{02} = 5.5$ GHz. The nominal design is $x^{(0)} = [30.47\ 11.60\ 19.20\ 0.47\ 2.46\ 1.30]^T$. The other two reference designs $x^{(1)} = [29.98\ 11.13\ 18.86\ 0.44\ 3.04\ 1.05]^T$ (maximum bandwidth) and $x^{(2)} = [29.72\ 11.03\ 19.04\ 0.27\ 3.01\ 0.67]^T$ (best reflection at the operating frequencies) were obtained by solving (9.11) and (9.12), respectively. The reference designs are generated using only 11 and 10 EM simulations, respectively.

Similarly as in Section 9.3.1, geometry parameter deviations are described by independent uniform probability distributions with the maximum deviation $d_{\max} = 0.05$ mm. The results of yield optimization using the considered algorithm and the benchmark methods have been gathered in Table 9.2. The size parameters $d_{c.k}$ were set to $2d_{\max}$ (training set size 53 samples, relative RMS error 0.8%). The surrogate model for the algorithm of Section 9.2.1 has been set up using 800 samples within the domain of size $10d_{\max}$ (relative RMS error 1.3%). The surrogate models for the algorithm of Section 9.2.2 have been set with 50 samples within the domain of size $3d_{\max}$ (relative RMS error 0.9% at the first domain centered at $x^{(0)}$). The optimum design produced by the considered approach is $x^* = [30.38\ 11.58\ 19.22\ 0.47\ 2.46\ 1.28]^T$. Figure 9.6 shows visualization of the EM-based Monte Carlo analysis (using 500 random samples) at the nominal design and at the yield-optimized design x^*.

The results obtained for this example are consistent with those discussed in Section 9.4.1. The considered approach outperforms both reference algorithms in terms of the computational efficiency while providing the results of similar quality. It should be noted that the predictive powers of the surrogate models are not as good as for the case of Section 9.4.1, therefore, slight discrepancies between the

Expedited Yield-Driven Design of High-Frequency Structures

Table 9.2. Yield optimization of the dual-band antenna of Fig. 9.5.

Optimization algorithm	Initial yield Estimated by surrogate model (%)	Initial yield EM-based (%)	Optimized yield Estimated by surrogate model (%)	Optimized yield EM-based (%)	CPU cost[a]
Algorithm of Section 9.2.1	64	65	95	94	800
Algorithm of Section 9.2.2	64	65	93	92	150[b]
Algorithm of Section 9.3.1	67	65	94	92	74[c]

Notes: [a]Optimization cost in number of EM analyses of the antenna structure.
[b]The algorithm convergence after three iterations (surrogate setup cost 50 training samples per iteration).
[c]The cost includes training data acquisition (53 EM analyses) and generation of the reference designs $x^{(1)}$ and $x^{(2)}$ (21 EM simulations in total).

surrogate-based and EM-based yield estimations can be observed. The computational cost of yield optimization using the algorithm of Section 9.3.1 is reduced by about 50% compared to the algorithm of Section 9.2.1, and it is significantly lower than for the algorithm of Section 9.2.2.

9.4.3 Case III: Triple-band uniplanar dipole antenna

The third example is a triple-band uniplanar dipole antenna shown in Fig. 9.7. The structure is based on the design of (Chen et al., 2006) and implemented on RO4350 substrate. The design variables are $x = [l_1\ l_2\ l_3\ l_4\ l_5\ w_1\ w_2\ w_3\ w_4\ w_5]^T$; other parameters are fixed: $l_0 = 30$, $w_0 = 3$, $s_0 = 0.15$ and $o = 5$ (all dimensions in mm). The EM-simulation model of the antenna is implemented in CST Microwave Studio (∼200,000 cells; simulation time 110 s).

The design objective is to minimize the in-band reflection within 4% symmetric bandwidths centered at the frequencies $f_{01} = 2.45$ GHz, $F_{02} = 3.6$ GHz, and $f_{03} = 5.3$ GHz. The nominal design is $x^{(0)} = [35.42\ 11.54\ 26.07\ 8.09\ 17.14\ 0.60\ 0.99\ 1.44\ 0.78\ 1.17]^T$. The other two reference designs $x^{(1)} = [35.38\ 11.50\ 25.72\ 6.60\ 15.87\ 0.52\ 0.90\ 1.38\ 0.81\ 1.09]^T$ (maximum bandwidth) and $x^{(2)} = [35.54$

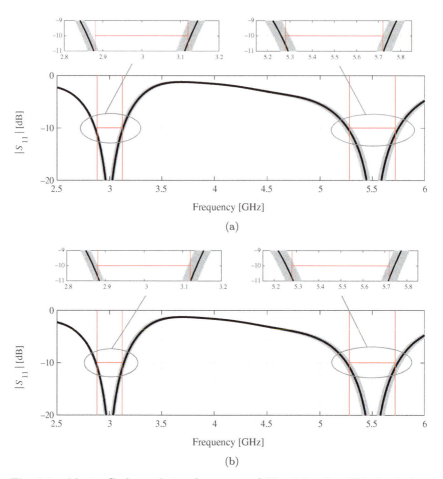

Fig. 9.6. Monte Carlo analysis of antenna of Fig. 9.5 using EM simulations (gray plots): (a) nominal design, (b) yield-optimized design obtained using the algorithm of Section 9.3.1. Black plots show the antenna response at the nominal and optimized designs, respectively.

11.69 26.53 6.63 15.76 0.70 1.36 1.90 0.48 0.67$]^T$ (best reflection at the operating frequencies) were obtained by solving (9.11) and (9.12), respectively. The computational cost of obtaining $\boldsymbol{x}^{(1)}$ and $\boldsymbol{x}^{(2)}$ is 16 EM antenna simulations in both cases.

Deviations of geometry parameters are described by the independent uniform probability distributions with the maximum deviation $d_{\max} = 0.05$ mm. The results of yield optimization using the

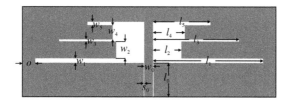

Fig. 9.7. Geometry of a triple-band uniplanar dipole antenna.

Table 9.3. Yield optimization of the triple-band antenna of Fig 9.7.

Optimization algorithm	Initial yield — Estimated by surrogate model (%)	EM-based (%)	Optimized yield — Estimated by surrogate model (%)	EM-based (%)	CPU cost[a]
Algorithm of Section 9.2.1	63	58	75	66	1,600
Algorithm of Section 9.2.2	62	58	72	69	400[b]
Algorithm of Section 9.3.1	60	58	72	69	132[c]

Notes: [a]Optimization cost in number of EM analyses of the antenna structure.
[b]The algorithm convergence after four iterations (surrogate setup cost 100 training samples per iteration).
[c]The cost includes training data acquisition (100 EM analyses) and generation of the reference designs $x^{(1)}$ and $x^{(2)}$ (32 EM simulations in total).

algorithm of Section 9.3 and the benchmark methods have been gathered in Table 9.3. The surrogate model setup for all methods is the same as for the previous cases. The size parameters $d_{c.k}$ were set to $2d_{\max}$ (training set size 100 samples, relative RMS error 1.3%). The surrogate model for the algorithm of Section 9.2.1 has been set using 1,600 samples within the domain of size $10d_{\max}$ (relative RMS error 2.8%). The surrogate models for the algorithm of Section 9.2.2 have been set with 100 samples within the domain of size $3d_{\max}$ (relative RMS error 2.1% at the first domain centered at $x^{(0)}$). The algorithm of Section 9.3.2 rendered the yield-optimized design $x^* = [35.37 \ 11.52 \ 26.0 \ 8.09 \ 17.14 \ 0.59 \ 1.01 \ 1.43 \ 0.80 \ 1.22]^T$.

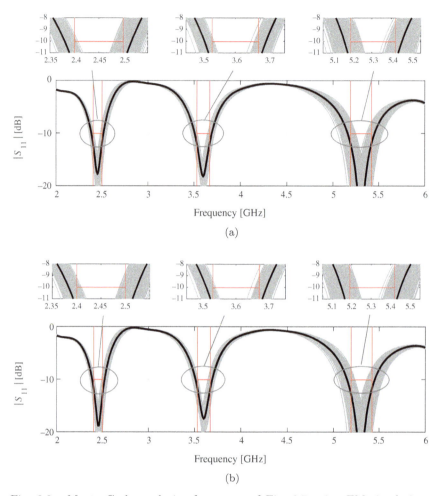

Fig. 9.8. Monte Carlo analysis of antenna of Fig. 9.7 using EM simulations (gray plots): (a) nominal design, (b) yield-optimized design obtained using the algorithm considered in this chapter. Black plots show the antenna response at the nominal and optimized designs, respectively.

Figure 9.8 shows visualization of the EM-based Monte Carlo analysis at the nominal design and at x^*.

Also in this case, the algorithm of Section 9.3.1 outperforms the benchmark ones and the overall results are consistent with those obtained for the previous examples. Due to higher dimensionality of the parameter space and nonlinearity of the antenna response (three resonances), in this case, the numbers of training samples necessary to construct the surrogate models are noticeably larger.

Notwithstanding, the algorithm of Section 9.2.2 (sequential approximate optimization) is not capable of finding as good design as those identified using the algorithm of Section 9.2.1 and the considered method (Section 9.3.2). This is most likely due to restricted size of the domain of the local surrogate. Furthermore, reliability of the algorithm of Section 9.2.1 is not as good as for the previous examples because of limited predictive power of the surrogate (2.8%). Finally, this verification example pronounces even more the benefits of the approach considered in this chapter, i.e., the capability of rendering high-quality design at the low computational cost.

9.4.4 Case IV: Compact microstrip rat-race coupler

As the last example, consider a compact microstrip rat-race coupler (RRC) shown in Fig. 9.9 (Koziel et al., 2015). The structure is implemented on Taconic RF-35 substrate ($\varepsilon_r = 3.50$, $h = 0.762$ mm). The transmission lines forming the circuit are folded to the inside of the structure to reduce the footprint area. The designable parameters are $\boldsymbol{x} = [l_1\ l_2\ l_3\ d\ w\ w_1]^T$, $d_1 = d + |w - w_1|$, the input line width and length, $w_0 = 1.7$ and $l_0 = 15$, are fixed (all in mm).

The design objectives for the RRC are to maintain equal power split at the operating frequency $f_0 = 1$ GHz, and to minimize the circuit matching and isolation within the frequency range from 0.9 GHz to 1.1 GHz. The same frequency range is also the minimum acceptable -20 dB bandwidth. Section 9.2.1 provides a rigorous formulation of these objectives. The nominal design $\boldsymbol{x}^{(0)} = [4.50\ 11.08\ 21.80\ 0.65\ 0.94\ 0.85]^T$ is obtained by solving the problem (9.6) (Section 9.2.1).

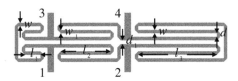

Fig. 9.9. Layout of the miniaturized folded rat-race coupler (Koziel et al., 2015).

For the purpose of yield estimation and optimization, deviations of the RRC geometry parameters are described by independent uniform probability distributions with the maximum deviation $d_{\max} = 0.05$ mm. Numerical experiments have been conducted using the following setup. For the methodology considered in this chapter, the first step was to generate the reference designs $\boldsymbol{x}^{(1)}$ and $\boldsymbol{x}^{(2)}$. These are $\boldsymbol{x}^{(1)} = [4.07\ 11.56\ 21.48\ 0.65\ 0.95\ 0.86]^T$, and $\boldsymbol{x}^{(2)} = [4.60\ 11.18\ 21.91\ 0.66\ 0.95\ 0.85]^T$, and were obtained by solving (9.18) and (9.19), respectively. The size parameters $d_{c.k}$ defining the surrogate model domain X_S were set to $2d_{\max}$, and the model itself has been constructed using 72 training samples (relative RMS error 2.3%).

The surrogate model for the algorithm of Section 9.2.1 has been set using 400 samples within the domain of size $10d_{\max}$ (relative RMS error 3.4%). The surrogate models for the algorithm of Section 9.2.2 have been set with 50 samples within the domain of size $3d_{\max}$ (relative RMS error 2.2% at the first domain centered at $\boldsymbol{x}^{(0)}$). In both cases, the objective was to render the models of similar predictive power with a relative error of less than 1%. The latter is sufficient to ensure yield estimation reliability. The results of yield estimation at the initial design, obtained yield improvements, as well as yield estimation at the final designs, have been gathered in Table 9.4. The considered methodology has been compared to

Table 9.4. Yield optimization of the compact coupler of Fig 9.9.

Optimization algorithm	Initial yield Estimated by surrogate model (%)	Initial yield EM-based (%)	Optimized yield Estimated by surrogate model (%)	Optimized yield EM-based (%)	CPU cost[a]
Algorithm of Section 9.2.1	67	62	90	83	400
Algorithm of Section 9.2.2	66	62	86	83	200[b]
Algorithm of Section 9.3.2	63	62	84	82	72

Notes: [a]Optimization cost in number of EM analyses of the antenna structure.
[b]The algorithm convergence after three iterations (surrogate setup cost 50 training samples per iteration).

Expedited Yield-Driven Design of High-Frequency Structures 319

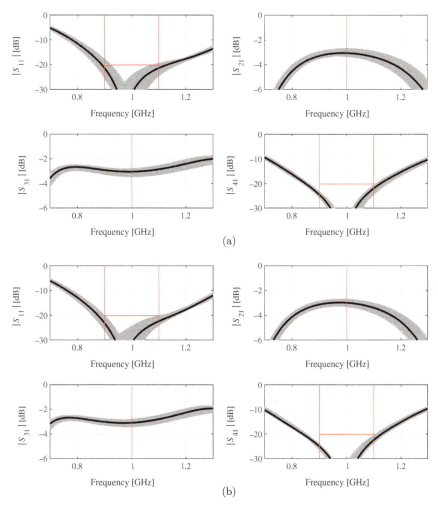

Fig. 9.10. Monte Carlo analysis of the coupler of Fig. 9.9 using EM simulations (gray plots): (a) nominal design, (b) yield-optimized design obtained using the algorithm of Section 9.3.2. Black plots show the antenna response at the nominal and the optimized designs, respectively.

the surrogate-assisted algorithms of Sections 9.2.1 and 9.2.2, respectively. The optimum design produced by the considered approach is $x^* = [4.65\ 11.10\ 21.87\ 0.71\ 0.95\ 0.81]^T$.

Figure 9.10 shows a visualization of the Monte Carlo analysis at the nominal design and at x^*. In each case, Monte Carlo analysis was performed using 500 samples generated according to the uniform probability distributions with a maximum deviation of 0.05 mm. It can be observed that the reliability of yield estimation is very good for the considered method and for the algorithm of Section 9.2.2, which is due to ensuring sufficient predictive power of the respective surrogates. Noticeable discrepancies between surrogate-predicted and EM-based yield for the algorithm of Section 9.2.1 are due to higher modeling error of the surrogate constructed in a larger domain.

A brief analysis of the results of Table 9.4 indicates that incorporating the concept of domain confinement into the yield optimization framework allows for combining the advantages of both algorithms considered as the benchmark. On the one hand, because the domain extends sufficiently into the most relevant directions determined by the vectors v_1 and v_2, there is no need to iterate the process of constructing and optimizing the surrogate (as in the algorithm of Section 9.2.1). On the other hand, because of maintaining the overall small size of the domain, the cost of setting up the surrogate is low. The overall computational overhead of the yield optimization process using the algorithm of Section 9.3 is reduced by 82% and 64%, as compared to the algorithms of Section 9.2.1 and 9.2.2, respectively. These savings are achieved without compromising the design quality (the final yield values are comparable for all considered methods) and reliability (the agreement with EM-based Monte Carlo analysis is acceptable).

9.5 Summary and Discussion

This chapter discussed a surrogate-assisted technique for statistical design of high-frequency structures. The presented approach incorporates the concept of performance-driven modeling to rendering a fast and reliable surrogate at a low computational cost. The latter is possible by spanning the domain of the model along the most relevant directions of the parameter space, i.e., corresponding to maximum variability of the system responses. At the same time, the dimensions in orthogonal directions are small, which allows for maintaining the low overall volume of the domain. This arrangement enables efficient yield optimization without the necessity of relocating the domain and re-constructing the surrogate (as in the sequential approximation optimization frameworks, SAO).

For the considered verification examples, the yield-optimized design is obtained at the average cost of less than one hundred of EM analyses of the structure at hand, which leads to a significant, over 80% and over 60% speedup as compared to the benchmark surrogate-based methods, a one-shot approach and the SAO-based algorithm, respectively. The reliability of the procedure is confirmed by comparisons with EM-based Monte Carlo analysis at the nominal and the optimized designs. The discussed approach may be considered a step towards the development of a generalized methodology, applicable to other types of microwave and antenna components, as well as the implementation of a generic yield optimization framework involving performance-driven surrogates.

Acknowledgments

This work was supported in part by the Icelandic Centre for Research (RANNIS) Grant 206606 and by National Science Centre of Poland Grant 2018/31/B/ST7/02369.

References

Abdel-Malek, H.L., Hassan, A.S.O., Soliman, E.A., Dakroury, S.A. (2006). The ellipsoidal technique for design centering of microwave circuits exploiting space-mapping interpolating surrogates, *IEEE Trans. Microwave Theory Techn.*, vol. 54, no. 10, pp. 3731–3738.

Biernacki, R., Chen, S., Estep, G., Rousset, J., Sifri, J. (2012). Statistical analysis and yield optimization in practical RF and microwave systems, *IEEE MTT-S Int. Microw. Symp. Dig.*, Montreal, pp. 1–3.

Broyden, C.G. (1965). A class of methods for solving nonlinear simultaneous equations, *Math. Comp.*, vol. 19, pp. 577–593.

Budimir, D., Goussetis, G. (2003). Design of asymmetrical RF and microwave bandpass filters by computer optimization, *IEEE Trans. Microwave Theory Techn.*, vol. 51, no. 4, pp. 1174–1178.

Chen, Y.-C., Chen, S.-Y., Hsu, P. (2006). Dual-band slot dipole antenna fed by a coplanar waveguide, *IEEE Int. Symp. Ant. Prop.*, pp. 3589–3592.

Conn, A.R., Gould, N.I.M., Toint, P.L. (2000). *Trust Region Methods*, MPS-SIAM Series on Optimization.

Du, J., Roblin, C. (2017). Statistical modeling of disturbed antennas based on the polynomial chaos expansion, *IEEE Ant. Wireless Prop. Lett.*, vol. 16, pp. 1843–1847.

Hassan, A.S.O., Abdel-Malek, H.L., Mohamed, A.S.A., Abuelfadl, T.M., Elqenawy, A.E. (2015). Statistical design centering of RF cavity linear accelerator via non-derivative trust region optimization, *IEEE Int. Conf. Numerical EM Multiphysics Modeling Opt. (NEMO)*, pp. 1–3.

Jin, H., Zhou, Y., Huang, Y.M., Ding, S., Wu, K. (2017). Miniaturized broadband coupler made of slow-wave half-mode substrate integrated waveguide, *IEEE Microwave Wireless Comp. Lett.*, vol. 27, no. 2, pp. 132–134.

Kennedy, M.C., O'Hagan, A. (2000). Predicting the output from complex computer code when fast approximations are available, *Biometrika*, vol. 87, pp. 1–13.

Kim, D.W., Choi, N.S., Lee, C.U., Kim, D.H. (2015). Assessment of statistical moments of a performance function for robust design of electromagnetic devices, *IEEE Trans. Magn.*, vol. 51, no. 3, paper no. 7205104.

Kouassi, A., Nguyen-Trong, N., Kaufmann, T., Lallechere, S., Bonnet, P., Fumeaux, C. (2016). Reliability-aware optimization of a wideband antenna, *IEEE Trans. Ant. Prop.*, vol. 64, no. 2, pp. 450–460.

Koziel, S., Bandler, J., Mohamed, A., Madsen, K. (2005). Enhanced surrogate models for statistical design exploiting space mapping technology, in *IEEE MTT-S Int. Microw. Symp. Dig.*, Long Beach, CA, pp. 1–4.

Koziel, S. (2017). Low-cost data-driven surrogate modeling of antenna structures by constrained sampling, *IEEE Antennas Wireless Prop. Lett.*, vol. 16, pp. 461–464.

Koziel, S., Bandler, J.W. (2015). Rapid yield estimation and optimization of microwave structures exploiting feature-based statistical analysis, *IEEE Trans. Microwave Theory Tech.*, vol. 63, no., 1, pp. 107–114.

Koziel, S., Bekasiewicz, A., Kurgan, P., Bandler, J.W. (2015). Expedited multi-objective design optimization of miniaturized microwave structures using physics-based surrogates, *IEEE MTT-S Int. Microwave Symp.*, pp. 1–3.

Koziel, S., Bekasiewicz, A. (2016). Low-cost surrogate-assisted statistical analysis of miniaturized microstrip couplers, *J. Electromagnetic Waves Appl.*, vol. 30, no. 10, pp. 1345–1353.

Koziel, S., Bekasiewicz, A. (2017). On reduced-cost design-oriented constrained surrogate modeling of antenna structures, *IEEE Ant. Wireless Prop. Lett.*, vol. 16, pp. 1618–1621.

Koziel, S., Bekasiewicz., A. (2018a). Sequential approximate optimization for statistical analysis and yield optimization of circularly polarized antennas, *IET Microwaves Ant. Prop.*, vol. 12, no. 13, pp. 2060–2064.

Koziel, S., Bekasiewicz., A. (2018b). Reduced-cost surrogate modeling of input characteristics and design optimization of dual-band antennas using response features, *Int. J. RF & Microwave CAE*, vol. 28, no. 2, article no. e21194, pp. 1–6.

Koziel, S., Pietrenko-Dabrowska, A. (2019). Performance-based nested surrogate modeling of antenna input characteristics, *IEEE Trans. Ant. Prop.*, vol. 67, no. 5, pp. 2904–2912.

Leifsson, L., Du, X., Koziel, S. (2020). Efficient yield estimation of multiband patch antennas by polynomial chaos-based kriging, *Int. J. Numerical Modeling*, vol. 33, no. 6, article no. e2722, pp. 1–10.

Li, X., Zhou, J., Duan, B., Yang, Y., Zhang, Y., Fang, J. (2015). Performance of planar arrays for microwave power transmission with position errors, *IEEE Ant. Wireless Prop. Lett.*, vol. 14, pp. 1794–1797.

Ma, B., Lei, G., Liu, C., Zhu, J., Guo, Y. (2018). Robust tolerance design optimization of a PM claw pole motor with soft magnetic composite cores, *IEEE Trans. Magn.*, vol. 54, no. 3, paper no. 8102404.

Matoglu, E., Pham, N., De Araujo, D., Cases, M., Swaminathan, M. (2004). Statistical signal integrity analysis and diagnosis methodology for high-speed systems, *IEEE Trans. Adv. Packaging*, vol. 27, no. 4, pp. 611–629.

Ochoa, J.S., Cangellaris, A.C. (2013). Random-space dimensionality reduction for expedient yield estimation of passive microwave structures, *IEEE Trans. Microwave Theory Techn.*, vol. 61, no. 12, pp. 4313–4321.

Queipo, N.V., Haftka, R.T., Shyy, W., Goel, T., Vaidynathan, R., Tucker, P.K. (2005). Surrogate-based analysis and optimization, *Progress in Aerospace Sciences*, vol. 41, no. 1, pp. 1–28.

Petrocchi, A., Kaintura, A., Avolio., G., Spina, D., Dhaene, T., Raffo, A., Schreurs, D.M. (2017). Measurement uncertainty propagation in transistor model parameters via polynomial chaos expansion, *IEEE Microwave Wireless Comp. Lett.*, vol. 27, no. 6, pp. 572–574.

Prasad, A.K., Ahadi, M., Roy, S. (2016). Multidimensional uncertainty quantification of microwave/RF networks using linear regression and optimal design of experiments, *IEEE Trans. Microwave Theory Techn.*, vol. 64, no. 8, pp. 2433–2446.

Rayas-Sanchez, J.E., Gutierrez-Ayala, V. (2006a). EM-based Monte Carlo analysis and yield prediction of microwave circuits using linear-input neural-output space mapping, *IEEE Trans. Microwave Theory Techn.*, vol. 54, no. 12, pp. 4528–4537.

Rayas-Sanchez, J.E., Gutierrez-Ayala, V. (2006b). EM-based statistical analysis and yield estimation using linear-input and neural-output space mapping, *IEEE MTT-S Int. Microwave Symp. Digest (IMS)*, pp. 1597–1600.

Ren, Z., He, S., Zhang, D., Zhang, Y., Koh, C.S. (2016). A possibility-based robust optimal design algorithm in preliminary design state of electromagnetic devices, *IEEE Trans. Magn.*, vol. 52, no. 3, paper No. 7001504.

Rossi, M., Dierck, A., Rogier, H., Vande Ginste, D. (2014). A stochastic framework for the variability analysis of textile antennas, *IEEE Trans. Ant. Prop.*, vol. 62, no. 16, pp. 6510–6514.

Scotti, G., Tommasino, P., Trifiletti, A. (2005). MMIC yield optimization by design centering and off-chip controllers, *IET Proceedings — Circuits, Devices and Systems*, vol. 152, no. 1, pp. 54–60.

Sengupta, M., Saxena, S., Daldoss, L., Kramer, G., Minehane, S., Cheng, J. (2005). Application-specific worst case corners using response surfaces and statistical models, *IEEE Trans. Comput.-Aided Design Integr. Circuits Syst.*, vol. 24, no. 9, pp. 1372–1380.

Simpson, T.W., Pelplinski, J.D., Koch, P.N., Allen, J.K. (2001). Metamodels for computer-based engineering design: Survey and recommendations, *Engineering with Computers*, vol. 17, pp. 129–150.

Spina, D., Ferranti, F., Antonini, G., Dhaene, T., Knockaert, L. (2014). Efficient variability analysis of electromagnetic systems via polynomial chaos and model order reduction, *IEEE Trans. Comp. Packaging Manufacturing Techn.*, vol. 4, no. 6, pp. 1038–1051.

Syrytsin, I., Zhang, S., Pedersen, G.F., Zhao, K., Bolin, T., Ying, Z. (2017). Statistical investigation of the user effects on mobile terminal antennas for 5G applications, *IEEE Trans. Ant. Prop.*, vol. 65, no. 12, pp. 6596–6605.

Vidal, B., Corral, J.L., Marti, J. (2005). Statistical analysis of WDM photonic microwave filters with random errors, *IEEE Trans. Microwave Theory Techn.*, vol. 53, no. 8, pp. 2600–2603.

Wu, H.W., Chiu, C.T. (2016). Design of compact multi-layered quad-band bandpass filter, *IEEE Microwave Wireless Comp. Lett.*, vol. 26, no. 11, pp. 879–881.

Zhang, L., Zhang, Q.J., Wood, J. (2008). Statistical neuro-space mapping technique for large-signal modeling of nonlinear devices, *IEEE Trans. Microwave Theory Techn.*, vol. 56, no. 11, pp. 2453–2467.

Zhang, J., Zhang, C., Feng, F., Zhang, W., Ma, J., Zhang, Q.J. (2018). Polynomial chaos-based approach to yield-driven EM optimization, *IEEE Trans. Microwave Theory Tech.*, vol. 66, no. 7, pp. 3186–3199.

© 2022 World Scientific Publishing Europe Ltd.
https://doi.org/10.1142/9781800610750_0010

Chapter 10

Solving the Inverse Problem Through Optimization — Applications to Analog/RF IC Design

Yi Wang and Paul Franzon

Abstract

RF circuit optimization is generally formed as a multi-objective optimization function. Weight setting is important can be guided by a principal component analysis (PCA) step that can also be used to screen parameters. This is illustrated on a circuit example. Two surrogate model-based optimization approaches are presented, one based on Bayesian Optimization, the second on a candidate search method. Bayesian optimization is best employed on all continuous domain problems while candidate search can be used on mixed integer-continuous problems. We implement co-Kriging in the candidate search algorithm to permit mixed fidelity electromagnetic evaluation. These algorithms are illustrated with several circuit examples.

Keywords: RF circuit optimization, principal component analysis, surrogate model-based optimization, Bayesian optimization, candidate search algorithm

10.1 Introduction

Multi-objective RF circuit optimization is a difficult and computationally expensive problem. In this paper, we focus on solutions based on surrogate-based optimization methodologies. We first give an overview of the flow, with a discussion of parameter screening, and present a summary of Bayesian optimization and candidate point search. We then give an example in which principal component analysis (PCA) is used to determine the weight settings in a multi-objective optimization problem using a mixer as an example. This is followed by an example of using Bayesian optimization in a circuit design. We then show how space mapping (SM) can be incorporated into a modified Candidate Search algorithm and finish with some examples employing that combination.

10.2 Overview of Proposed Design Flow

We propose a computationally effective methodology flow for analog/RF design that is aimed at implementing analog synthesis for fixed topology. It includes design analysis, and optimization as shown in Fig. 10.1.

10.2.1 *Design space analysis*

As the number of design parameters increases, the whole design space grows exponentially. Therefore, it is critical to identify the most relevant inputs in order to reduce complexity. Usually this is done with experienced analog designers' handwork. In this work, we apply the

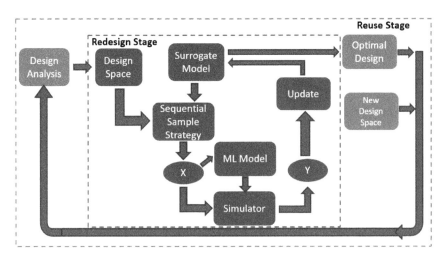

Fig. 10.1. Flowchart of proposed flow.

numerical dimensionality reduction method singular value decomposition (SVD) (Jolliffe, 2011) to help identify the key components in the circuit. We will illustrate this below in the context of weight setting.

10.2.2 Surrogate modeling method

The goal of surrogate modeling is to train a mapping function with a few samples from costly simulation. This mapping function can then be used as a surrogate for predicting the outcome from other unsampled inputs in the design space (Jin, 2011). A statistical model can then be trained to capture the variance trend between input variables and output performance. This cheap model can be utilized to explore the design space instead of running a SPICE/EM simulation. The circuit optimization process is built on the exploitation of the trained model and exploration of unknown design space. Generally, surrogate model construction consists of four steps:

(1) An appropriate initial sample plan is applied to cover the design space;
(2) A model is trained to fit the observation samples;
(3) A sample strategy is applied to add more samples in the region where either there might exist an optimal solution or it is not explored;
(4) The model is evaluated and validated. Figure 10.2 shows more detailed steps in surrogate modeling.

In this work, the model builder is set to a mixture of kriging, Gaussian correlation functions and cubic radial basis functions. The adaptive sample methods will be discussed in Section 10.2.3.

10.2.3 Adaptive sample strategy

In surrogate modeling, one-shot samples from the initial sampling plan are usually not enough to construct the globally feasible model due to many unknown nonlinearities and discontinuities in the target mapping function. An adaptive sampling strategy results in a search for new samples whose locations are not yet explored by analyzing the current response surface.

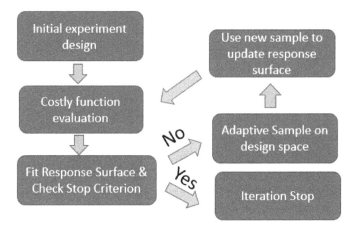

Fig. 10.2. Flowchart of surrogate model.

In this work, two different sampling methods are implemented in the proposed design flow, namely expected improvement from Bayesian optimization (Mockus, 2012) (the benchmark method) and a proposed candidate point search solution.

10.3 Bayesian Optimization Framework

10.3.1 *Overview of Bayesian optimization*

Standard Bayesian optimization (BO) consists of two components: a Gaussian process as the surrogate to model the black box function; an acquisition function as the sampling method to decide where to sample next to approach the optima.

Algorithm 1 demonstrates how the Bayesian optimization flow is executed to achieve the optima. BO first builds a GP surrogate with a pre-generated initial dataset. Then the surrogate is updated iteratively with the sample created from the acquisition function and its corresponding simulation value.

10.3.1.1 *Acquisition function*

An acquisition function is used to evaluate the next point to sample by choosing that point that maximizes the function. The most common acquisition functions applied in BO are probability of

Algorithm 1 Bayesian Optimization

Input: Initial samples & function evaluations (x_0, R_0);
Maximum function simulation iteration N; Expensive function $F(x)$
Output: Optimal function value y_{best}.

1: **procedure** BO$(N, (x_0, R_0))$
2:     **for** $i \leftarrow 0$ to $N-1$ **do**
3:         Build GP posterior estimation $f(x^*|(x_i, R_i))$
4:         $x_{acq} \leftarrow$ argmin acquisition function $a(x)$
5:         $y_{acq} = F(x_{acq})$
6:         $x_{i+1} = (x_i, x_{acq})$
7:         $R_{i+1} = (R_i, y_{acq})$
8:     **end for**
9:     $y_{best} \leftarrow \min(R_N)$
10:     **return** y_{best}
11: **end procedure**

improvement (PI), expected improvement (EI), and lower confidence bound (LCB). In BO, the essence of PI and EI can be interpreted as an expected loss calculation based on the surrogate function value $f(x)$ at x. x is then selected by minimizing the expected loss function. In this work, we will adopt all three of these methods to fulfill the EM simulation driven optimization tasks and compare them with the proposed modified candidate algorithm, MFSMO-CS.

PI & EI: The utility function of PI (Shahriari et al., 2015) is

$$u(x) = \begin{cases} 0 & f(\mathrm{x}) \geq f' \\ 1 & f(\mathrm{x}) < f' \end{cases}.$$

PI would receive a unit reward when $f(x)$ is smaller than the current optimal f'. Otherwise, PI rewards nothing. The PI acquisition function is expressed as

$$a_{PI} = E[u(x)|x, D] = \Phi(f'; \mu(x), \sigma^2(x)), \quad (10.1)$$

where Φ is the cumulative distribution function. The point with largest probability of improvement value is selected as the next sample in the BO loop. Based on the description of PI's utility function, it actually doesn't take the precise improvement into consideration. Similarly, an alternative utility function called expected improvement (EI) that accounts for the decremental change between function value $f(x)$ and f' is developed as

$$u(x) = \max(0, (f' - f(x)). \tag{10.2}$$

Then EI's acquisition function is shown as

$$\begin{aligned}a_{\text{EI}} = E[u(x)|x, D] = (f' - \mu(x))\Phi \\ \times (f'; \mu(x), \sigma^2(x)) + \sigma^2(x)\phi(f'; \times \mu(x), \sigma^2(x)),\end{aligned} \tag{10.3}$$

where Φ and ϕ are the cumulative distribution function and the probability density function of the standard normal distribution.

LCB: The LCB acquisition function is shown as

$$a_{LCB} = \mu(x) - \beta\sigma(x). \tag{10.4}$$

LCB (Snoek et al., 2012) does not follow the way of computing the expectation of utility function. However, all three acquisition functions described earlier contain the model exploitation term ($\mu(x)$) and design space exploration term ($\sigma^2(x)$).

Consequently, each acquisition function provides different degree of balance in exploitation and exploration.

10.4 Candidate Point Search

Mixed-integer optimization problems exist in many engineering applications. In general, this category of problems is NP-hard and difficult to solve because of the non-convex nature of the response surfaces. Traditionally, branch and bound methods, genetic or evolutionary strategies are applied to these problems after reformulation (Van Roy and Wolsey, 1987). But in real engineering applications, due to their complex nature, systems are often treated as black boxes and only inputs and outputs are known to designers. Meanwhile, evolutionary based methods like genetic algorithms require a large number of function evaluations to find the global optimum.

Thus, a new sample strategy, candidate point search is applied in Analog/RF circuit design in an attempt to mitigate drawbacks of existing methods. The candidate point search method not only solves the concerns mentioned earlier, but also can be used in large-dimensional design spaces without the issue of computation cost that exists in Bayesian optimization. It is adapted and improved from (MüLler *et al.*, 2013), and the steps of the candidate point search algorithm are given below:

Step 1: Find variables to perturb. (1) Evaluate objective function at initial points, and update counter, and (2) Group 1: Perturb only continuous variables of best point by randomly selecting from D variables for inclusion through probability calculation $P = 10/\text{data}_{\text{dim}}$ when data dimension > 10 and 1 when data dimension < 10; Randomly select perturbing rate. Group2: Perturb only discrete variables of best point. Group3: Perturb continuous and discrete variables of best point. Group4: Generate a random sample in the design space. Then compute function value based on existing response surface for four groups.

Step 2: Compute scoring criteria. Determine the distance of each candidate point to the set of already sampled points and do min–max normalization and discard points that are close to sampled points.

Step 3: Find best candidate & perform function evaluation. Normalize predicted function values and generate candidate total value for weighted sum of scaled distance and scaled function values.

Step 4: Find index of best candidate and perform expensive function evaluation at selected point. Go to Step 1 until the criterion is met.

As explained in the steps of the CAND algorithm, unlike Bayesian optimization, CAND only samples four groups to find the next sample and then updates the response surface. If not satisfied, more groups can be processed simultaneously in batch mode using parallelization. What's more, Bayesian optimization APIs currently on the market only support discrete-only, continuous-only, and/or categorical-only problems (Martinez-Cantin, 2014).

Next we illustrate the use of PCA in weight setting and give an example of employing Bayesian optimization. Then we show an improved Candidate algorithm based on the use of multi-fidelity optimization, often called SM.

10.5 Design Analysis: Weight Setting

A perennial problem in circuit optimization is how to set weights so as to solve multi-objective optimization problems. Here we present a methodology based on Principal Component Analysis and illustrate its use with a circuit application.

The simplified benchmark schematic of a mixer is illustrated in Fig. 10.3. A more detailed circuit is omitted for IP protection (Fujibayashi *et al.*, 2017). We will illustrate weight setting for multi-objective optimization using this example. Table. 10.1 shows the design input parameters and target performance outputs. In this section, we mainly focus on the output performance metrics analysis.

From a designer's perspective, the order of performance ranking is: VG > iP1dB > S11 > S22 > Lo_RF_Feedthrough > NF. That is to say, designers should focus on VG optimization more than iP1db, etc. This general scheme works well for the scenario that the first few PCs contribute similar variance in the newly transformed space,

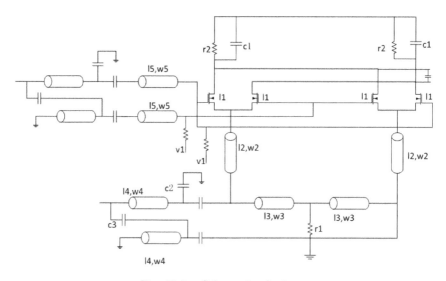

Fig. 10.3. Schematic of mixer.

Table 10.1. Mixer design summary.

Physical design parameter
X = l1,l2,l3,l4,l5,w2,w3,w4,w5,r1,r2,c1,c2,v1
Target electrical parameter
Y1 = Voltage gain
Y2 = Noise figure
Y3 = LO RF feedthrough
Y4 = S11(< −5dB)
Y5 = S22(< −3dB)
Y6 = iP1dB

Table 10.2. SVD result1.

	VG	NF	LO RF Feedth.	S11	S22	iP1dB
PC1	✓					✓
PC2				✓		✓
PC3				✓	✓	
PC4			✓		✓	
PC5			✓			✓
PC6	✓	✓				

Table 10.3. PC explained variance.

PC1	PC2	PC3	PC4	PC5	PC6
0.4920	0.2209	0.1572	0.074	0.048	0.006

especially for a high-dimensional space. In the mixer design case (six input dimensions), we discover that the normalized variance for each principal component is as shown in Tables 10.2 and 10.3. PC1 contributes nearly 50% of the total variance in the dataset. One strategy to select the most important features is sorting the first principal component and ranking the features according to the variance in one PC. The resulting ranking is VG > iP1dB > NF > S22 > S11 > Lo_RF_Feedthrough.

10.5.1 Weight setting and optimization

Now we turn to optimization. For the mixer design case, we first demonstrate the feasibility of applying PCA to weight setting in the multi-objective analog design problem. The idea behind this methodology is that we hope to set up proper initial multi-objective weights through PCA analysis. Then minor modifications on weights are made to meet the performance requirements through multi-objective optimization.

Based on the result from Section 10.5, a second strategy is to adopt only the features of explained variance in the first principal component (Table 10.4). The initial weights set for voltage gain, noise figure, LO RF feed through and iP1dB are –1, 1, 0.02 and –1, respectively. Thus, the weighted sum objective function is shown below in Equation (10.5) whereas a penalty function value of 50 is given if S11 is larger than –5 and S22 is larger than –3.

$$\text{objective function} = -VG + NF + LO/50 - iP1db. \quad (10.5)$$

The resulting optimal result is displayed in Table 10.5.

The optimized design variables are 99.38u, 14.243u, 285.98u, 14.243u, 1.368, 99.99, 908.24, 2.405p, 47.842u, 5.552u, 107.8u, 8.112u, 22.336f, 5.297u.

Table 10.4. SVD result2.

	VG	NF	LO RF Feedth.	S11	S22	iP1dB
PC1	−0.5741	0.5625	0.018	−0.095	0.1459	−0.5684

Table 10.5. Mixer multi-objective optimization result with PCA.

	8HP	8XP manual design
Voltage gain	7.201	18.28
Noise figure	12.019	9.39
LO RF feedthrough	−51.440	−27.59
S11	−9.166	−4.6 (not met)
S22	−6.792	−3.3
iP1dB	2.242	−4.67

Table 10.6. Mixer optimization result2 with PCA.

	8HP	8XP manual design
Voltage gain	16.31	18.28
Noise figure	7.43	9.39
LO RF feedthrough	−40.12	−27.59
S11	−6.3	−4.6 (not met)
S22	−11	−3.3
iP1dB	−9.79	−4.67

After the first round of optimization, we found out that noise figure was worse than manual design. Thus we increased the weight of noise figure as indicated in the second objective function 10.6:

$$\text{objective function} = -\text{VG} + 1.5\text{NF} + \text{LO}/50 - \text{iP1db}. \quad (10.6)$$

Through another round of optimization process, we obtain the result as shown in Table 10.6.

The most challenging part in this mixer design is to set a proper weight for iP1dB. iP1db can be positive or negative. After second round optimization, we found iP1db got worse so we reduced the weight of it assuming negative iP1dB. Meanwhile, we found Lo RF feed through had room for performance trade-off so we lowered its weight a little bit. In the third round of optimization, a new objective function (Equation (10.7)) was formulated:

$$\text{objective function} = -\text{VG} + 1.5\text{NF} + \text{LO}/60 - 1.3\text{iP1db}. \quad (10.7)$$

Table 10.7 shows the final solution to the mixer redesign problem. After three rounds of guided optimization, we achieved better voltage gain, noise figure, Lo RF feedthrough, S11 and S22. However, iP1db was not well treated since it is challenging to set up perfect weights that optimize all parameters.

10.6 Example of Optimization and Design Reuse with Bayesian Optimization

This section illustrates the use of Bayesian Optimization to design a mixer circuit and do so in two different technology nodes, i.e., reusing the design (IP or Intellectual Property) built in one node in another.

Table 10.7. Mixer multi-objective optimization result 3 with PCA.

	8HP	8XP manual design
Voltage gain	19.101	18.28
Noise figure	8.730	9.39
LO RF feedthrough	−31.12	−27.59
S11	−7.404	−4.6 (not met)
S22	−7.314	−3.3
iP1dB	−12.52	−4.67

Table 10.8. Mixer reuse result.

	22FDX	8XP	8XP manual design
Voltage gain	7.04	22.8	18.28
Noise figure	16.13	10.92	9.39
LO RF feedthrough	−36.84	−34.54	−27.59
S11	−6.8	−6.1	−4.6 (not met)
S22	−4.79	−3.13	−3.3
iP1dB	−0.694	−8.26	−4.67

We also used the mixer shown in Fig. 10.3 to test the feasibility of transferring the design between two semiconductor processes — from IBM 8XP to GF 22FDX. Transferring a design like this is an example of Intellectual Property (IP) reuse. The result is shown in Table 10.8. Although it is a big change from BJT technology to CMOS technology, our proposed design flow still works as long as the system is open-looped. The IP topology designed in 130 nm BJT was successfully transferred to 22 nm CMOS and the result was still within acceptance due to significant transistor and other device differences. More evaluation iterations could be applied to improve the performance.

In Table 10.8, it is shown that favorable voltage gain and phase noise values can not be attained. This is due to the S11 and S22 constraints set in the objective function. That is, a large penalty is applied if S11 or S22 does not meet its requirement. To further improve voltage gain and noise figure, we should expand the ranges for S11 and S22 constraints, i.e., we should make some trade-off on

S11 and S22 in order to get better voltage gain and phase noise. From another perspective, our experiment shows 22FDX may not a good technology for the mixer design used here.

10.7 Surrogate Model Extension in Physical Design: Multi-Fidelity Optimization for Electromagnetic Simulation Acceleration

As circuit speeds and frequency increase, fast and accurate capture of the details of the parasitics in metal structures, such as inductors and clock trees, becomes more critical. However, conducting high-fidelity 3D electromagnetic simulations within the design loop is very time consuming and computationally expensive. To address this issue, we propose a surrogate-based optimization methodology flow, namely multi-fidelity surrogate-based optimization with candidate search (MFSBO-CS), which integrates the concept of multi-fidelity to reduce the full wave EM simulation cost in analog/RF simulation-based optimization problems. To do so, a statistical co-kriging model is adapted as the surrogate to model the response surface and a parallelizable perturbation-based adaptive sampling method is used to find the optima. Within the proposed method, low-fidelity fast RC parasitic extraction tools and high-fidelity full-wave EM solvers are used together to model the target design and then guide the proposed adaptive sample method to obtain the final optimal design parameters. The sampling method in this work not only delivers additional coverage of design space but also helps increase the accuracy of the surrogate model efficiently by updating multiple samples within each iteration. Moreover, a novel modeling technique is developed to further improve the multi-fidelity surrogate model at an acceptable additional computation cost. The effectiveness of the proposed technique is validated by mathematical proofs and numerical test function demonstration. In this section, MFSBO-CS has been applied to two design cases and the result shows that the proposed methodology offers a cost-efficient solution for analog/RF design problems involving EM simulation. For the two design cases, MFSBO-CS either performs comparably or outperforms the optimization result from various Bayesian Optimization methods with only approximately one third to two thirds of the computation cost.

10.7.1 *Space mapping*

As the operating frequency of analog/RF circuit blocks increases, detailed electromagnetic modeling of passive wire structure becomes more difficult and increases time to market. In some high speed applications such as flash ADCs, the skew introduced by signal coupling and crosstalk in clock and data distribution paths has a significant effect on performance (Shahramian *et al.*, 2009). Circuits are often added to implement skew calibration and mismatch correction at the expense of added power and circuitry (Poulton *et al.*, 2003; Schvan *et al.*, 2008). Redesigning the physical layout of the interconnect to correct these issues is an alternative that requires expensive iterative electromagnetic simulations to ensure rigorous performance. From this perspective, a method to eliminate the high computation cost of multiple full-wave EM simulations would be a priority in the design process. The SM technique (Bandler *et al.*, 1994) was one of the earliest approaches used to reduce this cost. With this technique, coarse and fine EM simulation techniques are combined to reduce overall simulation time. In this technique, a linear mapping is found between coarse and fine model input parameter spaces by using an iterative process to evaluate the error between fine response surface (RS) and coarse RS. Later, variants of SM have merged to enhance the mapping function (Koziel *et al.*, 2005; Koziel and Bandler, 2007a, 2007b). However, SM techniques still require sufficiently faithful coarse models to assure good results and more sample data may not lead to improvement in the model accuracy (Koziel and Bandler, 2012). Beyond that, most work related to SM techniques focuses on the microwave device modeling rather than a systematic optimization methodology to acquire optimal design parameters, e.g., for circuit design. Recently, Bayesian optimization (BO) (Snoek *et al.*, 2012), a surrogate-based optimization technique has been shown to be useful in the investigation of simulation-based analog/RF ICs design automation where a large number of expensive SPICE simulations have to be executed in order to obtain the desired design performance (Lyu *et al.*, 2018; Qi, 2017; Zhang *et al.*, 2019). Compared with the traditional direct optimization methods like gradient-based or evolutionary-based ones (Vural and Yildirim, 2012), Bayesian optimization requires many fewer samples to find the proper optima. However it is still worth investigating other algorithms to further reduce the total computation cost and to be capable of handling high-dimensional design problems.

This proposed work emphasizes a cost-effective solution towards the EM simulation-based optimization in analog/RF design problems. Rather than simple utilization of costly fine EM evaluations, we employ EM simulation data at two different levels of sophistication, i.e., a large number of low-fidelity simulations and a small number of high-fidelity simulations. These are fitted to a stochastic model for predicting the uncovered area in the design space. The concept of multi-fidelity modeling is frequently implemented in the area of aerospace engineering since it is common to take over one day for producing a single low-resolution plot using physical simulators (Forrester et al., 2007; Geiselhart et al., 2011; Ghoreyshi et al., 2008; Giunta et al., 1995). The utilization of low-fidelity emulation data could assist the system to quickly capture the approximate high-fidelity response surface by evaluating large amount of low-fidelity samples at relatively low computation cost. In electromagnetic simulation, we define the high-fidelity simulators as 3D or 2.5D EM solvers whose algorithms rely on method of moments (MoM), finite element method (FEM) or finite difference time domain (FDTD) (Vandenbosch and Vasylchenko, 2011). Even within each specific high-fidelity simulator, various settings such as edge mesh, thickness, 3D model, ration, etc., could lead to different levels of fidelity at the trade-off between simulation accuracy and computation cost. This provides a wider selection of multi-fidelity simulators for EM simulation acceleration. In this work, the low-fidelity simulators adopted are ASITIC (Niknejad, 2000) and a commercial parasitics extraction tool for the two separate design cases. ASITIC relies on the simple Pi model to derive S parameters to a circuit representation. The basic algorithm behind the commercial parasitics extraction tool calculates the parasitics by geometrically analyzing conductors in 3D regions.

Apart from employing a multi-fidelity model as the surrogate, our proposed method leverages the cheap evaluation cost of the surrogate model by conducting a large number of surrogate evaluations. The current optimum and its corresponding design inputs are obtained first. Then the surrogate evaluations are executed by varying the inputs near the current optimum and within entire design space. Contrary to traditional Bayesian optimizations, there is no extra effort in balancing current surrogate model exploitation and uncovered design space exploration through optimization of the acquisition function. The model exploration is simply fulfilled with the perturbation of input variables in the whole design space due to the low evaluation cost of the surrogate model. Specifically, in the sequential sample method applied, we generate multiple groups of samples that have different perturbation properties in parallel and the variables to be perturbed are selected following a dropout style (Hinton *et al.*, 2012). The surrogate determines the best sample point so far in each group and then runs the expensive EM simulation on these points. Once the improvement from expensive 3D EM solver (high-fidelity) simulation is detected, not only would the high-fidelity data be used to update the surrogate, but also the system would re-evaluate this improved point in the low-fidelity simulator and update the low-fidelity dataset to further acquire a higher accuracy and resolution in this area of interest.

10.7.2 *Overview of EM simulation acceleration*

10.7.2.1 *Overview of multi-fidelity surrogate-based optimization with candidate search*

In this section, we decompose the whole framework of the proposed method into several pieces. The complete overview of the proposed MFSBO-CS algorithm for clock tree optimization is illustrated in Fig. 10.4. There is a slight difference in representing the flow chart for the inductor design case considering that inductor design only involves multi-fidelity EM simulations. From a high-level perspective, the proposed approach consists of four major components: (1) design exploration, (2) surrogate model construction, (3) adaptive sampling, and (4) model evaluation. The core of our approach is to utilize a statistical surrogate model trained from multi-fidelity data to search for the global optima with the fewest samples. A surrogate model is a model fitted to evaluations of a detailed model (that often takes a long time to evaluate). It thus serves as a "surrogate" for the detailed model. It is sometimes called a response surface. Conventionally a surrogate model is fitted to evaluations of only one detailed or fine model. In this work it is fitted to a combination of

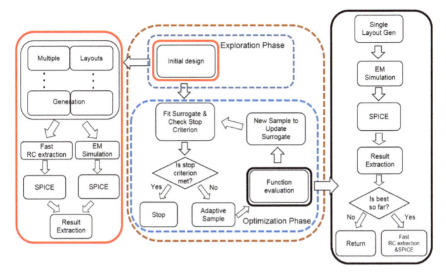

Fig. 10.4. Overview of MFSBO-CS.

fine model evaluations and (faster) coarse model evaluations — hence the term "multi-fidelity").

10.7.3 *Multi-fidelity surrogate-based optimization with candidate search flow*

Algorithm 2 shows the detailed MFSBO-CS flow that we developed. The flow is initialized with a multi-fidelity dataset obtained from the design of an experiment. Then we build a multi-fidelity statistical model based on the dataset acquired from the previous step. This surrogate model is updated iteratively as coded from steps 14 to 42. The prediction based upon the surrogate model is utilized as one of the criteria for adaptive sampling.

10.7.3.1 *Design exploration*

In traditional surrogate-based optimization, the aim of the design exploration stage is to seek a better initial point for the next optimization stage by evaluating a number of designed points with only high-fidelity simulations while enabling the model fitted on these data to provide a relatively accurate prediction over the entire design space. Generally, the layout structure EM simulation problem is

Algorithm 2 MFSBO-CS

Input: LF simu. budget L; Total simu. budget T; Problem config. File config; LF simulator $F_c(x_c)$; HF simulator $F_e(x_e)$; Design of Experiments **DoE**; Perturbation budget for each group in CS N.
Output: Optimal Result x_{best}

1: **procedure** MFSBO-CS(L, T, config, $F_c, F_e,$ **DoE**)
2:     $\epsilon \leftarrow$ tolerance, $x_c \leftarrow$ **DoE**(L), $x_e \leftarrow$ **DoE**(10)
3:     **if** integer variable in **x then**
4:        $G \leftarrow 4$ ▷ Define no. of groups for CS
5:     **else**
6:        $G \leftarrow 2$
7:     **end if**
8:     **for** $i \leftarrow 1$ to L **do**
9:        $y_{c,i} \leftarrow F_c(x_{c,i}), R_{c,i} \leftarrow \{R_{c,i-1}, y_{c,i}\}$
10:    **end for**
11:    **for** $i \leftarrow 1$ to 10 **do**
12:       $y_{e,i} \leftarrow F_e(x_{e,i}), R_{e,i} \leftarrow \{R_{e,i-1}, y_{e,i}\}$
13:    **end for**
14:    Build statistical surrogate model $P(\hat{y}_e|\hat{x}_e, (x_c, R_c), (x_e, R_e))$
15:    **for** $i \leftarrow 1$ to $(T - 10 - L)/G$ **do**
16:       **for** $j \leftarrow 1$ to G **do**
17:          $X_j, D_j \leftarrow CS(x_{best}, G, N)$
18:          **for** x in X_j **do**
19:             **if** $|x - x_e| < \epsilon$ **then**   ▷ check distance
20:                delete x, update D_j
21:             **end if**
22:          **end for**
23:          $x_{e,j} \leftarrow \min(D_j), y_{e,j} \leftarrow F_e(x_{e,j})$
24:          **if** $y_{e,j} < y_{best}$ **then**
25:             $y_{best} \leftarrow y_{e,j}, x_{best} \leftarrow x_{e,j}, x_{c,new} \leftarrow x_{e,j}$
26:          **end if**
27:          $x_e \leftarrow \{x_e, x_{e,j}\}, R_e \leftarrow \{R_e, y_{e,j}\}$
28:       **end for**
29:       **if** $x_{c,new}$ exist **then**   ▷ add LF data for training
30:          $y_{c,new} \leftarrow F_c(x_{c,new}), x_c \leftarrow \{x_c, x_{c,new}\}, R_c \leftarrow \{R_c, y_{c,new}\}$, delete $x_{c,new}$
31:       **end if**
32:       Update surrogate model $P(\hat{y}_e|\hat{x}_e, (x_c, R_c), (x_e, R_e))$
33:    **end for**
34:    **return** x_{best}
35: **end procedure**

highly nonlinear and non-convex so we don't have much prior information about how to sample this black box. The most economic way to find the approximate optima is to sample the design space with uniform coverage and then evaluate the samples using expensive simulations. Among the existing sampling methods, stratified Latin hypercube sampling (LHS) (McKay et al., 2000) is the simplest and most effective. This method ensures uniformly spread projections of all points on all axes. Even if an effective initial sampling method is used, it is obvious that the more data we have, the more accurate the surrogate model will be. Therefore it encapsulates the dilemma that training a good surrogate requires more costly EM simulations.

For the multi-fidelity surrogate-based optimization framework developed in this paper, we adopted LHS as the design exploration method for both high- and low-fidelity simulations (line 3 of Algorithm 2). Design exploration is also known as "design of experiments" (DoE) in some of the literature and is so named in line 3.

In Table 10.9, we also summarize the average simulation cost for two sets of multi-fidelity simulators used in this paper. The EM Solvers' are the high-fidelity solvers, while ASITIC and RC extraction simulator are low-fidelity ones. Between lines 9 and 13 of Algorithm 2, the samples generated from DoE will be evaluated with low-fidelity simulator $F_c(x_c)$ at x_c and high-fidelity simulator $F_e(x_e)$ at x_e.

In Algorithm 1, the initial surrogate is built with L evaluations of the low-fidelity model at x_c and 10 evaluations of the high-fidelity model at x_e. The number of low-fidelity model evaluations L should be larger than $2*(D+1)$ where D is the dimension of the design.

Table 10.9. Low and high-fidelity average simulation cost (in s).

Simulator	Simulation time cost	Comment
ASITIC	0.125	For inductor case
EM Solver 1	76.583	For inductor case
RC extraction simulator	$386e^{-0.64n} + 55.16$	For clock tree case
EM Solver 2	138.012	For clock tree case

Note: The "EM Solvers" are the high-fidelity solvers, the others are low-fidelity ones.

The requirement for the number of high-fidelity evaluations needs to be larger than D since the inverse of the covariance matrix can not be calculated if the number of high-fidelity samples is smaller than D.

For the inductor layout design example, the low-fidelity simulator is approximately 620 times faster than the high-fidelity EM solver, thus making the multi-fidelity model building very effective.

We follow the assumption, also used in Kennedy and O'Hagan (2000), that the high-fidelity simulators are trustworthy and all errors occur in the low-fidelity simulators. A multi-fidelity model is constructed based upon the samples obtained from the design exploration stage as indicated in line 15 of Algorithm 2. The optimum of the high-fidelity samples from the exploration phase will be selected as the starting point of the following optimization phase. The number of high-fidelity simulation samples in the design exploration stage is required to be larger than the problem dimension since we utilize the Gaussian process as our surrogate. The inversion of the covariance matrix can't be calculated when the dimension is larger than the number of samples.

Now that we have proper cheap and expensive simulators set up to evaluate EM effects in our layout design, we introduce the stochastic multi-fidelity model to bridge the low-fidelity and high-fidelity data in Section 10.7.3.2.

10.7.3.2 Statistical surrogate model

Co-kriging (Kennedy and O'Hagan, 2000), a statistical method to construct a mapping function in high-fidelity which is enhanced by low-fidelity data is used in this work. We first define the **n** samples of low-fidelity data as x_c, its corresponding function evaluation result set R_c, high-fidelity data as **m** samples of x_e, and its corresponding function evaluation result set R_e. Then the high-fidelity simulator can be expressed in a format of a scaled Gaussian process which denotes the low-fidelity simulator plus another Gaussian process that represents the difference between high-fidelity simulator and the scaled low-fidelity simulator as follows:

$$Z_e = \rho Z_c + Z_d. \qquad (10.8)$$

We can approximate the covariance matrix of the high-fidelity GP to be

$$V = \begin{pmatrix} \sigma_c^2 \Phi_c(x_c, x_c) & \rho \sigma_c^2 \Phi_c(x_c, x_e) \\ \rho \sigma_c^2 \Phi_c(x_e, x_c) & \rho^2 \sigma_c^2 \Phi_c(x_e, x_e) + \sigma_d^2 \Phi_d(x_e, x_e) \end{pmatrix}, \quad (10.9)$$

where Φ_c and Φ_d denote covariance matrix of Z_c and Z_d, respectively. In order to fit the hyperparameters θ_c, θ_d (unknown scaling factors in the covariance function), μ_c, μ_d and ρ, we perform a maximizing likelihood estimate (MLE) on both Z_c and Z_d.

Thus, the final posterior mean of Z_e given a new sample x is

$$\mu_e(x) = q(x)^T \alpha + k(x)^T V^{-1}(y - B\alpha), \quad (10.10)$$

where

$$q(x) = (\rho b(x)^T, b(x)^T), \quad (10.11)$$

$$\alpha = (B^T V^{-1} B)^{-1} B^T V^{-1} y, \quad (10.12)$$

$$k(x)^T = (\rho \sigma_c^2 \Phi_c(x, x_c), \rho^2 \sigma_c^2 \Phi_c(x, x_e) + \sigma_d^2 \Phi_d(x, x_e)), \quad (10.13)$$

$$B = \begin{pmatrix} b(x_1^c)^T & 0 \\ \cdots & \cdots \\ b(x_n^c)^T & 0 \\ \rho b(x_1^e)^T & b(x_1^e)^T \\ \cdots & \cdots \\ \rho b(x_m^e)^T & b(x_m^e)^T \end{pmatrix}. \quad (10.14)$$

The $b(x)$ above is a vector of polynomial regression functions fitted during the training process that describes Z_c roughly over the entire design space. In ordinary kriging, the model is in the format of a regression function $f(x)$ plus a Gaussian process Z. The regression $f(x)$ is constructed to describe the rough trend (largest variance) of the response surface and a Gaussian process Z is built to gap the residuals. In this case, $f(x)$ is modeled as a weighted sum of polynomials $\sum a_i b_i(x)$. $\Phi_c(x, x_c)$ represents the correlation vector between the point to be predicted, namely x, and the sample data points in low-fidelity x_c. It is a vector whose entries are calculated by the low-fidelity covariance function. $\Phi_c(x, x_e)$ denotes the correlation vector between the point to be predicted and the sample data points in high fidelity. The low-fidelity covariance function is applied as well to obtain the entries in the vector.

To visualize the modeling performance of co-kriging, a 1D example is used as an illustration in this section. Assume the high-fidelity function and the low-fidelity function as

$$F_e(x) = (3x - 2)^2 \sin(15x - 2), x \in (0, 1), \tag{10.15}$$

$$F_c(x) = 0.8(3x - 2)^2 \sin(15x - 2) + 5(x - 0.5) - 6,$$
$$x \in (0, 1). \tag{10.16}$$

We execute the flow from lines 3 to 15 with 20 low-fidelity and five high-fidelity samples. The green dots depicted in Fig. 10.5 are 20 LHS sampled points in range of (0,1) with values obtained from the low-fidelity simulator Equation (10.10). Similarly, a high-fidelity dataset of five points is generated and depicted as the blue dots in the figure. The co-kriging model is built based on these 25 samples using the mean estimator shown in Equation (10.4). The red line in the figure is plotted by sampling the trained co-kriging model in the interval of 0.01. The blue dashed line is generated by sampling $F_e(x)$ at the step of 0.01.

The co-kriging model performance shown in Fig. 10.5 has a leave-one-out cross validation error of 0.011999 and integrated mean square error of 0.005919.

10.7.3.3 *Adaptive sampling with dropout*

One-shot samples from the initial sampling plan are usually not enough to construct a globally feasible model due to many unknown nonlinearities and discontinuities in the target mapping function. Adaptive sampling allows searching for new samples whose locations are not yet explored by analyzing the current response surface. The adaptive sampling method developed in this work is adapted and improved from MüLler *et al.* (2013).

(a) Mixed integer optimization
Unlike Bayesian optimization, our adaptive sampling method spares no effort to optimize the acquisition function. When dealing with

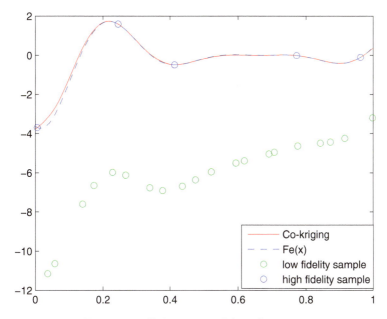

Fig. 10.5. Co-kriging model performance.

optimization tasks that involve continuous and discrete variables, a common solution of Bayesian optimization is to apply binning or to replace the continuous acquisition function optimization value with the closest integer. This could lead to serious mismatch between the points where the acquisition takes high values, and where the actual evaluation is performed. This issue can be avoided by finding the optimal next sample point with the proposed adaptive sample strategy in this work. Rather than running an optimization over a function with the posterior mean and covariance estimation information, the samples are generated in perturbation and evaluated with the existing surrogate model directly. The best will be selected based on the posterior mean estimation, which skips the binning operation and makes it fit for mixed integer optimization (Floudas, 1995). Algorithm 3 illustrates the details of proposed sampling method which is embedded in Algorithm 2 at line 22.

(b) Dropout for high dimension problem

A key feature of Algorithm 3 is the application of the concept of dropout, which is a widely used technique for training deep neural nets (Warde-Farley *et al.*, 2013). In that context, the concept

Algorithm 3 Candidate Search (CS)

Input: Optimal point so far x_{best}; Perturbation budget for single group N; Group size G
Output: Adaptive samples X_i; Adaptive sampling surrogate evaluation result D_i

1: **procedure** CS(x_{best}, G, N)
2:   **if** G = 4 **then**
3:     indint ← integer design variable index table
4:     indcont ← continuous design variable index table
5:     SM_{co} ← current surrogate model
6:     **for** $m \leftarrow 1$ to N **do**
7:       indexint ← random select (indint)
8:       indexcont ← random select (indcont)
9:       $X_1 \leftarrow$ mutate(x_{best}[indexint])
10:      $X_2 \leftarrow$ mutate(x_{best}[indexcont])
11:      $X_3 \leftarrow$ mutate(x_{best})
12:    **end for**
13:    $X_4[1 : w \times N] \leftarrow$ Sobol sample in design space
14:    $X_4[w \times N + 1 : N] \leftarrow$ random sample in design space
15:   **else**
16:    **for** $m \leftarrow 1$ to N **do**
17:      indexcont ← random select (indcont)
18:      $X_1 \leftarrow$ mutate(x_{best}[indexcont])
19:    **end for**
20:    $X_2[1 : w \times N] \leftarrow$ Sobol sample in design space
21:    $X_2[w \times N + 1 : N] \leftarrow$ random sample in design space
22:   **end if**
23:   **for** $i = 1$ to G **do**   ▷ run in parallel
24:     $D_i \leftarrow SM_{co}$.predict(X_i)
25:   **end for**
26:   **return** X_i, D_i
27: **end procedure**

is used to probabilistically remove (or "dropout") neurons in the network. Here it is used to remove dimensions, so as to reduce the modeling cost. The variables to be perturbed will be randomly selected when the number of the dimensions in the design space is high, i.e., the dimensions not selected are dropped, which is similar to the random dropout of a certain percentage of neurons in one layer. If both discrete and continuous variables are detected in the problem configuration file, an internal variable, group size G would be set to 4, otherwise, if only continuous variables are detected, G equals to 2 (Algorithm 2, lines 4–8). The input of proposed Candidate Search (CS) sampling method are the current optimal sample x_{best}, group size G and perturbation budget N for each group (Algorithm 2, line 18). The group size G denotes how many groups of sequential samples with different location properties are generated in parallel. Perturbation budget N for each group means that N samples will be generated for each group. For $G = 4$ case, we will have a group ($G1$) of N samples only performing perturbation on continuous variables of the current best point by randomly selecting continuous variables from all continuous variables. This generates the set $X1$ of new points to sample. Only "indexint" of the "indent" dimensions are sampled, and that set is randomly selected (i.e., The unsampled dimensions are dropped). The second group ($G2$) has N samples based only on perturbations on randomly chosen discrete variables of the current best point. This generates the set $X2$. The third group ($G3$) is formed by perturbing the randomly selected continuous and discrete variables of the current best point. This generates the set $X3$. The fourth group ($G4$) contains the N points sampled from the whole design space, generating set $X4$. The perturbation rates for groups $G1$–$G3$ are stored in a table for each chosen variable. The perturbation rate for each chosen variable is randomly selected in that table. The essence of the sampling method makes it a good candidate for high-dimensional problem optimization because only a subset of the design variables is used in the process.

Specially, when it comes to generating the four group of samples, part of the perturbation budget will assign to generate random samples and the rest is designed to be a low discrepancy sequence samples to ensure sufficient coverage of the design space. The low discrepancy sample sequence adopted in this work is *Sobol sequence sampling* (Sobol', 1967). The Sobol sequence was chosen because it produces very uniformly distributed samples. The weight w between the random sampling count and Sobol sampling count was adjustable and 0.2 was found suitable in the experiments described in this paper (Algorithm 3, lines 20, 21). All of these points X1, X2, X3, X4 are evaluated with the current Surrogate Model SM (Algorithm 3, line 24). These points and evaluations then become the set of points, Xi, and the set of results, Di, referred to in line 18 of Algorithm 2.

10.7.3.4 *Adaptive samples filtering*

After completing the adaptive sampling method CS, a distance checking is performed as described at lines 23 to 27 in Algorithm 2. The distance between the newly generated samples and existing samples that have already been stored in x_e is calculated with k-nearest neighbors (KNNs). Once the distance of a newly sampled point to the nearest existing training sample is shorter than the preset ϵ, it will be discarded. If no points are found nearer than the preset, it is highly likely that the potential minimum of D_i would aggregate at the location where the current known optimal lies. As a result, the optimization result can not be further improved and converges at a bad value. As the stopping criteria isn't met, the system would keep computing those repetitive samples with a costly high-fidelity simulator which leads to much waste in computation.

10.7.3.5 *Low-fidelity dataset update*

A new technique is introduced to improve the surrogate model accuracy at acceptable additional cost as described from lines 31 to 44 in Algorithm 2.

10.7.3.6 *Sample generation for model rebuild*

As depicted at line 23 in Algorithm 3, the surrogate is updated with G samples in every iteration until the stop criterion is met. The mechanism that generates multiple high-fidelity samples simultaneously to update the surrogate not only accelerates modeling convergence but also naturally leads to a balance in exploring local area near the current optimal and the entire design space. As long as the best point in $G4$ doesn't violate the distance checking requirement, at least one point generated globally would be added into the training dataset to update the surrogate model.

10.8 Experimental Results

To demonstrate the effectiveness of the proposed MFSBO-CS methodology in an EM simulation-based optimization task, two design examples are explored. A comparison is drawn between the proposed method and the state of the art Bayesian optimization with different acquisition function configurations. MFSBO-CS is implemented in Python and Matlab. The experiments are simulated with an Intel Xeon(R) W3503 2.40 GHz CPU and 8 GB of memory.

10.8.1 *Inductor design*

The layout of the square inductor is illustrated in Fig. 10.6. It contains four tuning variables: space, width, length and turn count. This test case could be categorized as a mixed integer optimization problem since the turn count of the inductor is discrete while the rest of the variables are all continuous. Table 10.10 summarizes the inductor design. The low-fidelity simulation is implemented in ASITIC (Niknejad, 2000) whose simulation relies on the estimation of a simple Pi model. We implement a subflow to generate ASITIC simulation configuration files and execute them in a batch mode.

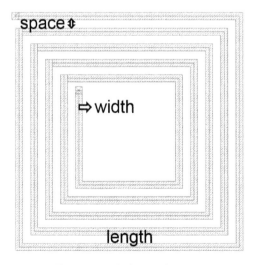

Fig. 10.6. Inductor layout.

Table 10.10. Inductor design summary.

Design parameter	Design range
Length	250–500
Width	5–12
Space	4–15
Turns (int)	2–5
Performance	Desired value
L	2.5 nH
Q	12

The high-fidelity simulation requires a 2.5D EM solver and X11-GUITest (Paulsen, 2009) is applied for EM simulation automation. The S-parameters generated by the EM solver are converted to Z-parameters which can be used to calculate inductor value and quality factor.

There are two inverse problems that need to be solved in the inductor case: (1) find the design variables' values for the inductance of 2.5 nH, and (2) find the design variables' values for the quality factor of 12. So the target functions are defined as $||L - 2.5\,\text{nH}||$ and $||Q - 12||$, respectively.

10.8.1.1 Optimization result comparison for inductor

Table 10.11 summarizes the experimental results for the two specifications: inductor value and quality factor at 2 GHz. Figures 10.7 and 10.8 demonstrate the optimization convergence plots for L and Q under different algorithms. In the baseline experiments, the budget of 160 EM simulations is set for each Bayesian Optimization method. For the L problem, the result shows that optimal values are found at 45th, 9th and 158th iteration within 160 iterations for PI, EI and LCB, respectively. All Bayesian optimization methods deliver similar final designs given the simulation configuration shown in Table 10.11. The results indicate that BO methods get stuck at the local optima [500, 5, 4, 5]. A total budget of 100 low-fidelity simulations and 60 high-fidelity simulations is set For MFSBO-CS. Due to the candidate point group 4 (G4) in the proposed adaptive sampling method

Table 10.11. Optimization results for inductor design.

Spec	Target	Method	Init. sim. config.	Opt. iter. in HF	Opt. value	Elapsed time (s)
L	2.5 nH	BO(PI)	160 HF	45	4.428nH	15029.23
		BO(EI)	160 HF	9	4.428nH	13378.02
		BO(LCB)	160 HF	158	4.428nH	16256.46
		MFSBO-CS	100 LF + 60 HF	29	2.501nH	5001.54
Q	12	BO(PI)	160 HF	135	12.0001	14420.16
		BO(EI)	160 HF	160	12.0024	14411.55
		BO(LCB)	160 HF	151	12.0112	14288.65
		MFSBO-CS	100 LF + 60 HF	9	12.0992	5442.17

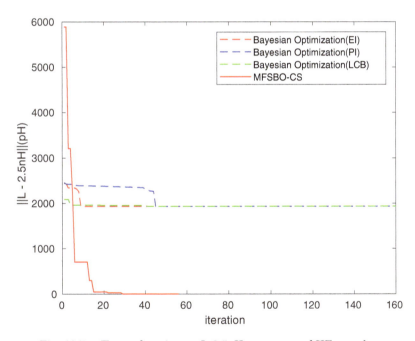

Fig. 10.7. Target function — L-2.5nH — vs. no. of HF samples.

and a surrogate with higher model accuracy, MFSBO-CS yields a much better and faster solution at the 29th EM simulation iteration out of 60. The total elapsed simulation time is around one third that of Bayesian optimization.

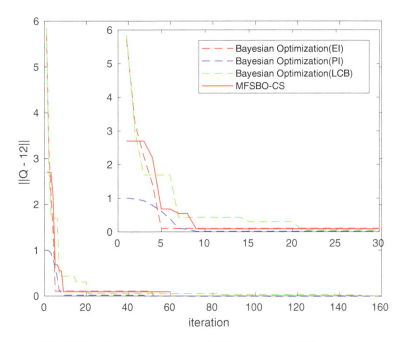

Fig. 10.8. Target function — Q-12 — vs. no. of HF samples.

In the experiment for finding design variable values for a given quality factor, MFSBO-CS delivers final solutions comparable to BO methods but it only requires nine high-fidelity EM simulations after 100 low-fidelity simulations (and thus is approximately 17 times faster than EI).

10.8.1.2 *IP redesign example — VCO*

The simplified benchmark schematic of the VCO is illustrated in Fig. 10.9 (Wang *et al.*, 2014). The design summary is included in Table 10.12. With this VCO, we focus on demonstrating IP redesign in our flow and the result is shown in Table. 10.13.

For the first and second experiments shown in Table 10.13, we are trying to establish a figure of merit (FoM) for the VCO. Since the second performance goal is set to be a range of center frequencies, we first simply set the FoM to be (Equation 10.17)

$$\text{FoM} = \mathcal{L}(\Delta f) - 20\log(f_o/10^{10}). \qquad (10.17)$$

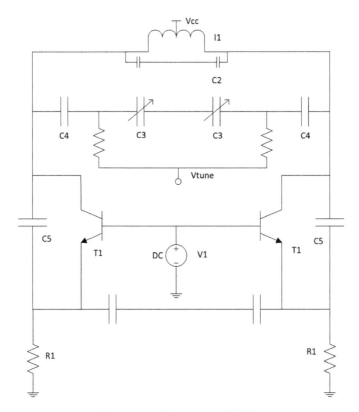

Fig. 10.9. Schematic of VCO.

Table 10.12. VCO design summary.

Physical design parameter

$X = l1, C1, C2, C3, C4, C5, T1, R1$
Target electrical parameter
$Y1$ = Phase noise @ 1 MHz offset
$Y2$ = Center frequency (desired at 9.5 G–10.5 G)

The first term in Equation (10.17) is the phase noise at an offset frequency Δf, and the second term is associated with a normalized center frequency. The final optimized result appears to be approximately as good as the original multiple-objective optimization, which proves the proposed FoM is feasible. Because BO is unable to

Table 10.13. VCO redesign result.

	PN @ 1M offset	Center freq (G)
CAND mixed integer opt.	−123.45	9.605
CAND mixed integer opt. (single objective)	−123.1	9.546
CAND continuous opt.	−122.91	10.3
Bayesian opt.	−118.8	9.95
Manual design	−122.2	10

process mixed-integer problems, we must first transform our design to contain only continuous variables in the third and fourth experiments. The result demonstrates that the CAND algorithm performs better than BO with the same maximum number of function evaluations — set to 200 in this case — with less processing time: (BO: 6 h; CAND: 5 h). The optimized result is shown to be far better than manual human design. This work was implemented using the IBM 9HP (90 nm) process.

In the previous discussion, a VCO FoM was developed based upon the specific RF designer's requirement, which is to minimize the phase noise given the fixed tuning range. To further validate our methodology, we test the flow with a generalized VCO FoM shown in Equation (10.18) (Kinget, 1999):

$$\text{FoM} = \mathcal{L}(\Delta f) + 10\log\left(\frac{P_{DC}}{1\text{mW}}\right) + 20\log\left(\frac{\Delta f}{f_0}\right). \quad (10.18)$$

The first term of Equation (10.18) is the phase noise measured at an offset frequency Δf. In this case, 1 MHz is chosen for the calculation in Table 10.7. The second term is the normalized DC power P_{DC}. The third term in Equation (10.18) is related to the center frequency of the design. A VCO design with a higher center frequency tends to have a better FoM but the other two terms could be worse. Generally a well-designed VCO should generate less phase noise and consume less power. By executing the flow on the FoM, we obtain the result in Table 10.14.

Table 10.14. VCO redesign with FoM (10.18).

	PN @ 1 MHz	Center freq	Tuning range	Power	FoM (10.18)
Our method	−120.90	9.6 G	6.99%	0.051 W	−182.1
Human design	−122	10 G	13.7%	N/A	−187

10.9 Conclusions

With the increasing importance of electromagnetic simulation in analog/RF system design, the computation cost of EM evaluation becomes a fundamental issue in analog/RF design automation. We have presented a novel flow to accelerate EM simulation in analog/RF circuit design problems. This flow automates the process of layout design generation and optimizes the layout based on EM simulation results. In the optimization loop, we leverage the multi-fidelity EM simulation data along with the proposed parallel adaptive sampling method, sample filtering and low-fidelity update mechanism to assist the system to quickly achieve the global optima with low computation cost, i.e., a shorter simulation time. The experimental results show the proposed method either performs comparably or outperforms the optimization result from various Bayesian optimization methods with only approximately one third to two thirds of the computation cost.

Acknowledgments

This material is funded in part by the National Science Foundation under Grant No. CNS 16-24811 and the industry members of the CAEML IUCRC.

References

Bandler, J. W., Biernacki, R. M., Chen, S. H., Grobelny, P. A., and Hemmers, R. H. (1994). Space mapping technique for electromagnetic optimization, *IEEE Transactions on Microwave Theory and Techniques*, vol. 42, no. 12, pp. 2536–2544.

Floudas, C. A. (1995). *Nonlinear and Mixed-Integer Optimization: Fundamentals and Applications*, Oxford University Press.

Forrester, A. I., Sóbester, A., and Keane, A. J. (2007). Multi-fidelity optimization via surrogate modelling, in *Proceedings of the Royal Society of London A: Mathematical, Physical and Engineering Sciences*, vol. 463 (The Royal Society), pp. 3251–3269.

Fujibayashi, T., Takeda, Y., Wang, W., Yeh, Y.-S., Stapelbroek, W., Takeuchi, S., and Floyd, B. (2017). A 76-to 81-ghz multi-channel radar transceiver, *IEEE Journal of Solid-State Circuits*, vol. 52, no. 9, pp. 2226–2241.

Geiselhart, K., Ozoroski, L., Fenbert, J., Shields, E., and Li, W. (2011). Integration of multifidelity multidisciplinary computer codes for design and analysis of supersonic aircraft, in *49th AIAA Aerospace Sciences Meeting Including the New Horizons Forum and Aerospace Exposition*, p. 465.

Ghoreyshi, M., Badcock, K., and Woodgate, M. (2008). Integration of multifidelity methods for generating an aerodynamic model for flight simulation, in *46th AIAA Aerospace Sciences Meeting and Exhibit*, p. 197.

Giunta, A., Narducci, R., Burgee, S., Grossman, B., Mason, W., Watson, L., and Haftka, R. (1995). Variable-complexity response surface aerodynamic design of an hsct wing, in *13th Applied Aerodynamics Conference*, p. 1886.

Hinton, G. E., Srivastava, N., Krizhevsky, A., Sutskever, I., and Salakhutdinov, R. R. (2012). Improving neural networks by preventing co-adaptation of feature detectors, arXiv preprint arXiv:1207.0580.

Jin, Y. (2011). Surrogate-assisted evolutionary computation: Recent advances and future challenges, *Swarm and Evolutionary Computation*, vol. 1, no. 2, pp. 61–70.

Jolliffe, I. (2011). *Principal Component Analysis*, Springer.

Kennedy, M. C. and O'Hagan, A. (2000). Predicting the output from a complex computer code when fast approximations are available, *Biometrika*, vol. 87, no. 1, pp. 1–13.

Kinget, P. (1999). Integrated ghz voltage controlled oscillators, in *Analog Circuit Design*, Springer, pp. 353–381.

Koziel, S. and Bandler, J. W. (2007a). Microwave device modeling using space-mapping and radial basis functions, in *Microwave Symposium, 2007. IEEE/MTT-S International (IEEE)*, pp. 799–802.

Koziel, S. and Bandler, J. W. (2007b). A space-mapping approach to microwave device modeling exploiting fuzzy systems, *IEEE Transactions on Microwave Theory and Techniques*, vol. 55, no. 12, pp. 2539–2547.

Koziel, S. and Bandler, J. W. (2012). Accurate modeling of microwave devices using kriging-corrected space mapping surrogates, *International Journal of Numerical Modelling: Electronic Networks, Devices and Fields*, vol. 25, no. 1, pp. 1–14.

Koziel, S., Bandler, J. W., Mohamed, A. S., and Madsen, K. (2005). Enhanced surrogate models for statistical design exploiting space mapping technology, in *Microwave Symposium Digest, 2005 IEEE MTT-S International (IEEE)*, pp. 4–pp.

Lyu, W., Xue, P., Yang, F., Yan, C., Hong, Z., Zeng, X., and Zhou, D. (2018). An efficient bayesian optimization approach for automated optimization of analog circuits, *IEEE Transactions on Circuits and Systems I: Regular Papers*, vol. 65, no. 6, pp. 1954–1967.

Martinez-Cantin, R. (2014). Bayesopt: A bayesian optimization library for nonlinear optimization, experimental design and bandits, *The Journal of Machine Learning Research*, vol. 15, no. 1, pp. 3735–3739.

McKay, M. D., Beckman, R. J., and Conover, W. J. (2000). A comparison of three methods for selecting values of input variables in the analysis of output from a computer code, *Technometrics*, vol. 42, no. 1, pp. 55–61.

Mockus, J. (2012). *Bayesian Approach to Global Optimization: Theory and Applications*, vol. 37, Springer Science & Business Media.

Müller, J., Shoemaker, C. A., and Piché, R. (2013). So-mi: A surrogate model algorithm for computationally expensive nonlinear mixed-integer black-box global optimization problems, *Computers & Operations Research*, vol. 40, no. 5, pp. 1383–1400.

Niknejad, A. M. (2000). *Analysis, Simulation, and Applications of Passive Devices on Conductive Substrates*.

Paulsen, D. K. (2009). X11::GUITest, http://sourceforge.net/projects/x11guitest (Dec. 2009).

Poulton, K., Neff, R., Setterberg, B., Wuppermann, B., Kopley, T., Jewett, R., Pernillo, J., Tan, C., and Montijo, A. (2003). A 20 gs/s 8 b adc with a 1 mb memory in 0.18/spl mu/m cmos, in *Solid-State Circuits Conference, 2003. Digest of Technical Papers. ISSCC. 2003 IEEE International (IEEE)*, pp. 318–496.

Qi, W. (2017). *IC Design Analysis, Optimization and Reuse via Machine Learning* (North Carolina State University).

Schvan, P., Bach, J., Falt, C., Flemke, P., Gibbins, R., Greshishchev, Y. M., Ben-Hamida, N., Pollex, D., Sitch, J., Wang, S.-C., et al. (2008). A 24gs/s 6b adc in 90nm cmos. in *ISSCC*, vol. 544, p. 81.

Shahramian, S., Voinigescu, S. P., and Carusone, A. C. (2009). A 35-gs/s, 4-bit flash adc with active data and clock distribution trees, *IEEE Journal of Solid-State Circuits*, vol. 44, no. 6, pp. 1709–1720.

Shahriari, B., Swersky, K., Wang, Z., Adams, R. P., and De Freitas, N. (2015). Taking the human out of the loop: A review of bayesian optimization, *Proceedings of the IEEE*, vol. 104, no. 1, pp. 148–175.

Snoek, J., Larochelle, H., and Adams, R. P. (2012). Practical bayesian optimization of machine learning algorithms, in *Advances in Neural Information Processing Systems*, pp. 2951–2959.

Sobol', I. M. (1967). On the distribution of points in a cube and the approximate evaluation of integrals, *Zhurnal Vychislitel'noi Matematiki i Matematicheskoi Fiziki*, vol. 7, no. 4, pp. 784–802.

Van Roy, T. J. and Wolsey, L. A. (1987). Solving mixed integer programming problems using automatic reformulation, *Operations Research*, vol. 35, no. 1, pp. 45–57.

Vandenbosch, G. A. and Vasylchenko, A. (2011). A practical guide to 3D electromagnetic software tools, in *Microstrip Antennas* (InTech).

Vural, R. and Yildirim, T. (2012). Analog circuit sizing via swarm intelligence, *AEU-International journal of Electronics and Communications*, vol. 66, no. 9, pp. 732–740.

Wang, W., Takeda, Y., Yeh, Y.-s., and Floyd, B. (2014). A 20ghz vco and frequency doubler for w-band fmcw radar applications, in *2014 IEEE 14th Topical Meeting on Silicon Monolithic Integrated Circuits in Rf Systems (IEEE)*, pp. 104–106.

Warde-Farley, D., Goodfellow, I. J., Courville, A., and Bengio, Y. (2013). An empirical analysis of dropout in piecewise linear networks, arXiv preprint arXiv:1312.6197.

Zhang, S., Lyu, W., Yang, F., Yan, C., Zhou, D., and Zeng, X. (2019). Bayesian optimization approach for analog circuit synthesis using neural network, in *2019 Design, Automation & Test in Europe Conference & Exhibition (DATE) (IEEE)*, pp. 1463–1468.

© 2022 World Scientific Publishing Europe Ltd.
https://doi.org/10.1142/9781800610750_0011

Chapter 11

An Automated and Adaptive Calibration of Passive Tuners Using an Advanced Modeling Technique

Maral Zyari, Francesco Ferranti, and Yves Rolain

Abstract

In this chapter, we describe an advanced modeling method oriented towards the development of an automated user-friendly tuner calibration procedure based on adaptive sampling and modeling techniques. The potential shown by the proposed technique can be a stepping stone towards further developments that could be also introduced into commercial products. Pertinent numerical results are presented to validate the proposed method.

Keywords: Microwave tuners, modeling, adaptive sampling

11.1 Introduction

The main purpose of an impedance tuner is to control the impedance at the ports of a nonlinear active component to characterize or optimize the circuit and the system level of the design. Impedance tuners are thus used in many high frequency applications, including passive source and load pull measurements for characterization and design (Berghoff *et al.*, 1998; Bossche, 2003; Ghannouchi and Bosisio, 1992; Suh *et al.*, 2007; Teppati and Bolognesi, 2012; Verbeyst and Bossche, 2005), and noise measurements (Axelsson *et al.*, 2012; McIntosh *et al.*, 1999). They can also be used in automatic impedance-matching devices like Automatic Antenna Tuning Units (Sun and Fidler, 1995;

De Mingo et al., 2004) and many other broadband wireless devices (Whatley et al., 2006). Synthesizing the desired port impedance at a desired frequency or over a frequency band of interest is the task of a tuner in all these applications. An accurate model that can characterize the tuner is very important to achieve the desired reflection coefficient at the frequency values of interest. To minimize the uncertainty of the synthesized impedance that is achieved by the tuner in all the mentioned applications, the tuner needs to be calibrated accurately.

Accuracy of the calibration of the tuners is an essential step as it assures repeatable and precise measurements. Moreover, the inaccuracy in the tuner calibration leads to errors in the measurements. The existing calibration procedures that are offered by the software provided by the manufacturers of the tuners can be time consuming, especially if the highest accuracy is desired. On top of that, defining the accuracy in commercial tuner software can also be an issue. In order to choose the accuracy of the desired calibration one has to choose between accuracy levels that are qualitative: for example, very low, low, high, and very high resolutions, which are not quantitative measures of the calibration accuracy. Therefore, the user has an ambiguous idea of accuracy, while choosing among the options. For all these reasons, the need of developing an automated, simple, user-friendly and accurate calibration procedure is very relevant.

Two main types of tuners are available, passive and active tuners. Each class of tuners has corresponding strengths and drawbacks. In this chapter, we focus on mechanical passive tuners that are often used in several application. The state-of-the-art maximally accurate calibration of these tuners requires a very large number of positions of the probes to be characterized for the desired frequencies prior to using the tuners in the measurement setups. This leads to a very time-consuming task: moving the probes to a high number of different positions and measuring the corresponding scattering parameters (S-parameters) at the ports of the tuner with a vector network analyzer (VNA). This calibration process has a very significant impact on the time needed for the total measurement procedure.

In the literature, a technique has been proposed to reduce the calibration time of mechanical passive tuners (Roff et al., 2008). It uses a 2D interpolation method to reduce the time needed for tuner calibration in comparison to a standard calibration. However, the choice of the probe position samples is not adaptive and there is no control on the achieved calibration accuracy. Undersampling/oversampling phenomena, which generate inaccuracies/waste of measurement time, can very likely exist when using only an initial fixed sample distribution to build a model without any adaptive refinement. Key questions like how the samples should be distributed and how many samples are needed are not addressed in Roff *et al.* (2008).

In this chapter, we describe a method that can be used to automate the calibration process of passive tuners based on mechanical probes. It is based on two main ideas: adaptive sampling and decomposing the horizontal and vertical movements of the probes. We show that using adaptive sampling techniques (Simpson *et al.*, 2004; Wang and Shan, 2007) can be very effective to achieve accuracy-controlled calibrations in an efficient time frame and for a reduced number of samples (measurements). The decomposition of the effects of the horizontal and vertical movements of the probes on the S-parameters is exploited by decomposing the 2D modeling problem to two 1D modeling problems. A very preliminary version of this work was discussed in Zyari *et al.* (2018).

We note that the aim of this chapter is to describe an advanced method oriented towards the development of an automated user-friendly tuner calibration procedure based on adaptive sampling and modeling techniques. We do not intend to present our technique as a fully ready alternative to commercial software proposed by tuner vendors. The potential shown by the proposed technique can be a stepping stone towards further developments that could be also introduced into commercial products. The proposed algorithm uses a software to control the position of the tuner probes, a VNA to measure the S-parameters at the ports of the tuner and a MATLAB routine to implement the modeling and sampling method. A detailed pseudocode of the different steps of the proposed technique in provided in this chapter.

The remainder of this chapter is organized as follows. In Section 11.2, the passive tuner behavior is described in more detail. Section 11.3 describes the proposed technique for the calibration of a passive tuner. Different numerical results based on measurements are shown in Section 11.4. Conclusions are given in Section 11.5.

11.2 Behavior of a Passive Mechanical Tuner

The passive tuners are also known as slide-screw tuners. This name also describes their functionality. The principle of mechanical passive tuners is simple: they consist of a slab-line with two grounded parallel plates, a center conductor and one or more adjustable probes that can move horizontally (X) and vertically (Y). The vertical movement of the probes changes the magnitude of the reflection coefficient (S_{11}) and the horizontal movement changes the phase.

With a vertical movement of the probe towards the slab-line, the electrical field distribution of the line is perturbed and part of the signal gets reflected back: the magnitude of the reflection measured at the tuner ports increases. Instead, the horizontal movement of the probe modifies the distance between the probe and the measurement ports, therefore it changes the phase of the reflection coefficient. The synthesis of an arbitrary passive impedance at the measurement ports is enabled by these probes movements (Simpson, 2015; Tsironis, 2004). An example of the S_{11} outcome of the movements of the probe on the Smith chart is shown in Fig. 11.1.

Passive tuners can be either manual or electro-mechanical. In the latter case, the probe movements are controlled with stepper motors that will assure more accurate positioning of the probes compared to movements that are controlled manually. The S-parameters measured at the ports of the tuner depend on the probes position. We describe an automated modeling technique able to represent the relation between the S-parameters data of the tuner and the physical horizontal and vertical movements of the probes at different frequencies. In our work, we use software positioning control of the probes in electro-mechanical tuners to move the probes.

Multi-harmonic tuners allow tuning at multiple frequencies (or harmonic frequencies) independently at the same time, as each probe covers a certain frequency band (Tsironis, 2006). In this chapter, we target the modeling of one probe of a passive electro-mechanical tuner in order to show the capability of our method. We can safely assume that all the probes of a tuner behave similarly.

We note that the concept of decomposing magnitude and phase variations by vertical and horizontal probe movements might become

An Automated and Adaptive Calibration of Passive Tuners 365

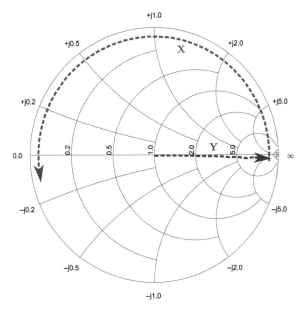

Fig. 11.1. Vertical and horizontal movements of the tuner's probe on the Smith chart.

less valid for high values of the reflection coefficient (S_{11} close to one in magnitude), therefore close to the boundary of the Smith Chart. This does not invalidate our technique, since it is adaptive and more samples will be located where the functions to be modeled are more dynamic. Numerical results will show cases where the reflection coefficient is high and will confirm accuracy of our method also in those cases.

11.3 The Proposed Algorithm

In the proposed method, we exploit the decomposition of the probe movements into two different directions: along vertical (Y) and horizontal (X) axis, which sets the magnitude and the phase of the reflection coefficient, respectively. An adaptive sampling technique is used to sample the 2D space ($X - Y$) and obtain high accuracy models with a reasonably low number of samples (measurements).

The variation of the position of the probes along the X (horizontal) and Y (vertical) directions is limited to a certain range: $[X_{\min}, X_{\max}]$ for the horizontal movement and $[Y_{\min}, Y_{\max}]$ for the vertical movement. In our method, instead of sampling the 2D space at once, we first sample the Y axis and then the X axis. The adaptive sampling procedure first generates a set of initial samples in the (X, Y) space. Then, based on the information obtained from these initial samples, final samples are generated. In the last step, a model that is able to predict the tuner behavior over the complete (X, Y) space at a set of desired frequencies will be built based on these samples.

The interaction of the user is limited to setting some desired accuracy thresholds ($\text{Err}_{\text{Desired}}$, $\text{Err}_{\text{DesiredX}}$, $\text{Err}_{\text{DesiredY}}$), defining the movement limits of the probes of the tuner under test ($X_{\min}, X_{\max}, Y_{\min}, Y_{\max}$), the number of initial samples for each axis (N_{init}) and the following parameters, which will be explained in detail in what follows: $N_{\text{dense}}, K_X, K_Y$. The proposed algorithm will ensure that $\text{Err}_{\text{DesiredY}} \leq \text{Err}_{\text{Desired}}$, $\text{Err}_{\text{DesiredX}} \leq \text{Err}_{\text{Desired}}$. The procedure of the proposed method is explained in further detail in what follows. A pseudocode is provided in Fig. 11.3.

11.3.1 Adaptive sampling technique

The samples distribution affects the accuracy and time needed for the calibration of the tuner. An adaptive sampling algorithm can result in reducing the calibration time, while providing an accurate calibration.

The adaptive sampling technique starts from a set of initial samples and provides a set of final samples. Initial samples are spread uniformly. Then final samples are selected sequentially in the (X, Y) space in order to well describe the functions of interest (S-parameters of the tuner in our case). Usually in adaptive sampling algorithms, the number of samples used as initial samples is small as the majority of the samples is chosen adaptively to catch the most of the dynamic behavior of the functions of interest. The sampling scheme is illustrated in Fig. 11.2 and is explained in more detail in what follows. This is just an illustration, the final location of samples depends on the function to be modeled as it will be shown in Section 11.4.

The sampling procedure starts by first sampling along the Y axis, while the value of X is fixed to $X_{\text{mid}} = \frac{X_{\min} + X_{\max}}{2}$. The number

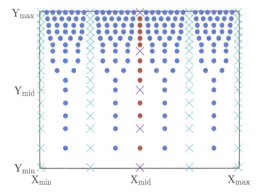

Fig. 11.2. Illustration of the sampling scheme to generate the initial and final samples. The initial samples are shown with crosses and the final generated samples are shown with dots. The difference in color between the two sets of crosses/dots represents the two steps of sampling the Y and X axes, which is described in Section 11.3.

Algorithm
1: PROPOSED TECHNIQUE($X_{min}, X_{max}, Y_{min}, Y_{max}, Err_{desired,Y}, Err_{desired,X}, K_Y, K_X$)

$\begin{cases} X_{mid} = \frac{X_{min}+X_{max}}{2}; \\ \text{Adaptive sampling } Y \text{ axis till } Err_{sampling,Y} < Err_{desired,Y}/K_Y; \\ A \text{ set of } (X_{mid}, Y_n) \text{ samples}, n = 1, ..., N_Y \text{ is obtained}; \\ \textbf{for } n \leftarrow 1 \textbf{ to } N_Y \\ \quad \textbf{do } \begin{cases} \text{Adaptive sampling } X \text{ axis till } Err_{sampling,X} < Err_{desired,X}/K_X; \\ A \text{ set of } (X_{p_n,n}, Y_n) \text{ samples}, p_n = 1, ..., N_{X,n} \text{ is obtained}; \end{cases} \\ A \text{ set of } (X_{p_n,n}, Y_n) \text{ samples}, n = 1, ..., N_Y, p_n = 1, ..., N_{X,n} \text{ is obtained} \end{cases}$

Fig. 11.3. Pseudocode of the proposed technique.

of initial samples is denoted as N_{init}. This initial set of uniformly distributed Y samples is defined as the starting points of the sampling procedure. An adaptive sampling then decides the location of the samples along Y axis with the fixed position for the horizontal axis X_{mid}. As it is explained later in the chapter, this step is repeated for the X axis as soon as the final Y samples are selected. When these initial Y samples are chosen, the adaptive sampling decides about the location of the new samples over the axis by adaptively building two interpolation-based models (Model$_1$, Model$_2$) and comparing them with each other. During the iterations of the adaptive sampling step for the Y axis, two proposed interpolation models are evaluated over a very dense set of samples (N_{dense}) first over the Y axis. Then, the Y sample (position in the Y axis) where these two models exhibit a maximum deviation

$$\text{Err}_{\text{sampling},Y} = \max_{Y_{\text{dense}}} \left(\text{Err}(\text{Model}_1, \text{Model}_2) \right), \quad (11.1)$$

becomes a new sample to be measured. Evaluating these two analytical interpolation models over a dense set of samples is computationally efficient. This sampling iteration continues until the required accuracy is satisfied:

$$\text{Err}_{\text{Sampling},Y} < \text{Err}_{\text{Threshold},Y}. \quad (11.2)$$

We take a margin in defining the error threshold, such as

$$\text{Err}_{\text{Threshold},Y} = \frac{\text{Err}_{\text{Desired},Y}}{K_Y}; \quad K_Y > 1, \quad (11.3)$$

because two models are being compared to evaluate $\text{Err}_{\text{Sampling},Y}$.

In this work, two model classes are used: Piecewise cubic hermite interpolating polynomial (PCHIP) interpolation and SPLINES interpolation (de Boor, 2001) for Model_1 and Model_2, respectively. The user can utilize different model classes as the method is not limited to the choices made here. Our choice on the model classes is based on the fact that PCHIP and SPLINES are well-established and well-performing. A set of samples $(X_{\text{mid}}, \mathbf{Y}_n)$ with $n = 1, \ldots, N_Y$ is obtained at the end of this step. A pseudocode for this step is provided in Fig. 11.4.

In the next step, for each of the $\mathbf{Y}_n$ samples previously generated, initial uniformly distributed samples are generated along the X axis. Then, adaptive sampling is used to decide the location of the final samples along the X axis, for each of the $\mathbf{Y}_n$ samples. As previously for the Y axis, an adaptive sampling loop is carried out to select a set of sampled X values at each $\mathbf{Y}_n$ value with the accuracy iteration criterion:

$$\text{Err}_{\text{sampling},X} = \max_{X_{\text{dense}}} \left(\text{Err}(\text{Model}_1, \text{Model}_2) \right), \quad (11.4)$$

$$\text{Err}_{\text{Sampling},X} < \text{Err}_{\text{Threshold},X}, \quad (11.5)$$

$$\text{Err}_{\text{Threshold},X} = \frac{\text{Err}_{\text{Desired},X}}{K_X}; \quad K_X > 1. \quad (11.6)$$

Algorithm
2: ADAPTIVE SAMPLING $Y(X_{fixed}, Y_{min}, Y_{max}, Err_{sampling,Y}, Err_{threshold,Y}, N_{init}, N_{dense})$

$\begin{cases} Y_{init} = linspace(Y_{min}, Y_{max}, N_{init}); \\ Y_{dense} = linspace(Y_{min}, Y_{max}, N_{dense}); \\ Gather\ Data_{init} = Data(X_{fixed}, Y_{init}); \\ Build\ Model_1\ and\ Model_2; \\ Evaluate\ Model_1\ and\ Model_2\ over\ Y_{dense}; \\ Err_{sampling,Y} = \max_{Y_{dense}}(Err(Model_1, Model_2)); \\ \textbf{if}\ Err_{sampling,Y} < Err_{threshold,Y} \\ \quad A\ set\ of\ (X_{fixed}, Y_n)\ samples, n = 1, ..., N_{init}\ is\ obtained; \\ \textbf{else} \\ \quad \textbf{while}\ Err_{sampling,Y} \geq Err_{threshold,Y} \\ \quad \textbf{do} \begin{cases} Y_{max,err} = \widetilde{Y}_{dense}\ for\ which\ Err(Model_1, Model_2) = \\ \quad = \max_{Y_{dense}}(Err(Model_1, Model_2)); \\ Y_{init} = Y_{init} \bigcup Y_{max,err}; \\ Gather\ Data_{max,err} = Data(X_{fixed}, Y_{max,err}); \\ Data_{init} = Data_{init} \bigcup Data_{max,err}; \\ Build\ Model_1\ and\ Model_2; \\ Evaluate\ Model_1\ and\ Model_2\ over\ Y_{dense}; \\ Err_{sampling,Y} = \max_{Y_{dense}}(\text{Err}(Model_1, Model_2)); \end{cases} \\ A\ set\ of\ (X_{mid}, Y_n)\ samples, n = 1, ..., N_Y\ is\ obtained. \end{cases}$

Fig. 11.4. Pseudocode of the adaptive sampling scheme along the Y axis.

A pseudocode for this step is provided in Fig. 11.5. Once the adaptive procedure is finalized, the full (X, Y) space is automatically and adaptively sampled. This leads to

- a set of $(\mathbf{X}_{pn,n}, \mathbf{Y_n})$ samples with $p_n = 1, \ldots, N_{X,n}$ for each $\mathbf{Y_n}$;
- a set of models $\mathbf{Models_{X,n}}(\mathbf{X}, \mathbf{Y_n}), n = 1, \ldots, N_Y$ for the $\mathbf{Y_n}$ samples.

11.3.2 Final model generation

A 2D model able to describe the tuner response for each sample in the space bounded by $([X_{\min}, X_{\max}], [Y_{\min}, Y_{\max}])$ can be achieved as explained in what follows. The data samples collected in the previous steps are used for this final model generation. Let us assume that we want to evaluate the reflection coefficient S_{11} of the tuner at a certain position pair $(\widetilde{X}, \widetilde{Y})$. Then, the different models $\mathbf{Models_{X,n}}(X, \mathbf{Y_n}), n = 1, \ldots, N_Y$ are evaluated at

**Algorithm
3:** ADAPTIVE SAMPLING $X(n, Y_n, X_{min}, X_{max}, Err_{sampling,X}, Err_{threshold,X}, N_{init}, N_{dense})$

$\begin{cases} X_{init} = linspace(X_{min}, X_{max}, N_{init}); \\ X_{dense} = linspace(X_{min}, X_{max}, N_{dense}); \\ Gather\ Data_{init} = Data(X_{init}, Y_n); \\ Build\ Model_1\ and\ Model_2; \\ Evaluate\ Model_1\ and\ Model_2\ over\ X_{dense}; \\ Err_{sampling,X} = \max_{X_{dense}}(Err(Model_1, Model_2)); \\ \textbf{if } Err_{sampling,X} < Err_{threshold,X} \\ \quad A\ set\ of\ (X_{p_n,n}, Y_n)\ samples, n = 1, ..., N_{X,n}\ is\ obtained; \\ \textbf{else} \\ \quad \textbf{while } Err_{sampling,X} \geq Err_{threshold,X} \\ \quad \textbf{do } \begin{cases} X_{max,err} = \widetilde{X}_{dense}\ for\ which\ Err(Model_1, Model_2) = \\ \quad = max_{X_{dense}}(Err(Model_1, Model_2)); \\ X_{init} = X_{init} \bigcup X_{max,err}; \\ Gather\ Data_{max,err} = Data(X_{max,err}, Y_n); \\ Data_{init} = Data_{init} \bigcup Data_{max,err}; \\ Build\ Model_1\ and\ Model_2; \\ Evaluate\ Model_1\ and\ Model_2\ over\ X_{dense}; \\ Err_{sampling,X} = max_{X_{dense}}(Err(Model_1, Model_2)); \end{cases} \\ A\ set\ of\ (X_{p_n,n}, Y_n)\ samples, n = 1, ..., N_{X,n}\ is\ obtained; \\ A\ Model_{X,n}(X, Y_n)\ is\ obtained\ from\ the\ latest\ Model_1. \end{cases}$

Fig. 11.5. Pseudocode of the adaptive sampling scheme along the X axis.

$\widetilde{X}$, which provides $S_{model,11}(\widetilde{X}, Y_n)$. Based on the data samples $S_{model,11}(\widetilde{X}, \mathbf{Y_n})$, another model can be built as a function of Y and evaluated at $\widetilde{Y}$, which finally provides $S_{model,11}(\widetilde{X}, \widetilde{Y})$.

We utilize the PCHIP model class to build this 2D model in the numerical results that we show in the corresponding section. We note that once the set of $(\mathbf{X_{p_n,n}}, \mathbf{Y_n})$ samples, $p_n = 1, \ldots, N_{X,n}, n = 1, \ldots, N_Y$ is obtained, any kind of 2D model can be built using these data points. In this work, the main idea of the 2D modeling/sampling decomposition into two 1D modeling/sampling procedures is used to build a final 2D model and to have control on the accuracy of the 2D model.

11.3.3 *Validation and error estimation*

A set of samples over the (X, Y) space is used to show validation results for the proposed technique in this chapter. In the following example, we choose 71 × 71 uniformly distributed samples. The S-parameters response is measured at all these samples and this dataset is then used to show validation results. This dataset is collected before starting the adaptive sampling procedure. In the proposed technique, we compare two models over a very dense set of samples (N_{dense}) to predict the location of the next samples during the adaptive sampling steps and we estimate the accuracy achieved by a final 2D model over the (X,Y) space by using $\text{Err}_{\text{Sampling},X}$ and $\text{Err}_{\text{Sampling},Y}$ in the adaptive sampling loops. We use the factors K_X, K_Y to have stricter error convergence criteria than just

$$\text{Err}_{\text{Sampling},Y} < \text{Err}_{\text{Desired},Y}, \text{Err}_{\text{Sampling},X} < \text{Err}_{\text{Desired},X},$$

and take into consideration that the error between two models is used for accuracy estimation. This is different from relying on a validation data set to assess the accuracy of a model.

11.4 Numerical Results Based on Measurements

The proposed method is applied on a Focus Microwave ©tuner MPT-1818-TC that is a multi-harmonic tuner with three probes. This tuner covers the frequency range [1.8–18] GHz. In this work, we consider the first probe of the tuner out of three. We keep the other two probes completely withdrawn from the center conductor, so that they do not have any influence on the behavior of the tuner. The VNA that is used to perform the measurements is an Agilent E8364B which is calibrated with an Electronic Calibration (ECal) module for the frequency band of interest. In this example, the number of initial values is set to $N_{\text{init}} = 4$ and $N_{\text{dense}} = 100,000$ to compare two interpolation models during the adaptive sampling. The tuner under test has the following limitation on the movements of its probe: the horizontal movement (X) is limited to the range [0–5913] and the vertical movement (Y) to the range [0–4960]. The resolution of the probe steps is 25 μm. The measurements are performed over the frequency band [1–8] GHz with a frequency resolution of 100 MHz. Three main accuracy levels are of interest, namely $\text{Err}_{\text{Desired}} = -20, -30, -40$ dB for three frequencies $1, 2, 4$ Hz. As the repeatability of the tuner is between $-40, -50$ dB (Focus Microwaves Group, 2017), we consider -40 dB as the last accuracy level of interest. The error settings for the sampling of the Y axis have been chosen as

$$\text{Err}_{\text{Sampling},Y} < \text{Err}_{\text{Threshold},Y},$$

with

$$\text{Err}_{\text{Threshold},Y} = \text{Err}_{\text{Desired},Y}/2,$$
$$K_Y = 2; \ \text{Err}_{\text{Desired},Y} = -50\,\text{dB},$$

and for the sampling of the X axis

$$\text{Err}_{\text{Sampling},X} < \text{Err}_{\text{Threshold},X},$$

with

$$\text{Err}_{\text{Threshold},X} = \text{Err}_{\text{Desired},X}/2,$$
$$K_X = 2; \ \text{Err}_{\text{Desired},X} = \text{Err}_{\text{Desired}}.$$

Concerning the sampling over the Y axis, we aim at high accuracy independently on the final accuracy desired ($\text{Err}_{\text{Desired}}$) over the complete (X, Y) space. This is because it is a fundamental step for the complete procedure. The proposed algorithm has been used to model the measured S_{11} of the tuner.

A set of validation data is used based on 71×71 samples in the (X, Y) space to show validation results for the accuracy of the proposed approach. We note that this validation data set is not used in any steps of the described adaptive sampling and modeling technique. Table 11.1 shows the desired accuracy, the obtained accuracy over the validation grid (71×71 samples), the number of samples and the CPU time needed to perform the proposed procedure for the three targeted accuracy levels and the three frequencies of interest. The CPU time indicated in Table 11.1 refers to the overall time composed of

- communication time between MATLAB and the tuner software to control the probe movement;
- time to move mechanically the probes and measure the S-parameters;
- execution time of the MATLAB code of the proposed modeling and sampling technique.

Table 11.1. Experimental results.

Freq (GHz)	Err$_{desired}$ (dB)	Err$_{validation}$ (dB)	Samples	CPU time (s)
1	−20	−27	67	869
2	−20	−26	84	1003
4	−20	−17	128	1453
1	−30	−36	84	1040
2	−30	−35	119	1302
4	−30	−30	188	1949
1	−40	−42	113	1337
2	−40	−40	169	1742
4	−40	−36	279	2803

The mechanical movement of the probe is the most significant contribution to the total CPU time.

The samples chosen by the proposed technique, the real and imaginary part of S_{11} computed by the final 2D model, and the absolute error of this final 2D model are illustrated in Figs. 11.6–11.8 for different accuracy and frequency values. The samples generated to achieve the desired accuracy of −40 dB at 4 GHz are shown in more detail in Figs. 11.9 and 11.10 for the real and imaginary part of S_{11}. The numerical results confirm that the sampling density becomes higher when the function to model becomes more dynamic and the accuracy requirement is more demanding. The proposed automated modeling and sampling procedure for calibration performs well concerning accuracy with a limited and adaptively chosen number of samples. The technique is able to efficiently and accurately reconstruct the tuner behavior for different frequencies and accuracy levels. The algorithm uses the error between two models for accuracy estimation. The targeted accuracy is achieved in almost all cases and in the cases where this does not happen the difference between the desired and achieved accuracy is limited. Additional error estimation schemes, such as cross validation or binary search validation, could be used during the adaptive sampling loops. However, the proposed algorithm has already shown good performances and we consider the refinement of this technique as future work.

374 *Surrogate Modeling for High-Frequency Design*

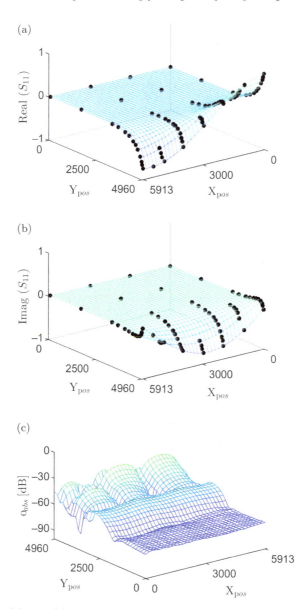

Fig. 11.6. (a) and (b) Real and imaginary parts of S_{11} for Freq = 1 GHz and −20 dB as desired accuracy based on the final 2D model. The black dots represent the samples chosen by the proposed algorithm. (c) Absolute error of the final 2D model over the validation grid.

An Automated and Adaptive Calibration of Passive Tuners

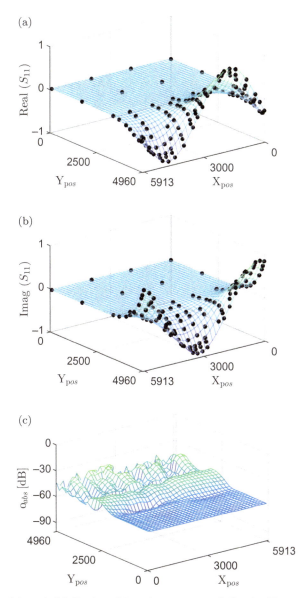

Fig. 11.7. (a) and (b) Real and imaginary parts of S_{11} for Freq = 2 GHz and −30 dB as desired accuracy based on the final 2D model. The black dots represent the samples chosen by the proposed algorithm. (c) Absolute error of the final 2D model over the validation grid.

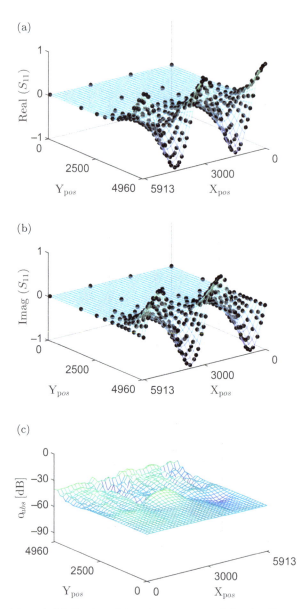

Fig. 11.8. (a) and (b) Real and imaginary parts of S_{11} for Freq = 4 GHz and −40 dB as desired accuracy based on the final 2D model. The black dots represent the samples chosen by the proposed algorithm. (c) Absolute error of the final 2D model over the validation grid.

An Automated and Adaptive Calibration of Passive Tuners 377

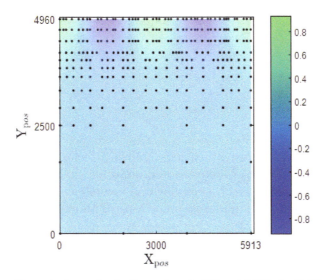

Fig. 11.9. Generated samples for the real part of S_{11} at 4 GHz (−40 dB as desired accuracy).

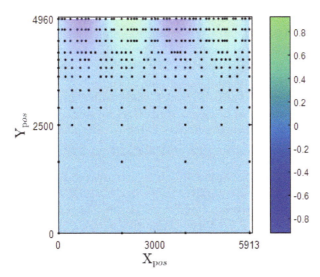

Fig. 11.10. Generated samples for the imaginary part of S_{11} at 4 GHz (−40 dB as desired accuracy).

As discussed in the Introduction, the potential shown by the proposed technique can be a stepping stone towards further developments that could be also introduced into commercial products.

11.5 Conclusion

We have described an advanced modeling technique that can be used for an automated, efficient and adaptive calibration of passive microwave tuners. The 2D sampling and modeling is decomposed into two 1D sampling and modeling procedures. This exploits the different effects of the vertical and horizontal movement of the mechanical probe on the S-parameters behavior of the tuner. Different numerical results have confirmed the accuracy and efficiency of the proposed automated approach. Choosing the samples with adaptive sampling decreases the calibration time while keeping the calibration accuracy at a desired level.

References

Axelsson, O., Thorsell, M., Andersson, K., Stenarson, J., and Rolain, Y. (2012). Noise temperature of an electronic tuner for noise parameter measurement systems, in *79th ARFTG Microwave Measurement Conference*, Montreal, Canada, pp. 1–2.

Berghoff, G., Bergeault, E., Huyart, B., and Jallet, L. (1998). Automated characterization of hf power transistors by source-pull and multiharmonic load-pull measurements based on six-port techniques, *IEEE Transactions on Microwave Theory and Techniques*, vol. 46, no. 12, pp. 2068–2073.

Bossche, M. V. (2003). Curiosities on accurate large-signal measurements in combination with passive tuners, in *62nd ARFTG Microwave Measurements Symposium*, Boulder, USA, pp. 171–180.

de Boor, C. (2001). *A Practical Guide to Splines, Applied Mathematical Sciences*, Springer-Verlag, New York.

De Mingo, J., Valdovinos, A., Crespo, A., Navarro, D., and Garcia, P. (2004). An rf electronically controlled impedance tuning network design and its application to an antenna input impedance automatic matching system, *IEEE Transactions on Microwave Theory and Techniques*, vol. 52, no. 2, pp. 489–497.

Focus Microwaves Group (2017). Multi-purpose tuner 1818, https://focusmicrowaves.com/wp-content/uploads/2017/04/DS-MPT-818-V02.pdf.

Ghannouchi, F. M. and Bosisio, R. G. (1992). Source-pull/load-pull oscillator measurements at microwave/mm wave frequencies, *IEEE Transactions on Instrumentation and Measurement*, vol. 41, no. 1, pp. 32–35.

McIntosh, C. E., Pollard, R. D., and Miles, R. E. (1999). Novel mmic source-impedance tuners for on-wafer microwave noise-parameter measurements, *IEEE Transactions on Microwave Theory and Techniques*, vol. 47, no. 2, pp. 125–131.

Roff, C., Graham, J., Sirois, J., and Noori, B. (2008). A new technique for decreasing the characterization time of passive load-pull tuners to maximize measurement throughput, in *72nd ARFTG Microwave Measurement Symposium*, Portland, USA, pp. 92–96.

Simpson, G. (2015). A beginner's guide to all things load pull, https://www.maurymw.com/pdf/datasheets/5A-062.pdf.

Simpson, T. W., Booker, A. J., Ghosh, D., Giunta, A. A., Koch, P. N., and Yang, R. (2004). Approximation methods in multidisciplinary analysis and optimization: A panel discussion, *Structural and Multidisciplinary Optimization*, vol. 27, no. 5, pp. 302–313.

Suh, I., Doo, S. J., Roblin, P., Cui, X., Kim, Y. G., Strahler, J., Bossche, M. V., Rojas, R., and Park, H. D. (2007). Negative input resistance and real-time active load-pull measurements of a 2.5ghz oscillator using a lsna, in *69th ARFTG Conference*, Honolulu, USA, pp. 1–6.

Sun, Y. and Fidler, J. (1995). High-speed automatic antenna tuning units, in *Ninth International Conference on Antennas and Propagation*, Eindhoven, Netherlands.

Teppati, V. and Bolognesi, C. (2012). Evaluation and reduction of calibration residual uncertainty in load-pull measurements at millimeter-wave frequencies, *IEEE Transactions on Instrumentation and Measurement*, vol. 61, no. 3, pp. 817–822.

Tsironis, C. (2004). Adaptable pre-matched tuner system and method, https://www.google.com/patents/US6674293, US Patent 6,674,293.

Tsironis, C. (2006). Triple probe automatic slide screw load pull tuner and method, https://www.google.com/patents/US7135941, US Patent 7,135,941.

Verbeyst, F. and Bossche, M. V. (2005). Measurement-based behavioral model under mismatched conditions: A new and easy approach for an accurate model, in *European Microwave Conference*, Paris, France, pp. 1–4.

Wang, G. G. and Shan, S. (2007). Review of metamodeling techniques in support of engineering design optimization, *J. Mech. Des.*, vol. 129, no. 4, pp. 370–380.

Whatley, R. B., Zhou, Z., and Melde, K. L. (2006). Reconfigurable rf impedance tuner for match control in broadband wireless devices, *IEEE Transactions on Antennas and Propagation*, vol. 54, no. 2, pp. 470–478.

Zyari, M., Ferranti, F., and Rolain, Y. (2018). An adaptive modeling method for the calibration of passive tuners, in *IEEE MTT-S International Conference on Numerical Electromagnetic and Multiphysics Modeling and Optimization (NEMO)*, Reykjavik, Iceland, pp. 1–4.

© 2022 World Scientific Publishing Europe Ltd.
https://doi.org/10.1142/9781800610750_0012

Chapter 12

Surrogate Modeling of High-Frequency Electronic Circuits

Xhesila Xhafa and Mustafa Berke Yelten

Abstract

This chapter focuses on applying surrogate models in a low noise amplifier (LNA), which is one of the main building blocks of modern communication circuits. In microelectronics, surrogate modeling has been shown to yield promising results to describe complex design surfaces. Using this feature, surrogate models can be established to extract specific design insights that enable the development of integrated circuits with higher performance and reliability. Examples of surrogate modeling in LNAs covered here explore the relationships between the voltage gain, noise figure, as well as input and output reflection coefficients. In particular, the variability of the voltage gain has been characterized through a surrogate model. Input parameters are composed of the gate, source, and drain inductors, along with their quality factors. Results confirm that highly accurate surrogate models for performance metrics such as voltage gain if an LNA can be built and employed in circuit design optimization.

Keywords: Low noise amplifier, voltage gain, noise figure, input reflection coefficient, output reflection coefficient, variability modeling, circuit optimization, quality factor, bondwires, inductance

12.1 Introduction

The rapid growth of microelectronics applications in our daily lives facilitated the growth of the economies and increased human living standards. Computers that fill a large room once upon a time can now be squeezed within a small chip. This unbelievable change was realized by transistor size scaling, which was projected by Gordon Moore in 1965. As microelectronics fabrication revolutionizes its capabilities, process engineers were able to develop smaller devices that operate at larger currents. The other key progress in microelectronics was the integration of transistors that increased both the system performance and complexity. Integrated transistors in the 1970s enabled the design of practical circuits that can achieve essential tasks in computing, health, and communications.

As transistor sizes continued to scale down, device transconductances went up, and transit frequencies have been elevated (Razavi, 2011). As a result, transistors can amplify at higher frequencies. Thus, integrated circuits could be introduced to implement various communications standards in hardware. Especially starting in the mid-1990s, radio frequency integrated circuits (RFICs) emerged as a promising field where several companies such as Qualcomm, Broadcom, Infineon, RFMD produced many high-performance RF circuit blocks and systems.

The term of high-frequency electronic circuits involves both RFICs, as well as microwave ICs and millimeter-wave ICs (MMICs). These circuits can be designed using complementary-metal-oxide-semiconductor (CMOS) field-effect-transistors, hetero-junction bipolar transistors (HBTs), or high-electron-mobility transistors (HEMTs). Although the latter two have a much larger transit frequency due to their current conduction mechanisms, CMOS-based systems can be integrated at a much less fabrication cost; hence, for mass-production, CMOS high-frequency integrated circuits have become the primary tools. With transistors having a channel length of 180 nm, CMOS RFICs can successfully capture the L- and S-band regimes (Çağlar and Yelten, 2019), whereas circuit made up by transistors of 65 nm channel length can reach to the V-band (Niknejad and Hashemi, 2008).

Fabrication of transistors at short channel lengths is laborious since the number of development steps increases dramatically. Although the sizes are shrinking, the fabrication equipment cannot keep up with the same tolerance; thus, variations of transistor characteristics are observed (Orshansky *et al.*, 2007). The level of these variations increases as the channel geometry of the transistors becomes smaller. More complex models describing the transistor behavior, both at DC and high frequencies, have been generated to deal with this challenge. Such models are based on a plethora of parameters, and during simulations, they lead to long convergence times. As the number of transistors grows, it can take days to hours for transient simulations to converge in short-channel technologies, particularly. Therefore, a need for a less sophisticated model has arisen that should be as faithful to the experimental results as possible.

Behavioral models can be a candidate to replace the complex and semi-empirical yet accurate transistor models (Dündar and Yelten, 2020). Such models try to capture, in essence, the mathematical relationship between the input and output by treating it as a black-box system. The number of input parameters can be decimated through techniques like principal component analysis or sensitivity analysis (Yelten, 2016). With less but more impactful input parameters, the black-box system behavior is regenerated through variable basis functions. Variables of these basis functions can be tuned for the best accuracy based on a given set of input–output pairs called training samples. The model success depends heavily on the number of parameters and the sensitivity of the output to them. Models based on insensitive parameters can yield unsatisfactory results, especially when validated by a set of test samples different than the training samples.

Another approach for behavioral modeling involves space mapping (Bandler *et al.*, 1994). This technique suggests maintaining a physics-based relationship between the input and output. For example, in a simple CMOS field-effect-transistor, the so-called "square law" governs the current–voltage relationship for long-channel devices that deviates at short channel technologies due to velocity saturation. However, engineers would like to use the "square law" since it yields a practical means between the transistor current and terminal voltages. Instead of modeling the actual transistor current–voltage relationship based on velocity saturation, space mapping suggests modifying the input (process parameters, terminal voltages) and output parameters (here, just the drain current) such that the "square law" holds. Variable basis functions can be employed that will be tuned for the best achievable accuracy.

One side note should be made about the variable basis functions. The optimization of the basis functions is, by itself, an important focus of research. Linear optimization would often be insufficient to accomplish the desired results since the design space, on which optimization will be performed, is generally nonlinear. Consequently, novel mathematical techniques for parameter-search are adopted. In recent studies, machine learning approaches such as artificial neural networks and support vector machine methods have been employed and shown to be promising (Yelten *et al.*, 2012a). Problem complexity can easily be scaled down in artificial neural networks where the number of neurons in its hidden layer can be commensurately increased.

Both behavioral modeling and space mapping can be grouped under the term surrogate modeling. As its name suggests, surrogate modeling aims to establish a representative model that will replace the complex, nonlinear relationship, which results in longer simulation times (Yelten *et al.*, 2012b). The surrogate model must provide fast, accurate results. One problem related to the model accuracy is overfitting. When the model is forced to follow the data at the training samples, the model may perform poorly with the test samples if the dimensionality is high, i.e., if the system is described by many input parameters.

Surrogate models have been utilized in the circuit design for various purposes. In one example, the drain current of a MOSFET has been described by a surrogate model in terms of terminal voltages, charge carrier mobility, temperature, channel doping, oxide thickness, channel length variation, intrinsic threshold voltage, series drain-source resistance, and the age of the transistor (Yelten *et al.*, 2011a). A total of 11 parameters has been employed in the model, reflecting the impact of process variations and the reliability degradation. A custom circuit simulator has been developed that utilizes these transistor surrogate models. It is shown that the mismatch of the transistors can lead to more severe performance drops when compared to process corner results. Furthermore, it is demonstrated that the aging degradation tolerance of circuit blocks can be ranked where the aging refers to the performance drops due to reliability deterioration based on long-term usage (Yelten *et al.*, 2011b).

This chapter describes the use of surrogate models in the analysis and optimization of high-frequency circuit blocks.

12.2 Surrogate Modeling as a Circuit Optimization Tool

In the circuit design, as the technology scales down, it becomes harder to express the circuit parameters in simple, conspicuous relations since the short-channel effects, process variations, and parasitics become more evident than in long-channel devices. Being so, simpler optimization methods that are purely based on data can be used to help the intuition of circuit designers. Through these methods, mathematical models that "describe" the behavior of the circuit components or of the circuit itself can be generated. Simulation-driven optimization methods are considered as fast-design exploration approaches and can be separated into categories such as macromodeling, intelligent algorithms, and metamodeling (surrogate modeling) (Garitselov et al., 2011).

Macromodeling is a broadly used method for optimization problems, yet since it is fully based on circuit simulations, these models can still be expensive. However, variations to conventional macromodeling such as sequential design space decomposition techniques (Ding and Vemuri, 2006) or structure-aware performance modeling (Sun et al., 2013), have been proven to ameliorate the modeling process. Moreover, intelligent algorithms such as swarm optimization (Fakhfakh et al., 2010; Thakker et al., 2009), artificial bee colony-inspired modeling (Bouyghf et al., 2017), and genetic algorithms (Zebulum et al., 2018) can be used in circuit optimizations, as well.

On the contrary, surrogate modeling is preferred as an alternative method due to its ability to simplify complicated systems while still being fast and accurate. Since the model is generated from the sampled data (or data obtained from mathematical functions (Gorissen et al., 2008)), any problem that requires fitting and optimization can be addressed via surrogate modeling. For instance, an application of surrogate modeling can be observed in the passive electromagnetics field. In Koziel (2017) and Koziel and Pietrenko-Dabrowska (2019), the authors demonstrate how the function of computationally expensive full-wave electromagnetic simulations can be replaced by surrogate modeling as a means of designing fast and reliable antenna structures.

Surrogate modeling can also be used in RF design at both circuit and device model levels. For instance, RF integrated inductors are among the most challenging devices to design since the non-idealities they introduce can considerably deteriorate the circuit performance. Generally, on-chip inductors are characterized via electromagnetic simulators, which are inefficient due to the time and memory they require. On-chip inductors have been modeled in the literature using both physical and surrogate models. In the physical modeling method, the inductor is commonly expressed as an entity composed of other circuit devices such as resistors and capacitors in conjunction with ideal inductors (π-model). These components act as variables that affect the overall inductor performance, i.e., the quality factor and the inductance, for which separate models can be established, as in Akso *et al.* (2018). Another method to tackle the problem of on-chip is inductors is to generate surrogate models using samples from the EM simulations to predict the behavior of the inductor in a constrained design space. According to Passos *et al.* (2015), the on-chip inductors used in a common RF block such as the LNA can be easily modeled using a surrogate modeling strategy for the optimization of the S-parameters. The latter approach has been shown to be more accurate compared to the physical model of on-chip inductors.

The accuracy of the constructed surrogate models highly depends on the amount of available data, the choice of sampling method, and the learning technique. Depending on the nature of the problem, different sampling and training methods can be more instrumental over each other. Surrogate modeling attempts to fit the given samples with high accuracy, such that it is able to correctly predict results for new, untested samples. Nevertheless, since there are a plethora of existing methods to fit sampled data, the choice becomes problem-specific. Therefore, former studies on the area of IC optimization using metamodels should be considered and evaluated *a priori*.

To demonstrate the need for surrogate modeling in the design of an RF circuit, we first have to understand the theory behind the circuit itself. For this reason, a tunable low noise amplifier (LNA) is used as an example.

The LNA is the first block in a conventional RF receiver; therefore, it should be meticulously designed. As wireless communication technology advances, the need for compact hardware devices arises. Amid the existence of numerous communication standards, new approaches in RF circuit design that are able to satisfy multiple standards simultaneously have been of interest to designers for the past decades (Tasic et al., 2006). In RF circuits, a plausible method to introduce the latter concept is through using tunable circuit components. The same idea has been applied to the LNA topology introduced in Fig. 12.1. The tunability is achieved through the use of two accumulation-mode voltage-controlled capacitors (varactors) in both input and output ports. The varactors enable the effective input and output capacitance values to change such that the respective impedance matching shifts in the frequency domain. This means that the operating frequency of the LNA can be tweaked within a certain bandwidth, thereby yielding significant savings in chip estate.

The LNA, as the name suggests, is an RF circuit block that aims to introduce minimal noise while amplifying the received signal. Here, the given topology is a conventional cascode structure with inductive degeneration, which operates within 2–3 GHz. Cascoding maintains high reverse isolation while also ensuring that the noise contribution of the common gate transistor is degraded by the output impedance of the common source transistor. Despite the latter noise reduction, the thermal noise contribution from the common source transistor

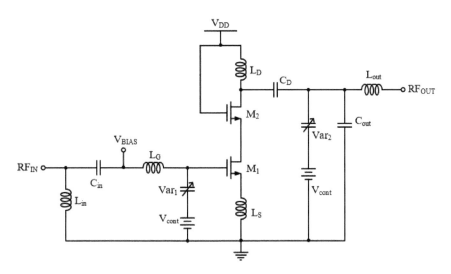

Fig. 12.1. The topology of a tunable LNA.

itself has to be mitigated. When the transistors operate in the saturation region, the transconductance (g_m) and the gate-to-source parasitic capacitance (C_{gs}) are the responsible parameters for the drain and gate current thermal noise (Tedja et al., 1994). Consequently, the minimum noise figure (NF$_{min}$) depends on the drain current density and the sizing of the common source transistor. NF$_{min}$ represents the noise value when the optimum source impedance is introduced; hence, the actual noise figure (NF) should be optimized such that it converges as close as possible to NF$_{min}$.

In the given LNA, two of the most dominant parameters on NF are the source and gate inductors, L_g and L_s. The values of these components should be optimized, such that NF falls at least below 3 dB. Conversely, since the topology is source-degenerated, L_s impacts the voltage gain as well, i.e., reduces the gain as its value increases. The voltage gain expression can be derived as

$$A_v = \frac{g_m R_L}{2\omega_0 R_s (C_{gs} + C_{var})}, \qquad (12.1)$$

where C_{var} is the capacitance introduced by the varactor and R_s is the real portion of the input impedance due to the source inductance, which can be expressed as

$$R_s = \frac{g_m L_s}{C_{gs}}. \qquad (12.2)$$

Moreover, aside from the gain and noise figure, an important specification of an LNA is the input power matching. The latter is realized when the input impedance Z_{in} is matched with the conventional source impedance of $Z_{in} = 50\,\Omega$. The input impedance given by

$$Z_{in} = j\omega L_g + j\omega L_s + \frac{1}{j\omega(C_{gs} + C_{var})} + \frac{g_m L_s}{(C_{gs} + C_{var})}, \qquad (12.3)$$

is a function of the source and gate inductors, as well. Thus, the optimization of their values is performed considering the noise figure, gain, and the input power matching, simultaneously.

As it can be seen in these equations, optimizing an LNA is not a straightforward task since there exist several distinct trade-offs between the design specifications. To quantify these trade-offs through conventional circuit simulations is an expensive and impractical endeavor. Hence, surrogate modeling (metamodeling) becomes an imperative tool for a circuit designer.

12.3 Variability Analysis of an LNA Using Surrogate Modeling

There are several steps that need to be followed in order to generate a surrogate model. Firstly, a set of data points are produced from the circuit simulator within the design space specified by the user. Then, a well-distributed initial sample set is selected for the training to begin. Different sampling techniques such as Monte Carlo sampling, Latin hypercube sampling (LHS), and Voronoi Tesselation have been described in Yelten et al. (2012b). The dimensionality of the samples is crucial when deciding the type of sampling methods since the model greatly depends on the effective coverage of the design space. In literature, sampling techniques are often compared since their choice can critically affect the performance of the model (Forrester et al., 2008). As concluded in Garitselov et al. (2011), uniform sampling techniques are preferred as they offer higher accuracy than the random sampling approaches. LHS is one of the methods that uniformly covers the design space while still ensuring some randomness when choosing samples from each Latin square.

However, the initial samples, more than often, are insufficient to satisfy the accuracy of the model. For this reason, the concept of sequential design methods is introduced. Sequential design is an iterative algorithm that chooses new samples based on the performance of the previously generated model. Given that there exists a variety of sequential design methods, it is important for the designer to attain a theoretical background about them. The choice of an appropriate method can lead to highly accurate surrogate models while keeping both the number of samples needed and the time required to meet the error requirements low.

In the study of Crombecq et al. (2009), a thorough comparison of sequential design methods has been conducted in the context of the optimization of a narrowband LNA. The authors introduce a novel sequential design method, LOLA-Voronoi, which is an amalgamation of exploitation-based and exploration-based sampling methods. This means that the algorithm is able to select data from highly nonlinear regions (exploitation) while exploring the design space as uniformly as possible (exploration).

Another frequently used sequential design sampling option is the model error-based method through which new samples are chosen from the largest error-producing design space locations. After producing an initial amount of samples, the fitting and modeling methodology should be specified. For this purpose, methods such as artificial neural networks (ANNs), Kriging functions, and radial basis functions can be employed (Koziel et al., 2011).

ANNs are commonly used to solve complex problems in various disciplines regarding optimization, classification, pattern recognition, and data fitting (Yelten et al., 2012a; Ngwar and Wight, 2015; Giordano et al., 2019). Artificial networks have been employed in circuit design, as well Gencer et al. (2020). However, ANNs are mostly implemented in software considering the extensive capabilities of modern computing. Since the circuit design involves the optimization of several parameters simultaneously, neural networks serve as a reliable method to generate surrogate models.

An alternative choice to ANNs is the Kriging interpolation method. Originally used in geostatistics, Kriging is a preferred method for uniformly-spaced samples as its predicted value is intimately based on the sample in the training set (McConaghy and Gielen, 2005). Due to the simplicity of its algorithm, Kriging is a fast fitting method. Nonetheless, depending on the complexity of the problem, the generation of a surrogate model may need a large number of training samples.

In general, the process of designing a surrogate model is accomplished when the model has reached the user-specified accuracy level. However, it can also be terminated if all of the available samples have been utilized or if the user has set a time limitation for the model generation. The accuracy of the constructed model is evaluated through an error function and an error measure. Error functions such as root-relative squared error (RRSE) given by

$$\text{RRSE} = \sqrt{\frac{\sum_{i=1}^{n}(y_i - \tilde{y}_i)^2}{\sum_{i=1}^{n}(y_i - \bar{y})^2}}, \quad (12.4)$$

and mean relative error (MRE),

$$\text{MRE} = \max \left| \frac{y_i - \tilde{y}_i}{y_i} \right|, \quad (12.5)$$

are commonly used in data fitting problems to determine how much the predicted outputs deviated from the actual outcomes of the samples. In these equations, y_i stands for the actual sample output, $\tilde{y}_i$

for the predicted output by the surrogate model, and $\bar{y}$ for the mean of y_i.

When it comes to the model validation, in literature, there are two preferred methods, namely the holdout and K-fold cross-validation. In the holdout case, the data is separated into the training and test data while in the cross-validation method, all the data is used in the training process, but the samples are separated into K batches, where $K - 1$ batches are used for training, and the left-over batch is used for validation. This process is then repeated until all the batches have been employed as both training and testing data sets.

To generate the surrogate models presented in this chapter, SUMO, a MATLAB-based toolbox (Gorissen et al., 2010), is used. This toolbox has been specifically designed to reduce the simulation time of complex problems and help the engineers better understand and analyze their products through surrogate modeling. The toolbox is an advanced software framework that comes with various plugins on model types, optimization algorithms, sequential sampling methods, design of experiments, and model evaluation algorithms. Through an XML file, the options above can be easily chosen while specifying the model accuracy percentage. The toolbox adaptively generates the models through a sequential sampling algorithm that allows the process to be smarter and faster. The toolbox includes a default sequential method, which will be utilized in the case studies below as the SUMO method. Its algorithm is a variation of the LOLA-Voronoi sequential sampling combined with the error-based sampling. After the best model has been simulated, the toolbox provides all the necessary information about the surrogate, as well as a detailed graphical user interface.

12.3.1 Case study I: Surrogate modeling of LNA performance parameters based on bondwire inductances

One of the manufacturing steps of an integrated circuit (IC) is the process of wire-bonding. When the IC is packaged, the bare die pads need to be electrically connected to the outside pins through golden or aluminum wires. Therefore, bondwires introduce a redundant inductance to the IC. Depending on the bondwire length and material, the expected performance can degrade or shift to unwanted design specifications in RF circuits. Consider the inductors in the given LNA topology. As seen in Fig. 12.1, L_g, L_s, L_d, L_{in}, and L_{out} are directly connected through bondwires to the bias voltage, supply voltage, and ground terminals. As a result, the effective inductance value at these terminals will increase, thereby altering the LNA performance. However, the noise performances of the LNA and its voltage gain depend on these inductors. Hence, if the effect of the bondwires is not properly taken into account, the projected LNA performance could be jeopardized. Moreover, the LNA is the first building block in a receiver system, so its performance degradation impacts the system as a whole.

The wire-bonding problem has been addressed in literature (Ishiko et al., 2006; Su and Chiang, 2003) as a separate issue from the design itself. Artificial neural networks and genetic algorithms have been used to predict the optimal wire positioning or the best choice of material such that the percentage of degradation is brought to a minimum. Here, it will be shown that the wire-bonding problem can be tackled before the circuit is manufactured, i.e., during the design process.

Estimation of the effective inductance contribution from the wire-bonding can be achieved through surrogate modeling. The first step in creating a surrogate model is to extract the data from a circuit simulator. Ranges of the input parameters are chosen by expecting a maximum of 25% increase in the inductance values. Also, the input inductor values have been normalized in the range of −1 to 1 for simplicity. Since the model needs to be evaluated with new data that are not part of the learning process, samples should be separated into the training and test sets. The ratio chosen for this case is 70/30, respectively.

The NF and the input return ratio S_{11} are used as metrics of noise performance evaluation, whereas S_{21} is employed to represent the gain. Both L_d and L_{out} impact the gain and the output matching of the LNA; hence, to investigate the trade-off between the two, S_{22} (the output return ratio) has been included as an output as well.

This case study has been separated into two subsections. We first model the S_{21} and S_{22} performances, while also conducting experiments on the type of model and the choice of the sequential design algorithm. Then, the surrogate model of NF and S_{11} will be generated and presented.

12.3.1.1 Surrogate modeling of S_{21} and S_{22}

Samples taken from circuit simulations have been used for model generation. To select the training samples, the LHS, which is a uniform sampling method, has been chosen. Different sequential sampling methods, such as LOLA-Voronoi, the error-based, and the LOLA-error hybrid, have been explored. As for the choice of the fitting method, Kriging, ANN, and the genetic ANN have been used to generate the surrogate models. In the literature, ANN and Kriging have been previously explored in the case of LNA optimization. Here, the genetic ANN has been chosen as an alternative method that combines both intelligent algorithms and ANNs. The genetic algorithm aids the pre-learning process of the ANNs by deciding the ANN structure (amount of layers, neurons, interconnections). The RRSE has been chosen as the error function, while the 5-fold cross-validation is chosen as the error measure method. The surrogate model has been terminated when the RRSE drops below 0.005. Then the model is tested with the test samples. The results for each fitting and sequential sampling method have been provided in Tables 12.1 and 12.2.

When analyzing the results, the first thing to note is the difference in the number of samples required to generate the models through the genetic ANN (applies for ANN as well) and the Kriging. As expected, ANN models can be generated using fewer samples. An interesting outcome can be seen in Fig. 12.2, in which the model is generated using only 24 samples with the genetic ANN method. The error performances, shown in Tables 12.1 and 12.2, indicate that

Table 12.1. The error performance of the S_{21} model.

	RRSE		
	Error-based	SUMO method	LOLA-Voronoi
Kriging	0.0352	0.0383	0.0356
Genetic ANN	0.0047	5.62E-04	9.16E-04
ANN	9.88E-04	0.0046	0.0014

Table 12.2. The error performance of the S_{22} model.

	RRSE		
	Error-based	SUMO method	LOLA-Voronoi
Kriging	0.0236	0.0207	0.0206
Genetic ANN	0.0352	0.0116	0.1073
ANN	0.0228	0.0178	0.0722

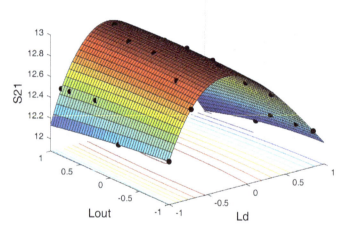

Fig. 12.2. The surrogate model of S_{21} using the genetic ANNs.

through ANNs, superior models can be generated in terms of accuracy. S_{21} performs better using genetic ANNs, while S_{22}, which has a more dynamic and complex design surface, pairs better with ANN when the sequential design setting based on the SUMO method is employed.

The developed models can be used toward finding the optimum values of L_d and L_{out}. They can also predict how the inductance change due to the wire-bonding will alter S_{21} and S_{22} performances. It can be seen in Figs. 12.2–12.5, both for the ANN and Kriging techniques, the best S_{21} and S_{22} performances are achieved with inductance values around the mean of their range. Finally, at the

Surrogate Modeling of High-Frequency Electronic Circuits 395

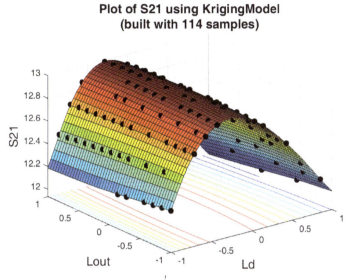

Fig. 12.3. The surrogate model of S_{21} using the Kriging technique.

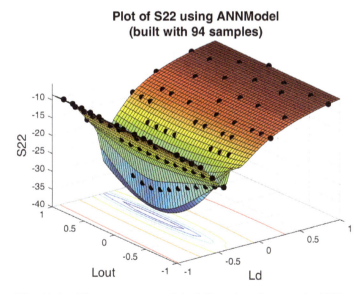

Fig. 12.4. The surrogate model of S_{22} using the genetic ANNs.

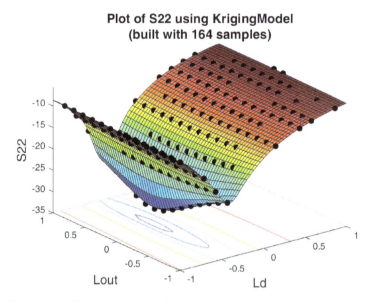

Fig. 12.5. The surrogate model of S_{22} using the Kriging technique.

corner values of L_d and L_{out}, S_{22} is around $-10\,\text{dB}$, which is near the boundary of an acceptable result regarding RF design principles.

12.3.1.2 Surrogate modeling of NF and S_{11}

An important application of surrogate models is using multiple input parameters simultaneously. In this section L_g, L_{in}, and L_s have been chosen as the inputs to generate two surrogate models for S_{11} and NF. The results have been shown in Figs. 12.6 and 12.7. The graphs demonstrate the relationship between L_g and L_{in} for the three values of L_s on terms of NF and S_{11}, provided in dB scale. It is observed that the source inductor L_s has a significant impact on the noise performance. Although the NF model is almost uniform regarding L_g and L_{in}, the performance degrades as L_g approaches to the normalized value of 1. Also, it is interesting to note the trade-off between the two models. The bottom slice of L_s on the NF model (best performance) corresponds to the upper slice (worst performance) in the S_{11} case. Based on the accuracy metrics shown in Table 12.3, it can be concluded that the outcomes of the model can yield invaluable intuition on the LNA performance to an RF circuit designer.

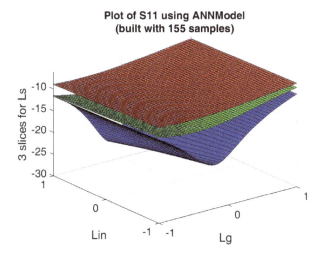

Fig. 12.6. The surrogate model of S_{11}.

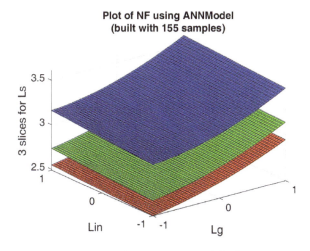

Fig. 12.7. The surrogate model of NF.

12.3.2 Case study II: Surrogate model of the LNA gain variability

The performance of on-chip inductors depends on the inductance they introduce, as well as the quality factor. A high value of the quality factor strives to the ideal performance. On-chip inductors are known for their parasitics; in particular, the quality degrades as the

Table 12.3. The error performance of the NF and S_{11} models.

	RRSE Training	RRSE Test	MRE Training	MRE Test
NF (@ **2 GHz**)	0.0011	0.0012	2.8E-04	3.4E-04
S_{11} (@ **2 GHz**)	0.0012	0.0038	0.001	0.0071

Table 12.4. Case study II: The error performance of the A_v model (Xhafa and Yelten, 2020).

	RRSE Training	RRSE Test	MRE Training	MRE Test
A_v (@ **2 GHz**)	4.37E-04	0.001	0.0083	0.025
A_v (@ **3 GHz**)	3.59E-04	0.0013	7.08E-04	0.0024

frequency increases due to the skin effect. Several studies have dealt with the optimization of on-chip inductors using surrogate modeling. Here, the dependency of the LNA gain-performance on the inductors and their quality factors is explored. Since the LNA is tunable and its performance changes with frequency, models have been generated for both corner frequencies of the range, 2 and 3 GHz.

Following the results of the surrogate models in the previous case study where NF and S_{11} were modeled with only three input parameters, the second case study aims to demonstrate a comprehensive variability analysis of the gain while incorporating six variables, L_g, L_s, and L_d, as well as their quality factors.

The procedure of the design is similar to the first case study. The main difference is the modeling approach chosen here as ANN since it yields superior results compared to other methods. The RRSE error tolerance is set as 0.005, and the sequential design algorithm is based on the SUMO toolbox method.

As shown in Table 12.4, overall, the model is successfully generated, and the RRSE is around the required error tolerance, except for the MSE of the 2 GHz case, where the error performance is worse than expected. The reason might be the fact that the case has been modeled to converge for the RRSE. The increased number of neurons and layers, coupled with a larger number of samples, serve as

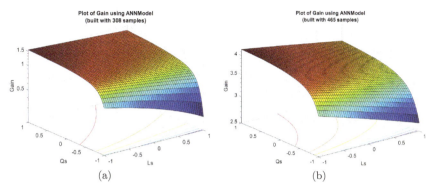

Fig. 12.8. A_v surrogate models when $Q = 2$ for $f = 2\,\text{GHz}$ (a) and $f = 3\,\text{GHz}$ (b) (Xhafa and Yelten, 2020).

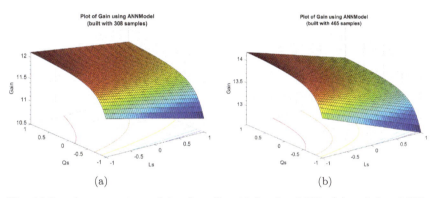

Fig. 12.9. A_v surrogate models when $Q = 10$ for $f = 2\,\text{GHz}$ (a) and $f = 3\,\text{GHz}$ (b) (Xhafa and Yelten, 2020).

proof for the augmented model complexity. For instance, building the model at 3 GHz, 465 samples were used, different and more than the 308 samples used for the model built at 2 GHz.

Six graphs have been plotted to showcase the importance of the quality factor (Q) in the RF inductor design (Xhafa and Yelten, 2020). The lowest, middle, and the highest unnormalized values of the quality factor are 2, 10, and 18, respectively. It should be emphasized that for the lowest Q values, the gain values would not be sufficient to fulfill the required LNA design specification, as seen in Fig. 12.8. Meanwhile, the average Q value results in Fig. 12.9 align with the desired gain value for an LNA circuit.

If better results are required, a quality factor of 18 should be sought, as in Fig. 12.10. It is important to notice that the error results

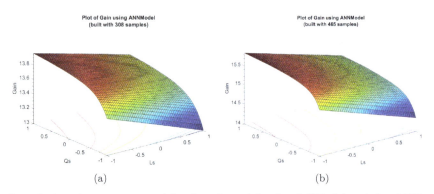

Fig. 12.10. A_v surrogate models when $Q = 18$ for $f = 2\,\text{GHz}$ (a) and $f = 3\,\text{GHz}$ (b) (Xhafa and Yelten, 2020).

have improved for the samples taken at 3 GHz, which was expected since the performance of the designed LNA is enhanced at 3 GHz. The generated models can be leveraged to provide an insight to the designer on the quality factor and the inductance values necessary to satisfy the LNA gain requirement.

12.4 Concluding Remarks

Surrogate modeling has become an important vehicle for understanding and representing complex engineering problems in the last few decades. Its impact increased even more, when its capabilities have been extended with machine learning techniques. As a result, the severity of the trade-off between accuracy and model complexity has decreased considerably. Surrogate modeling has also benefited from the advanced computational resources that are growing commensurate to technological developments. High-performance processors enable the complex models to be rendered within a few hours, in general. These factors add up to the widespread use of surrogate modeling in various technical challenges.

RFIC design involves many difficult simulation tasks given that the transistors at high frequencies have different secondary effects, which should be properly included in the circuit analysis. Even with the advanced processors, the circuit analysis can take hours to days to complete. Hence, surrogate modeling tools could be a significant help for those problems. As shown in this chapter, the insight provided by surrogate models can be instrumental in understanding the intricate relationships between design quantities. The knowledge extracted from the models can be leveraged to make critical design decisions specific to the circuits investigated.

Overall, surrogate modeling has a great potential in circuit design and optimization. In particular, high-frequency electronics may utilize the advantages of surrogate modeling to reduce the analysis complexity without giving up on the accuracy requirements. It can be concluded that in the near future, more high-frequency circuit applications will employ surrogate modeling tools, thereby reducing the computational cost and the time-to-market durations of products.

References

Akso, E., Soysal, İ.B., Yelten, M.B. (2018). Surrogate modeling and variability analysis of on-chip spiral inductors, *International Journal of Numerical Modelling: Electronic Networks, Devices and Fields*, vol. 31, no. 5, p. e2313.

Bandler, J.W., Biernacki, R.M., Shao Hua Chen, Grobelny, P.A., Hemmers, R.H. (1994). Space mapping technique for electromagnetic optimization, *IEEE Transactions on Microwave Theory and Techniques*, vol. 42, no. 12, pp. 2536–2544.

Bouyghf, H., Benhala, B., Raihani, A. (2017). Optimal design of RF CMOS circuits by means of an artificial bee colony technique, *Focus on Swarm Intelligence Research and Applications*, 221–246.

Çağlar, A., Yelten, M.B. (2019). Design of cryogenic LNAs for high linearity in space applications, *IEEE Transactions on Circuits and Systems I: Regular Papers*, vol. 66, no. 12, pp. 4619–4627.

Crombecq, K., De Tommasi, L., Gorissen, D., Dhaene, T. (2009). A novel sequential design strategy for global surrogate modeling, in *Proceedings of the 2009 winter simulation conference (WSC)*, pp. 731–742.

Ding, M., Vemuri, R. (2006). Efficient analog performance macromodeling via sequential design space decomposition, in *19th International Conference on VLSI Design held jointly with 5th International Conference on Embedded Systems Design (VLSID'06)*, p. 1–4.

Dündar, G., Yelten, M.B. (eds). (2020). *Modelling Methodologies in Analogue Integrated Circuit Design*, IET — Materials, Circuits and Devices. Stevenage, the United Kingdom.

Fakhfakh, M., Cooren, Y., Sallem, A., Loulou, M., Siarry, P. (2010). Analog circuit design optimization through the particle swarm optimization technique, *Analog Integrated Circuits and Signal Processing*, vol. 63, no. 1, pp. 71–82.

Forrester, A., Sobester, A., Keane, A. (2008). *Engineering Design via Surrogate Modelling: A Practical Guide*, John Wiley & Sons, West Sussex, the United Kingdom.

Garitselov, O., Mohanty, S.P., Kougianos, E. (2011). A comparative study of metamodels for fast and accurate simulation of nano-cmos circuits, *IEEE Transactions on Semiconductor Manufacturing*, vol. 25, no. 1, pp. 26–36.

Gencer, F.B., Xhafa, X., İnam, B.B., Yelten, M.B. (2020). Design and validation of an artificial neural network based on analog circuits, *Analog Integrated Circuits and Signal Processing*, pp. 1–9.

Giordano, M., Cristiano, G., Ishibashi, K., Ambrogio, S., Tsai, H., Burr, G.W., Narayanan, P. (2019). Analog-to-digital conversion with reconfigurable function mapping for neural networks activation function acceleration, *IEEE Journal on Emerging and Selected Topics in Circuits and Systems*, vol. 9, no. 2, pp. 367–376, doi:10.1109/JETCAS.2019.2911537.

Gorissen, D., De Tommasi, L., Hendrickx, W., Croon, J., Dhaene, T. (2008). Rf circuit block modeling via kriging surrogates, in *MIKON 2008- 17th International Conference on Microwaves, Radar and Wireless Communications*, pp. 1–4.

Gorissen, D., Couckuyt, I., Demeester, P., Dhaene, T., Crombecq, K. (2010). A surrogate modeling and adaptive sampling toolbox for computer based design, *Journal of Machine Learning Research*, vol. 11, pp. 2051–2055.

Ishiko, M., Usui, M., Ohuchi, T., Shirai, M. (2006). Design concept for wire-bonding reliability improvement by optimizing position in power devices, *Microelectronics Journal*, vol. 37, pp. 262–268, doi:10.1016/j.mejo.2005.09.015.

Koziel, S., Ciaurri, D.E., Leifsson, L. (2011). *Surrogate-Based Methods*, pp. 33–59, Springer, Berlin, Germany.

Koziel, S. (2017). Low-cost data-driven surrogate modeling of antenna structures by constrained sampling, *IEEE Antennas and Wireless Propagation Letters*, vol. 16, pp. 461–464.

Koziel, S., Pietrenko-Dabrowska, A. (2019). Performance-based nested surrogate modeling of antenna input characteristics, *IEEE Transactions on Antennas and Propagation*, vol. 67, no. 5, pp. 2904–2912.

McConaghy, T., Gielen, G. (2005). Analysis of simulation-driven numerical performance modeling techniques for application to analog circuit

optimization, in *2005 IEEE International Symposium on Circuits and Systems*, pp. 1298–1301.
Moore, G. (1965). Moore's law, *Electronics Magazine*, vol. 38, no. 8, p. 114.
Ngwar, M., Wight, J. (2015). A fully integrated analog neuron for dynamic multi-layer perceptron networks, in *2015 International Joint Conference on Neural Networks (IJCNN)*, pp. 1–8, doi:10.1109/IJCNN.2015.7280448.
Niknejad, A.M., Hashemi, H. (2008). *mm-Wave Silicon Technology: 60 GHz and Beyond*, Springer Science & Business Media, New York City, NY, USA.
Orshansky, M., Nassif, S., Boning, D. (2007). *Design for Manufacturability and Statistical Design: A Constructive Approach*, Springer Science & Business Media.
Passos, F., Kotti, M., González-Echevarría, R., Fino, M.H., Fakhfakh, M., Roca, E., Castro-López, R., Fernández, F.V. (2015). Physical vs. surrogate models of passive rf devices, in *2015 IEEE international symposium on circuits and systems (ISCAS)*, pp. 117–120.
Razavi, B. (2011). *RF Microelectronics (Prentice Hall Communications Engineering and Emerging Technologies Series)*, 2nd edn., USA, Prentice Hall Press.
Su, C.-T., Chiang, T.-L. (2003). Optimizing the IC wire bonding process using a neural networks/genetic algorithms approach, *Journal of Intelligent Manufacturing*, vol. 14, no. 2, pp. 229–238, doi: 10.1023/A:1022959631926, URL https://doi.org/10.1023/A:1022959 631926.
Sun, S., Li, X., Gu, C. (2013). Structure-aware high-dimensional performance modeling for analog and mixed-signal circuits, in *Proceedings of the IEEE 2013 Custom Integrated Circuits Conference*, pp. 1–4.
Tasic, A., Serdijn, W.A., Long, J.R. (2006). Adaptive multi-standard circuits and systems for wireless communications, *IEEE Circuits and Systems Magazine*, vol. 6, no. 1, pp. 29–37.
Tedja, S., Van der Spiegel, J., Williams, H.H. (1994). Analytical and experimental studies of thermal noise in mosfet's, *IEEE transactions on electron devices*, vol. 41, no. 11, pp. 2069–2075.
Thakker, R.A., Baghini, M.S., Patil, M.B. (2009). Low-power low-voltage analog circuit design using hierarchical particle swarm optimization, in *2009 22nd International Conference on VLSI Design*, pp. 427–432.
Xhafa, X., Yelten, M.B. (2020). Design of a tunable lna and its variability analysis through surrogate modeling, *International Journal of Numerical Modelling: Electronic Networks, Devices and Fields*, vol. 33, no. 6, p. e2724, doi:10.1002/jnm.2724.

Yelten, M.B., Franzon, P.D., Steer, M.B. (2011a). Surrogate-model-based analysis of analog circuits–part I: Variability analysis, *IEEE Transactions on Device and Materials Reliability*, vol. 11, no. 3, pp. 458–465.

Yelten, M.B., Franzon, P.D., Steer, M.B. (2011b). Surrogate-model-based analysis of analog circuits–part II: Reliability analysis, IEEE Transactions on Device and Materials Reliability, vol. 11, no. 3, pp. 466–473.

Yelten, M.B., Franzon, P.D., Steer, M.B. (2012a). Comparison of modeling techniques in circuit variability analysis, *International Journal of Numerical Modelling: Electronic Networks, Devices and Fields*, vol. 25, no. 3, pp. 288–302, doi:10.1002/jnm.836.

Yelten, M.B., Zhu, T., Koziel, S., Franzon, P.D., Steer, M.B. (2012b). Demystifying surrogate modeling for circuits and systems, *IEEE Circuits and Systems Magazine*, vol. 12, no. 1, pp. 45–63.

Yelten, M.B. (2016). A heuristic sensitivity analysis technique for high-dimensional systems, in *2016 IEEE International Conference on Electronics, Circuits and Systems (ICECS)*, pp. 181–184.

Zebulum, R.S., Pacheco, M.A., Vellasco, M.M.B. (2018). *Evolutionary electronics: Automatic Design of Electronic Circuits and Systems by Genetic Algorithms*, CRC Press, Boca Raton, FL, USA.

© 2022 World Scientific Publishing Europe Ltd.
https://doi.org/10.1142/9781800610750_0013

Chapter 13

Sensitivity Analysis and Optimal Design with PC-co-kriging

Leifur Leifsson and Jethro Nagawkar

Abstract

This chapter describes a recently developed multi-fidelity surrogate modeling method called polynomial chaos-based co-kriging (PC-co-kriging). PC-co-kriging is a multivariate extension of the PC-kriging method and its construction is similar to co-kriging. In particular, a PC-kriging model of a set of data points is created by using a polynomial chaos expansion (PCE) to model the global trend and, as in a regular kriging model, a Gaussian process to model the local deviations from the global trend. Two such PC-kriging models are created. One for the low-fidelity model data and another for the difference of the high- and low-fidelity model responses. The two models are combined using the same approach as in co-kriging. The PC-co-kriging modeling method is demonstrated on three different problems: response modeling of an analytical function, sensitivity analysis of ultrasonic testing measurement simulation model, and the optimal design under uncertainty of transonic airfoil shapes.

Keywords: Global sensitivity analysis, optimal design, surrogate modeling, multi-fidelity methods, Sobol' indices, PC-co-kriging

13.1 Introduction

Physics-based simulation models are important in engineering, design and uncertainty quantification. Examples of engineering areas where simulations play an important role include non-destructive testing (NDT) and aerodynamic shape optimization (ASO). NDT refers to the process of evaluating, inspecting and testing a mechanical part or a system without physically damaging it (Crawley, 2001). The outcome of the NDT measurements depend on uncertain parameters in the system, such as the probe angle and location as well as the location, size and orientation of defects. Some of these parameters have a higher effect on the output response. To determine to what degree each parameter affects the response, model-based sensitivity analysis can be performed (Lilburne and Tarantola, 2009; Castillos *et al.*, 2007). ASO involves the design of efficient aerodynamic shapes, such as airfoils, wings, and rotor blades, to be used in engineered systems, e.g., vehicles and turbomachinery (Leung and Zingg, 2012; Lyu *et al.*, 2015; Yu *et al.*, 2018).

Model-based NDT measurement simulations use finite element methods (Zeng *et al.*, 2009) and boundary element methods (Zhang and Gross, 2002). Unfortunately, for model-based sensitivity analysis, in order to propagate variability parameters to output responses, a large number of model evaluations may be needed, which can be impractical to evaluate if the physics-based simulations are time-consuming. Many deterministic ASO problems are readily solved using gradient-based search methods with adjoint sensitivity information (He *et al.*, 2019; Secco and Martins, 2019). ASO problems involving uncertainty, on the other hand, are more expensive than its deterministic counterparts owning to an increased number of model evaluations to allow for the propagation of input uncertainties to the model output responses (Shah *et al.*, 2015; Vuruskan and Hosder, 2019).

Surrogate models (Forrester *et al.*, 2008) can be used to alleviate the computational cost in time-consuming applications, such as sensitivity analysis and optimal design under uncertainty. Data-fit methods (Queipo *et al.*, 2005) and multi-fidelity methods (Peherstorfer *et al.*, 2018) are two classes of surrogate modeling methods. In data-fit methods, a response surface is fitted to high-fidelity model (a model that estimates the output with the necessary accuracy for the current task) evaluations. In multi-fidelity methods, low-fidelity model (models that estimate the output with a lower accuracy and cost than the high-fidelity one) evaluations can be used to enhance the prediction at a limited number of high-fidelity model evaluations.

This chapter describes the use of fast surrogate models for global sensitivity analysis with Sobol' indices (Sobol', 2001; Sobol' and Kucherekoand, 1993) and optimal design under uncertainty. In particular, the chapter is focused on the PC-co-kriging surrogate modeling method (Du and Leifsson, 2020). In PC-kriging (Schobi et al., 2015), the global polynomial trend function in the kriging model (Krige, 1951) is replaced with a polynomial chaos expansion (PCE) (Wiener, 1938). PCE captures the global trend well whereas the kriging model (Krige, 1951) can capture the local deviations. PC-co-kriging (Du and Leifsson, 2020) is a multivariate extension of the PC-kriging model (Schobi et al., 2015).

The remainder of this chapter is organized as follows. Section 13.2 describes the methods used to construct the PC-co-kriging surrogate models, as well as methods used for the sensitivity analysis Sobol' indices and optimization using the surrogate model. Section 13.3 describes the application of the PC-co-kriging surrogate model method to three different problems. The PC-co-kriging surrogate modeling method is compared to state-of-the-art surrogate modeling methods, namely, Kriging (Krige, 1951), PCE (Blatman, 2009), PC-kriging (Schobi et al., 2015) and co-kriging (Kennedy and O'Hagan, 2000). The chapter ends with conclusions and suggestions of future work.

13.2 Methods

In this section, the surrogate modeling methods are described. The section begins by outlining the workflow of the general surrogate modeling and analysis algorithm. This is followed by descriptions of the sampling plan, the approaches for constructing the surrogate models, and their validations. The use of the surrogate models for both sensitivity analysis as well as optimization are then described.

13.2.1 *Surrogate modeling and analysis workflow*

The surrogate-based modeling and analysis workflow is shown in Fig. 13.1. The input design/variability space first needs to be sampled, represented by $\mathbf{X}$ in Fig. 13.1, to generate the training data used to construct the surrogate model. Two separate sets of data ($\mathbf{x}_c$ and $\mathbf{x}_e$) are used to construct the low- and high-fidelity surrogate models by evaluating the responses ($\mathbf{y}_c$ and $\mathbf{y}_e$) from the low- and

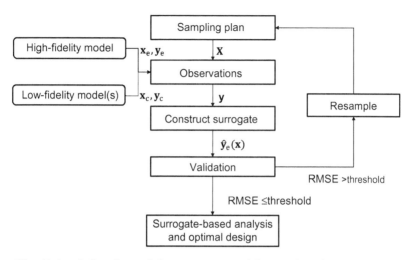

Fig. 13.1. A flowchart of the surrogate modeling and analysis workflow.

high-fidelity physics-based simulation models, respectively. These surrogate models are then combined to construct the multi-fidelity model. In Fig. 13.1, $\mathbf{y}$ represents the combined observations of $\mathbf{y}_c$ and $\mathbf{y}_e$, whereas $\hat{\mathbf{y}}_e$ is the surrogate model prediction. To validate the global accuracy of the surrogate model, a separate set of high-fidelity data, called the testing data, is used. The accuracy is measured using the root mean squared error (RMSE). This procedure may need to be repeated several times to satisfy the testing criteria, i.e., the threshold selected by the designer. Note that the amount of training data gradually increases due to resampling. On meeting the required global accuracy, this surrogate model can be used for either sensitivity analysis or to find the optimum design.

13.2.2 *Sampling plan*

The first step involved in constructing the surrogate model is sampling. Sampling is the process of selecting discrete samples in the parameter space (Forrester *et al.*, 2008). In this work, the training data is generated using the Latin hypercube sampling (LHS) (McKay *et al.*, 1979) method. The number of high-fidelity samples generated is typically lower than those of the low-fidelity samples due to the high computational cost required to evaluate the high-fidelity physics-based model. To generate the testing data, either the LHS method or the Monte Carlo sampling (MCS) (Shapiro, 2003) method is used.

13.2.3 Constructing the surrogate model

The PC-co-kriging model, developed by Du and Leifsson (2020), is generated by combing the PCE (Blatman, 2009) and co-kriging (Kennedy and O'Hagan, 2000) surrogate models. The modeling process starts with the kriging model (Forrester et al., 2008) which is given by

$$M^{\text{KR}}(\mathbf{X}) = \mathbf{g}^T(\mathbf{X})\gamma + \sigma^2 Z(\mathbf{X}), \tag{13.1}$$

where $\mathbf{X} \in \mathbb{R}^m$ is the vector of m-dimensional input variability parameters. The first term is the global trend term and the second term is the local deviation term. $\mathbf{g}^T(\mathbf{X})$ is a set of regression basis terms and γ is a constant. σ^2 is the constant variance of the Gaussian process $Z(\mathbf{X})$ with zero mean and unit variance. The Matern-5/2 (Gneiting et al., 2010) function is used as the correlation term in $Z(\mathbf{X})$, in this work.

The PC-kriging model (Schobi et al., 2015) uses the PCE surrogate model (Blatman, 2009) as the global trend term. The PCE surrogate model is given by

$$M^{\text{PCK}}(\mathbf{X}) = \boldsymbol{\alpha}^T \boldsymbol{\Phi}(\mathbf{X}) + \sigma^2 Z(\mathbf{X}). \tag{13.2}$$

Here, $\boldsymbol{\Phi}$ is a set of orthogonal polynomial basis function and $\boldsymbol{\alpha}$ is a constant. PC-kriging combines the advantage of the PCE model, which captures the global behavior of the computational model well and the kriging model, which captures the local variations well.

Co-kriging (Kennedy and O'Hagan, 2000) is a fusion-based model which combines information from multiple levels of fidelities to enhance prediction accuracy, especially in the presence of limited amount of high-fidelity data. In this work, information from two levels of fidelities are used. First, a kriging model ($M_{\text{LF}}^{\text{KR}}(\mathbf{X})$) is created using only low-fidelity data, followed by a second kriging model ($M_{\text{Diff}}^{\text{KR}}(\mathbf{X})$) on the difference of the high- and low-fidelity data. A generic form of the co-kriging model is (Kennedy and O'Hagan, 2000)

$$M^{\text{CoK}}(\mathbf{X}) = \rho M_{\text{LF}}^{\text{KR}}(\mathbf{X}) + M_{\text{Diff}}^{\text{KR}}(\mathbf{X}), \tag{13.3}$$

where ρ is a constant scaling factor.

The PC-co-kriging model (Du and Leifsson, 2020) is a multivariate version of the PC-kriging model and is constructed similar to the co-kriging model. The PC-co-kriging model is

$$M^{\text{PC-CoK}}(\mathbf{X}) = \rho M_{\text{LF}}^{\text{PCK}}(\mathbf{X}) + M_{\text{Diff}}^{\text{PCK}}(\mathbf{X}). \tag{13.4}$$

Here, $M_{\text{LF}}^{\text{PCK}}(\mathbf{X})$ and $M_{\text{Diff}}^{\text{PCK}}(\mathbf{X})$ are the PC-kriging model on the low-fidelity and the difference between the low and high-fidelity data, respectively.

13.2.4 *Validation*

The global accuracy of the surrogate model is measured in this work using the RMSE, which is given by

$$\text{RMSE} = \sqrt{\sum_{i=1}^{n_t} (\hat{y}_{\text{testing}}^{(i)} - y_{\text{testing}}^{(i)})^2 / n_t}, \tag{13.5}$$

where n_t is the total number of testing data, $\hat{y}_{\text{testing}}^{(i)}$ and $y_{\text{testing}}^{(i)}$ are the surrogate model estimation and high-fidelity model observation of the ith testing point, respectively. In this work, $1\%\sigma_{\text{testing}}$ (standard deviation of testing points) is considered as an acceptable global accuracy.

13.2.5 *Surrogate-based sensitivity analysis*

For the NDE case, the surrogate model is used to perform sensitivity analysis based on Sobol' indices (Sobol' and Kucherekoand, 1993). Sensitivity analysis is used to determine the effect of each variability parameter on the model response.

Consider a black box model given by

$$M(\mathbf{X}) = f(\mathbf{X}), \tag{13.6}$$

where $\mathbf{X}$ is a $m \times 1$ random variable input vector. Decomposing this equation gives (Sobol', 2001)

$$M(\mathbf{X}) = f_0 + \sum_{i=1}^{m} f_i(X_i) + \sum_{i<j}^{m} f_{i,j}(X_i, X_j)$$
$$+ \cdots + f_{1,2,\ldots,m}(X_1, X_2, \ldots, X_m), \tag{13.7}$$

where f_0 is a constant, and f_i is a function of X_i. These terms are orthogonal and can then be decomposed in terms of conditional expected values given by (Sobol', 2001)

$$f_0 = \mathbb{E}(M(\mathbf{X})), \tag{13.8}$$

$$f_i(X_i) = \mathbb{E}(M(\mathbf{X})|X_i) - f_0, \tag{13.9}$$

$$f_{i,j}(X_i, X_j) = \mathbb{E}(M|X_i, X_j) - f_0 - f_i(X_i) - f_j(X_j), \tag{13.10}$$

and so on. The variance of (13.7) is then (Sobol', 2001)

$$\mathbb{V}\mathrm{ar}(M(\mathbf{X})) = \sum_{i=1}^{m} V_i + \sum_{i<j}^{m} V_{i,j} + \cdots + V_{1,2,\ldots,m}, \tag{13.11}$$

where

$$V_i = \mathbb{V}\mathrm{ar}_{X_i}(\mathbb{E}_{\mathbf{X}_{\sim i}}(M(\mathbf{X})|X_i)), \tag{13.12}$$

$$V_{i,j} = \mathbb{V}\mathrm{ar}_{X_{i,j}}(\mathbb{E}_{\mathbf{X}_{\sim i,j}}(M(\mathbf{X})|X_i, X_j)) - V_i - V_j, \tag{13.13}$$

and so on, where the set of all variables except X_i is denoted by $\mathbf{X}_{\sim i}$.

The first-order Sobol' indices are given by (Sobol' and Kucherekoand, 1993)

$$S_i = \frac{V_i}{\mathbb{V}\mathrm{ar}(M(\mathbf{X}))}, \tag{13.14}$$

and the total-effect Sobol' indices are given by (Sobol' and Kucherekoand, 1993)

$$S_{T_i} = 1 - \frac{\mathbb{V}\mathrm{ar}_{\mathbf{X}_{\sim i}}(\mathbb{E}_{X_i}(M(\mathbf{X})|\mathbf{X}_{\sim i}))}{\mathbb{V}\mathrm{ar}(M(\mathbf{X}))}. \tag{13.15}$$

13.2.6 *Surrogate-based optimal design*

To find the optimal design based on the surrogate model, the sequential least squares programming (SLSQP) (Kraft, 1988) gradient-based optimizer, available in SciPy (Virtanen *et al.*, 2020), is used in this work. Forty LHS starting points are selected randomly in the space to start the SLSQP algorithm search.

13.3 Application Examples

The results of applying the PC-co-kriging surrogate modeling method to three different problems are presented in this section. The first problem involves the modeling of is an eight-parameter analytical function, called the borehole function. The second problem is an ultrasonic testing (UT) NDT benchmark case. The third problem is the optimal design under uncertainty of transonic airfoil shapes. The performance of the PC-co-kriging modeling algorithm is compared to kriging, PCE, PC-kriging and co-kriging.

13.3.1 *Surrogate modeling of the borehole function*

The high-fidelity borehole function (Harper and Gupta, 1983) is an eight-parameters problem used to model flow of water through a borehole and is given by

$$f_{\text{HF}}(\mathbf{x}) = \frac{2\pi T_u (H_u - H_l)}{\ln(r/r_w)\left(1 + \frac{2LT_u}{\ln(r/r_w)r_w^2 K_w} + \frac{T_u}{T_l}\right)}. \quad (13.16)$$

Each of the parameters and its distribution are described in Table 13.1.

The low-fidelity borehole model, developed by (Xiong et al., 2013), is given by

$$f_{\text{LF}}(\mathbf{x}) = \frac{5T_u (H_u - H_l)}{\ln(r/r_w)\left(1.5 + \frac{2LT_u}{\ln(r/r_w)r_w^2 K_w} + \frac{T_u}{T_l}\right)}. \quad (13.17)$$

Table 13.1. Variability parameters for the borehole function (Harper and Gupta, 1983).

Variability parameters	Distribution
Radius of borehole, $r_w(m)$	$N(0.1, 0.0161812^2)$
Radius of influence, $r(m)$	$LogN(7.71, 1.0056^2)$
Transmissivity of upper aquifer, $T_u(m^2/yr)$	$U(63070, 115600)$
Potentiometric head of upper aquifer, $H_u(m)$	$U(990, 1110)$
Transmissivity of lower aquifer, $T_l(m^2/yr)$	$U(63.1, 116)$
Potentiometric head of lower aquifer, $H_l(m)$	$U(700, 820)$
Length of borehole, $L(m)$	$U(1120, 1680)$
Hydraulic conductivity of borehole, $K_w(m/yr)$	$U(9855, 12045)$

Table 13.2. Borehole function surrogate modeling cost.

Surrogate model	HF sample cost
Kriging	300
PCE	100
PC-kriging	100
Co-kriging	200[a]
PC-co-kriging	3[a]

Note: [a]Plus 170 LF training points.

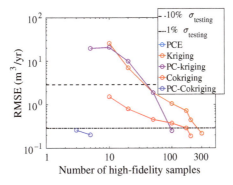

Fig. 13.2. Borehole function surrogate model construction.

Table 13.2 shows the number of high-fidelity samples required by each surrogate model to reach the global accuracy of $1\%\sigma_{testing}$. This accuracy is measure using 1,000 MCS generated testing points. Note that the multi-fidelity models use an additional 170 low-fidelity sample points. PC-co-kriging outperforms the other models requiring only 3 high-fidelity data points. PCE and PC-kriging follow the exact same trend (Fig. 13.2) and require 100 high-fidelity points to meet the 1% threshold, while co-kriging and kriging require 200 and 300, respectively. This case assumes that sampling the low-fidelity model is computationally efficient and its cost is negligible.

13.3.2 Model-based sensitivity analysis of ultrasonic testing

In this problem, the spherically-void-defect under planar transducer UT benchmark case developed by the World Federal Non-destructive

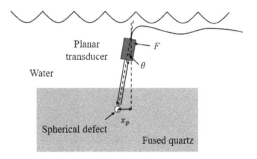

Fig. 13.3. Setup of the ultrasonic testing case.

Table 13.3. Variability parameters for the ultrasonic test case.

Variability parameters	Distributions
θ (deg)	$N(0, 0.5^2)$
x_p (mm)	$U(0, 1)$
F	$U(13, 15)$

Evaluation Center (Gurrala et al., 2017) is used. The five surrogate models are compared in terms of cost required to reach the global accuracy of $1\%\sigma_{\text{testing}}$. Sensitivity analysis using each of the surrogate models are also performed and the results are compared.

The setup for the UT benchmark case is shown in Fig. 13.3. The three variability parameters, namely, the probe angle (θ), the x location of the probe (x_p) and the F-number (F) along with their corresponding distributions are shown in Table 13.3. For this study, the Thompson–Grey analytical model (Thompson and Gray, 1983) is used as the high-fidelity model, while the Kirchhoff approximation (Darmon et al., 1970) is used as the low-fidelity one. The center frequency of the transducers are set to 5 MHz, while the density, the longitudinal and the shear wave speeds of the fused quartz block with spherical pore are $2,000\,\text{kg/m}^3$, $5,969.4\,\text{m/s}$ and $3,774.1\,\text{m/s}$, respectively.

Figure 13.4 shows the variation of the RMSE for all the surrogate models with increasing number of high-fidelity training points for the defect of size 0.5 mm. PC-co-kriging outperforms all the other surrogate models and requires 20 high-fidelity samples to reach the

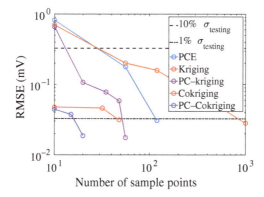

Fig. 13.4. Ultrasonic testing case surrogate model construction.

Table 13.4. Ultrasonic testing case surrogate modeling cost.

Surrogate model	HF sample cost
Kriging	1000
PCE	120
PC-kriging	56
Co-kriging	48[a]
PC-co-kriging	20[a]

Note: [a]Plus 1,000 LF training points.

1% threshold. The computational cost of all the models are shown in Table 13.4. 1,000 additional low-fidelity samples are used to construct the multi-fidelity models. For this case it is assumed the cost of observing the low-fidelity samples is negligible.

To verify the accuracy of the surrogate models for different defect sizes, the normalized root mean squared error (NRMSE) is calculated. The NRMSE is given by

$$\text{NRMSE} = \text{RMSE}/(\max(\mathbf{y}_{\text{testing}}) - \min(\mathbf{y}_{\text{testing}})). \quad (13.18)$$

Here, the maximum and minimum high-fidelity values from the testing data are $\max(\mathbf{y}_{\text{testing}})$ and $\min(\mathbf{y}_{\text{testing}})$, respectively.

Figure 13.5 shows the variation of the NRMSE with increasing defect size. The NRMSE is nearly constant with respect to defect size and within $1\%\sigma_{\text{testing}}$. Similar to the previous case, $1,000$ MCS

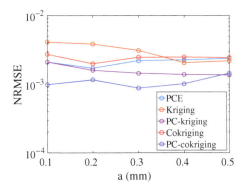

Fig. 13.5. Ultrasonic testing case surrogate model normalized root mean squared error (NRMSE) vs. the defect size.

generated testing points per defect size were used to measure the accuracy of the surrogate model.

Figures 13.6 and 13.7 show the first and total order Sobol' indices for the different surrogate models, respectively. The results from the surrogate models are compared the those obtaining from directly sampling the physics-based model. In all the models except PCE, 75,000 MCS were used to perform sensitivity analysis. In PCE, the coefficients are used to calculate these indices. The values of these indices match well for all the surrogate models. Surrogate model-based sensitivity analysis is highly efficient for this case as instead of evaluating the model response 75,000 times to get the Sobol's indices, the surrogate models output these responses at little to no cost. The F-number has negligible effect on the model response. This method can be used as a precursor to experimental measurements, where the number experiments can be reduced by keeping the value of the F-number constant.

13.3.3 *Optimal design of transonic airfoil shapes*

The final case is the optimal design of the RAE 2822 airfoil under uncertainty in the Mach number. This case is a modified version of a benchmark case developed by the AIAA Aerodynamic Optimization Design Discussion Group (ADODG). The five surrogate models are constructed and used to find the optimal design. The results from this optimal design under uncertainty are compared to the deterministic

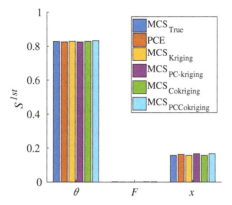

Fig. 13.6. First-order Sobol' indices for the ultrasonic testing case.

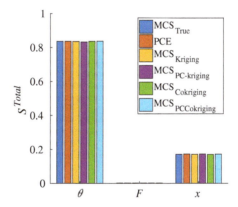

Fig. 13.7. Total-order Sobol' indices for the ultrasonic testing case.

optimal design (Nagawkar et al., 2020). Note that the type of parameterization (B-spline) as well as the computational fluid dynamics (CFD) setup and validation is the same those by Nagawkar et al. (2020). The CFD simulations are performed using the Stanford University Unstructured (Economon et al., 2015) and the mesh is generated using pyHyp[1] (He et al., 2019). For this study, however, the bounds of the design variables are increased from $(1 \pm 15\%)\mathbf{x}_0$ to $(1 \pm 25\%)\mathbf{x}_0$. $\mathbf{x}_0$ is the baseline design variable values.

---

[1]https://github.com/mdolab/pyhyp.

The objective is to minimize the drag coefficient (C_d) of a RAE 2822 airfoil in viscous flow with the freestream Mach number (M_∞) varying uniformly in the range 0.725 to 0.743, subject to a fixed lift coefficient (C_l) of 0.824 as well as pitching moment coefficient (C_m) and cross-sectional area constraints (A).

The optimization problem is formulated as

$$\min_{\mathbf{l} \leq \mathbf{x} \leq \mathbf{u}} (\mu(C_d) + \omega \cdot \sigma(C_d)), \tag{13.19}$$

subject to the equality constraint

$$C_l = 0.824, \tag{13.20}$$

and inequality constraints

$$\mu(C_m) \geq -0.092, \tag{13.21}$$

and

$$A \geq A_{\text{baseline}}. \tag{13.22}$$

$\mathbf{x}$ is the design variable vector, while $\mathbf{l}$ and $\mathbf{u}$ are the lower and upper bounds, respectively, of each design variable. μ and σ refer to the mean and standard deviation. ω is the weighting factor and is set to one for this case. A is the cross-sectional area of the airfoil non-dimensionalized with the square of the chord length (c). A_{baseline} is the baseline area of the airfoil with a value of $0.07787c^2$.

For this case, two different surrogate models were constructed for the objective function and the pitching moment constraint function. The variation of the RMSE with number of training points used to construct the surrogate models are shown in Figs. 13.8 and 13.9, respectively. The multi-fidelity model used an additional 1,604 low-fidelity training points for this case. 115 high-fidelity testing points were used to measure the RMSE. The multi-fidelity model shows significant higher accuracy for the drag coefficient surrogate models in presence of low amount of high-fidelity data as shown in Fig. 13.8. This is not the case for the pitching moment coefficient surrogate model (Fig. 13.9). For both the cases, increasing the number of high-fidelity training points, decreases the difference in accuracy between the multi-fidelity model and the data-fit surrogate models.

Figure 13.10 shows the variation of the drag coefficient with Mach number. The performance of the optimal design under uncertainty

Sensitivity Analysis and Optimal Design with PC-co-kriging 419

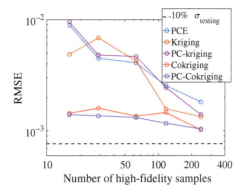

Fig. 13.8. Drag coefficient surrogate model construction.

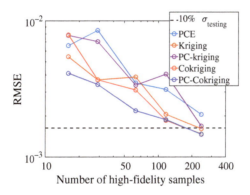

Fig. 13.9. Pitching moment surrogate model construction.

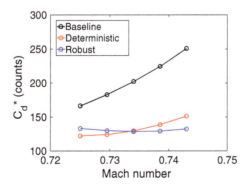

Fig. 13.10. Variation of the drag coefficient with respect to the Mach number.

Table 13.5. Performance of the baseline and optimized airfoil shapes.

Case	$\mu(C_d)$	$\sigma(C_d)$	$\Delta\mu(C_d)$	$\Delta\sigma(C_d)$	$\mu(C_m)$
Baseline	205.3	33.4	—	—	−0.099
Deterministic	133.0	12.1	72	21	−0.092
Robust	130.6	1.9	75	31	−0.090

Note: C_d is given in drag counts (one drag count is defined to be $1E-4$).

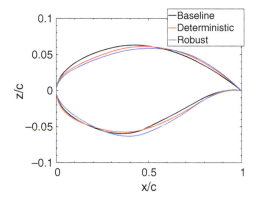

Fig. 13.11. Comparison of airfoil shapes.

shape is compared with the shapes for the deterministic optimal design as well as the baseline. Significant reduction in both the mean and standard deviation of the drag coefficient with respect to the Mach number is noticed (Table 13.5). For the optimal design under uncertainty the standard deviation of the drag coefficient is around 10 drag counts lower than the deterministic optimal design. This shows that optimal design under uncertainty results in an airfoil shape that is less sensitive to the Mach number.

The differences in airfoil shapes between the optimized and baseline shapes is shown in Fig. 13.11. Both the optimized shapes have a lower curvature on the suction side of the airfoil from the leading edge to around mid-chord. This shape reduces the acceleration of the flow over the suction side, resulting in a lower local Mach number, which in turn reduces the shock strength, thereby reducing the drag.

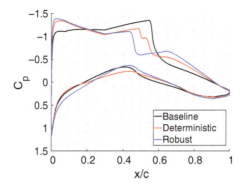

Fig. 13.12. Pressure coefficient distributions at a Mach number of 0.734.

Small differences in the shape has significant effect on the drag as well as pressure coefficient on the airfoil as seen in Fig. 13.12. The optimized shapes for both the cases have not eliminated the shock completely, but has reduced its strength significantly.

13.4 Conclusion

In this chapter, the polynomial chaos-based co-kriging (PC-co-kriging) surrogate modeling method has been described and applied to three different numerical problems. The first problem involved an eight-parameter nonlinear analytical function, another involved the sensitivity analysis of physics-based ultrasonic testing measurement simulations of a non-destructive testing system, and the last one the optimal design under uncertainty of physics-based fluid flow simulations past airfoil shapes at transonic speeds. All the problems involve uncertain parameters.

The results show that PC-co-kriging outperforms state-of-the-art surrogate modeling methods such as Kriging, polynomial chaos expansions (PCE), PC-kriging, and co-kriging, for the cases considered. The two main properties of PC-co-kriging allowing for efficient performance in dealing with cases involving uncertain parameters are the use of PCE to capture the global trend of the model response and the multivariate nature of co-kriging to utilize cheap data from low-fidelity model evaluations to reduce the need for data from the expensive high-fidelity model evaluations.

Future research will include the characterization of the PC-co-kriging method as the number of deterministic and non-deterministic problem parameters increase. Of particular interest is to understand how the convergence rate of the figures of merit change with the number of data points and when to terminate the outer-loop application.

Acknowledgments

This material is based upon work supported by the National Science Foundation under Grant Number (1846862). Any opinions, findings, and conclusions or recommendations expressed in this material are those of the authors and do not necessarily reflect the views of the National Science Foundation.

References

Blatman, G. (2009). *Adaptive Sparse Polynomial Chaos Expansion for Uncertainty Propagation and Sensitivity Analysis*, PhD Thesis, Blaise Pascal University, France.

Castillos, E., Conejo, A., Minguez, R., and Castillos, C. (2007). A closed formula for local sensitivity analysis in mathematical programming, *Engineering Optimization*, vol. 38, pp. 93–112.

Crawley, P. (2001). Non-destructive testing — current capabilities and future directions, *Journal of Material: Design and Applications*, vol. 215, no. 4, pp. 213–223.

Darmon, M., Leymarie, N., Chatillon, S., and Mahaut, S. (1970). *Modelling of Scattering of Ultrasounds by Flaws for NDT*, vol. 128, pp. 61–71, doi:10.1007/978-3-540-89105-5_6.

Du, X. and Leifsson, L. (2020). Multifidelity modeling by polynomial chaos-based cokriging to enable efficient model-based reliability analysis of NDT systems, *Journal of Nondestructive Evaluation*, vol. 39, no. 13, pp. 1–15. https://doi.org/10.1007/s10921-020-0656-8

Economon, T. D., Palacios, F., Copeland, S. R., Lukaczyk, T. W., and Alonso, J. J. (2015). SU2: An open-source suite for multiphysics simulation and design, *AIAA Journal*, vol. 54, no. 3, pp. 828–846.

Forrester, A. I. J., Sobester, A., and Keane, A. J. (2008). *Engineering Design via Surrogate Modelling: A Practical Guide*, John Wiley and Sons, Ltd, United Kingdom.

Gneiting, T., Kleiber, W., and Schlather, M. (2010). Matérn cross-covariance functions for multivariate random fields, *Journal of the American Statistical Association*, vol. 105, pp. 1167–1177.

Gurrala, P., Chen, K., Song, J., and Roberts, R. (2017). Full wave modeling of ultrasonic NDE benchmark problems using Nystrom method, *Review of Progress in Quantitative Nondestructive Evaluation*, vol. 36, no. 1, pp. 1–8.

Harper, W. V. and Gupta, K. S. (1983). Sensitivity/uncertainty analysis of a borehole scenario comparing Latin Hypercube Sampling and deterministic sensitivity approaches, *Technical Report*, Office of Nuclear Waste Isolation, Columbus, OH.

He, X., Li, J., Mader, C. A., Yildirim, A., and Martins, J. R. R. A. (2019). Robust aerodynamic shape optimization — from a circle to an airfoil, *Aerospace Science and Technology*, vol. 87, pp. 48–61.

Kennedy, M. C. and O'Hagan, A. (2000). Predicting the output from a complex computer code when fast approximations are available, *Biometrika Trust*, vol. 87, no. 1, pp. 1–13.

Kraft, D. (1988). A software package for sequential quadratic programming, *Technical Report DFVLR-FB 88-28, Institut für Dynamik der Flugsysteme, Oberpfaffenhofen*.

Krige, D. G. (1951). Statistical approach to some basic mine valuation problems on the witwatersrand, *Journal of the Chemical, Metallurgical and Mining Engineering Society of South Africa*, vol. 52, no. 6, pp. 119–139.

Leung, T. M. and Zingg, D. W. (2012). Aerodynamic shape optimization of wings using a parallel newton-krylov approach, *AIAA Journal*, vol. 50, no. 3, pp. 540–550.

Lilburne, L. and Tarantola, S. (2009). Sensitivity analysis of spatial models, *International Journal of Geographical Information Science*, vol. 23, pp. 151–168.

Lyu, Z., Kenway, G. K., and Martins, J. R. R. A. (2015). Aerodynamic shape optimization investigations of the common research model wing benchmark, *AIAA Journal*, vol. 53, pp. 968–985.

McKay, M. D., Beckman, R. J., and Conover, W. J. (1979). A comparison of three methods for selecting values of input variables in the analysis of output from a computer code, *Technometrics*, vol. 21, no. 2, pp. 239–245.

Nagawkar, J., Leifsson, L., and Du, X. (2020). Applications of polynomial chaos-based cokriging to aerodynamic design optimization benchmark problems, *AIAA Scitech 2020 Forum, 6-10 January, Orlando Florida*.

Peherstorfer, B., Wilcox, K., and Gunzburger, M. (2018). Survey of multifidelity methods in uncertainty propagation, inference, and optimization, *Society for Industrial and Applied Mathematics*, vol. 60, no. 3, pp. 550–591.

Queipo, N. V., Haftka, R. T., Shyy, W., Goel, T., Vaidyanathan, R., and Tucker, P. K. (2005). Surrogate-based analysis and optimization, *Progress in Aerospace Sciences*, vol. 21, no. 1, pp. 1–28.

Schobi, R., Sudret, B., and Wairt, J. (2015). Polynomial-chaos-based kriging, *International Journal of Uncertainty Quantification*, vol. 5, pp. 193–206.

Secco, N. R. and Martins, J. R. R. A. (2019). RANS-based aerodynamic shape optimization of a strut-braced wing with oversetmeshes, *Journal of Aircraft*, vol. 56, pp. 217–227.

Shah, H., Hosder, S., Koziel, S., Tesfahunegn, Y. A., and Leifsson, L. (2015). Multi-fidelity robust aerodynamic design optimization under mixed uncertainty, *Aerospace Science and Technology*, vol. 45, pp. 17–29.

Shapiro, A. (2003). Monte carlo sampling methods, *Handbooks in Operations Research and Management Science*, vol. 10, pp. 353–425.

Sobol', I. (2001). Global sensitivity indices for nonlinear mathematical models and their monte carlo estimates, *Mathematics and Computers in Simulation*, vol. 55, pp. 271–280.

Sobol', I. and Kucherekoand, S. (1993). Sensitivity estimates for nonlinear mathematical models, *Mathematical Modelling and Computational Experiments*, vol. 1, pp. 407–414.

Thompson, R. B. and Gray, T. A. (1983). Analytic Diffraction Corrections to Ultrasonic Scattering Measurements, *Library of Congress Cataloging in Publication Data*, Springer, vol. 2A.

Virtanen, P., Gommers, R., Oliphant, T. E., Haberland, M., Reddy, T., Cournapeau, D., Burovski, E., Peterson, P., Weckesser, W., and *et al.* (2020). Scipy 1.0: fundamental algorithms for scientific computing in python, *Nature Methods*, vol. 17, no. 3, pp. 261–272, doi:10.1038/s41592-019-0686-2.

Vuruskan, A. and Hosder, S. (2019). Impact of turbulence models and shape parameterization on robust aerodynamic shape optimization, *Journal of Aircraft*, vol. 56, no. 3, pp. 1099–1115, doi:10.2514/1.C035039.

Wiener, N. (1938). The Homogeneous Chaos, *American Journal of Mathematics*, vol. 60, no. 4, pp. 897–936, doi:10.2307/2371268.

Xiong, S., Qian, P. Z., and Wu, J. C. (2013). Sequential design and analysis of high-accuracy and low-accuracy computer codes, *Technometrics*, vol. 55, no. 1, pp. 37–46.

Yu, Y., Lyu, Z., Xu, Z., and Martins, J. R. R. A. (2018). On the influence of optimization algorithm and initial design on wing aerodynamic shape optimization, *Aerospace Science and Technology*, vol. 75, pp. 183–199.

Zeng, Z., Udpa, L., and Udpa, S. S. (2009). Finite-element model for simulation of ferrite-core eddy-current probe, *IEEE Transaction on Magnetics*, vol. 46, pp. 905–909.

Zhang, C. and Gross, D. (2002). A 2D hyper singular time-domain traction BEM for transient elastodynamic crack analysis, *Wave Motion*, vol. 35, pp. 17–40.

Index

A

absolute error, 25–26
accuracy, 58, 69, 362–363, 365–366, 368, 370–373, 378
accurate, 362, 364, 366
active learning, 7
adaptive hybrid sampling, 253
adaptive meshing, 43
adaptive response scaling (ARS), 70
adaptive sampling, 7, 245–246, 327, 337, 340, 346–347, 350, 352, 357, 363, 365–368, 371, 373, 378
adaptive sampling and modeling, 361, 372
adaptive sampling region, 265
adaptive sampling region updating strategy, 247, 257–258
adaptively adjusted design specifications (AADS), 70
adjoint neuro-TF model, 98, 100
adjoint sensitivities, 215
adjoint sensitivity analysis, 114
aerodynamic shape optimization, 405
aerospace engineering, 339
aggressive SM (ASM), 196, 199
aleatory, 294
algorithm convergence, 70
AMG algorithm, 83
analog/RF circuit design, 330, 338, 357
analog/RF design, 326, 339, 357
analog/RF IC design, 325
analog/RF ICs design, 338
analog/RF simulation-based optimization, 337
analog/RF system design, 357
analytical formulas, 58
anisotropy of substrates, 44
ANN, 393–394, 398
ANN models, 82
antenna design automation, 123, 125
antenna input characteristics, 65, 228
antenna reflection, 71
antenna resonance, 65
antenna response, 66, 68
antenna systems, 213
antennas, 2
approximation capability, 3
approximation error, 13, 23, 25
approximation layer, 58
approximation model management optimization (AMMO), 70
approximation models, 2
approximation surrogates, 2
approximation-based modeling, 1
artificial intelligence (AI), 126, 128
artificial neural networks (ANNs), 22, 215, 246, 295, 390, 395
ASITIC, 339, 343, 351
automated model generation (AMG), 82
automatic relevance determination (ARD), 21
average error, 69

average RMS error, 64
axial ratio, 45–46

B

back-propagation algorithm, 23
bandpass filter, 59, 63
bandstop microstrip filter, 253, 284–285
basis functions, 2, 10–11, 17–18, 24, 43, 383, 390
Bayesian model fusion, 24, 216
Bayesian optimization, 325–326, 328–329, 331, 335, 337–338, 346–347, 352–353, 357
Bayesian optimization (BO), 328–329, 338–339, 352, 354–356
bias error, 4
bit error ratio, 158
block design, 5
body-centric imaging applications, 134
bootstrapping, 28
borehole function, 412–413
Box Behnken, 162–164
box-Behnken design, 5
broadband patch antenna, 228, 231, 234
Broyden, 196, 201, 204

C

calibration, 362–363, 366, 373, 378
candidate point search, 326, 328, 330–331
candidate search, 325, 340, 348
cascode, 387
center frequency, 65
central composite design, 5
characteristic points, 61–62, 65–66, 297
circuit designer, 388
clock and data recovery (CDR), 158, 160–161, 163, 184, 198
CMOS, 336, 383
co-kriging, 13, 216, 296, 325, 344, 346–347, 405, 409, 412–413, 421

coarse mesh, 45
coarse model, 84, 155, 196–199, 202, 204, 258–260, 284
coarse-discretization, 40, 43, 46, 56
coarse-discretization EM models, 258
coarser model, 44, 46, 49
cognition-driven design, 70
compact microstrip rat-race coupler (RRC), 317
compact quadruple-band indoor base station antenna for 2G/3G/4G/5G systems, 141
compact slotted monopole antenna for ultra-wide band (UWB) body-centric imaging applications, 123
compact slotted monopole antenna for UWB, 134
compact slotted monopole antenna for UWB body-centric applications, 135, 148
compact slotted monopole antenna for UWB body-centric imaging applications, 127
complementary-metal-oxide-semiconductor (CMOS), 336, 382–383
complex pole/residues, 95
computational efficiency, 218
computational model, 68, 228, 232, 311
computational time, 44
Computer Simulation Technology–Microwave Studio (CST-MWS) trust region framework, 134
constrained modeling, 293
conventional model, 69
conventional sampling, 265
conventional sampling methods, 247
convergence, 50, 71
correction, 40
correction matrix, 70
correction method, 53
correction technique, 41
correction terms, 59

correlation function, 12
correlation matrix, 12
correlation parameters, 13
correlations, 217, 300
cost function, 19
coupling, 338
covariance function, 21
covariance matrix, 22
cross-coupling effects, 295
cross-validation, 11, 25, 28
crosstalk, 338
crossvalidation, 20
CST, 52, 71
CST Microwave Studio, 68, 71, 309, 312–313
CST MWS transient solver, 46
CST-MWS particle swarm optimization (PSO), 134–135, 138–139, 141, 144, 148
CST-MWS TRF, 135, 138–139, 141, 144, 148
CTLE, 163, 198
curse of dimensionality, 2, 40–41, 216

D

data acquisition, 4
data-driven, 39, 215
data-driven modeling, 2, 9, 20, 22
data-driven models, 1–2, 58–59, 67
data-driven surrogate modeling, 246
data-driven surrogates, 2–3, 59, 65, 215, 245
DE algorithm, 131, 256
decomposing, 363–364
decomposition, 363, 365, 370
deep neural nets, 347
deep neural networks, 83
Delaunay triangulation, 8, 224, 250–251
derivative free optimization (DFO), 276
derivative free TR optimization, 278
derivative free trust region (TR), 276
design center, 271–273, 275–277, 284
design centering combines, 282

design for test (DFT), 159, 174, 191, 197
design objective, 219
design of experiments (DoE), 1, 3–8, 41, 155, 162–165, 170, 174–178, 180, 182–183, 187, 189, 196, 198, 203, 222, 343
design of experiments procedure, 226
design optimality, 218
design optimization, 42, 44, 47, 69
design space, 2, 7, 60, 65, 67, 337, 343
design space dimensionality, 296
design specification, 298, 392, 399
design utility, 223, 229, 238
design variables, 60
design-ready surrogates, 303
determinant, 22
dielectric permittivity, 53–54, 68
dielectric resonator, 68
dielectric resonator antenna (DRA), 44–45, 52, 65–66, 68–69
differential evolution (DE), 128, 134, 256
dimensionality, 39, 384, 389
discretization density, 44, 47
distance matrix, 91
distribution, 22
DoE strategies, 41
domain confinement, 303, 320
domain definition, 225
domain thickness, 222–223
dropout, 347
dual-band bandpass filter, 48
dual-band dipole antenna, 228
dual-band microstrip dipole antenna, 228
dual-band microstrip filter, 48
dual-band uniplanar dipole antenna, 307, 311

E

effective pole/residues, 95
efficiency, 378
efficient, 363, 368, 378
electrical validation, 156

electro-mechanical tuners, 364
electromagnetic simulation, 337
EM analyses, 213, 295, 306, 315, 318, 321
EM antenna model, 227
EM antenna simulations, 314
EM cross-couplings, 49
EM model, 65
EM model responses, 65
EM sensitivity information, 90
EM simulation driven optimization, 329
EM simulation-based optimization, 339
EM simulations, 53, 55–56, 66, 294, 311, 314, 316, 319, 337
EM solver, 352
EM-based Monte Carlo, 311
EM-based Monte Carlo analysis, 312, 316, 320–321
EM-based Monte Carlo simulations, 297
EM-based statistical analysis, 301
EM-based stochastic optimization, 296
EM-based yield, 320
EM-based yield estimations, 313
EM-driven design, 214
EM-driven Monte Carlo, 308
EM-simulation model, 222, 313
empirical risk minimization, 19
engineering design, 293
enhanced space mapping, 59
epistemic, 294
EQ coefficients, 183, 188, 191, 194, 198, 201–202
equal-power split microstrip rat-race coupler, 297
equalization, 153, 158, 160, 170–171, 180, 182–183, 190, 202–203
equivalent circuit models, 40, 49, 53–54
equivalent network models, 214
error estimation, 25
error function, 24–25, 28, 87
error measures, 25

Ethernet, 154, 156
Euclidean distance, 250, 253
Euclidean distance-based sampling, 250
expedited design automation of antennas, 126
experimental design, 4
exploitation, 7–9, 330, 340
exploitative, 28
exploration, 7, 9, 330, 340–341, 344
eye diagram, 155, 158–160, 171–172, 181, 183–188, 191–192, 194, 201, 203

F

5G technology, 214
fabrication, 382
factorial designs, 5, 41
factorial DoEs, 5
fast replacement models, 295
feasible region, 274–276, 278, 289
feasible region (*FR*), 272
feature point, 65, 67–68
feature-based model, 69
feature-based modeling, 64–65
feature-based optimization (FBO), 70, 215, 229
feature-based surrogate, 68–69
feature-based technique, 69
feed-forward, 23
field-effect-transistors, 382
figure of merit, 298
figures of interest, 218, 223, 297
filter, 49, 55, 59, 63
fine model, 170, 172, 177–179, 182, 186–187, 196–199, 201–202, 204, 260–262, 284, 287
fine-discretization, 56
fine-model, 200
finer models, 46, 49
finite difference time domain (FDTD), 339
finite differentiation, 306
finite element method (FEM), 339
first-order consistency, 42, 51, 69, 71

forward modeling, 213
fractional factorial design, 5
frequency characteristics, 61, 64
frequency scaling, 55–56, 94
frequency shifts, 56
frequency-scaled surrogate, 56
full factorial design, 5, 247
full factorial sampling, 248
full wave EM simulation, 337
full-wave electromagnetic (EM) analysis, 293
full-wave electromagnetic (EM) simulation models, 295
full-wave electromagnetic (EM) simulations, 44, 213, 246
full-wave EM analysis, 214
full wave EM simulator, 276
full-wave EM simulation, 2, 43, 65, 338
full-wave EM solver, 284, 337
function approximation layer, 58
Fuzzy System surrogate, 59

G

gain coefficient, 89
Gaussian, 11–12, 16, 21–22, 300
Gaussian correlation, 327
Gaussian distribution, 301
Gaussian kernels, 20
Gaussian process (GP), 21, 128, 130, 328, 344–345
Gaussian process regression (GPR), 10, 20–22, 25, 216, 246
Gaussian quadrature, 17
generalization, 54, 60
generalization capability, 3, 19, 24, 28, 53
generalization error, 20, 25
generalization estimator, 25
generalization performance, 170–171, 173, 176–178, 180–181, 183
generalized regression neural networks (GRNN), 162, 166, 176–177, 181
generalized shape-preserving response prediction, 61

generalized space mapping (GSM), 271, 277–278, 282, 289
generalized shape-preserving response prediction (GSPRP), 61–64
genetic algorithm, 263–265, 330
geometrical approaches, 275
global optimization, 9, 13, 23
global sampling, 251, 253, 257
GP surrogate model, 131, 134
GP surrogate modeling, 129
gradient-based procedures, 214
granularity, 8
grid convergence studies, 41, 50
grid density, 50
grid sampling, 6
GSM technique, 271
GSPRP, 61
GSPRP model, 63–64

H

Halton sequence, 8
Hammersley sampling, 7
Hessian matrix, 280–281
high-dimensional model representation (HDMR), 3, 216
high-dimensional problem, 349
high-fidelity, 39–40, 42, 337, 340–341
high-fidelity data, 58–59
high-fidelity model, 10, 40–41, 43, 45–48, 50–55, 57–58, 61–62, 64–65, 69–72
high-fidelity models, 40, 56, 71
high-fidelity simulation, 4, 339
high-fidelity simulator, 350
high-frequency component, 246
high-frequency electronics, 18, 42–44
high-frequency structures, 1–2, 40, 43, 61, 293, 297, 320
high-frequency structures systems, 293
high-speed input/output (HSIO), 153–156, 158, 160–163, 169, 173–176, 180–181, 189, 196–198, 203–204

hybrid adaptive sampling, 254
hybrid differential evolution for antenna optimization, 123
hybrid sampling, 245–246, 251, 255–257, 265–266
hybridization, 296
hyperbolic truncation, 17
hyperparameters, 12–13, 21, 345

I

impedance tuner, 361
implicit SM, 55, 58
implicit space mapping, 54
importance sampling, 275
inductance, 386, 388, 391–392, 394, 397, 400
inductor, 351, 388, 396–399
inductor design, 351
inductor design summary, 352
infill criteria, 8–9
infill points, 8, 66
infill samples, 7–8
infill strategy, 8
infilling sample criterion, 257
infilling samples, 7, 251, 253
initial samples, 251
input return ratio, 392
input space mapping, 52–54, 261
intellectual property (IP), 335
internet of things (IoT), 214
inverse problem, 325
IP redesign, 354
IP reuse, 336
IP topology, 336

J

Jacobian, 71, 283, 306
Jacobian matrix, 304
jitter, 154, 158, 160–161, 187–191, 203–204
jitter tolerance (JTOL), 158, 160–161, 187–188, 190, 192–193, 195
joint probability density, 15

K

kernel approach, 19
kernel function, 19
knowledge-based model, 84
kriging, 1, 3, 10, 12–13, 20, 25, 58–59, 64, 68, 155, 162, 166–168, 176–177, 181, 187, 189, 198, 204, 215–216, 228–229, 327, 390, 393–396, 405, 407, 412–413, 421
kriging interpolation, 64, 69, 72, 215, 222, 234, 301, 305, 307
kriging interpolation model, 220–221
kriging interpolation surrogate, 224
kriging model, 276, 409
kriging surrogate, 64, 293
kriging, RBF, 217
Kronecker symbol, 15

L

Lagrange polynomial, 18
latin hypercube sampling (LHS), 6, 58, 64, 227, 245–247, 249–251, 255–257, 271, 275, 283–284, 287, 343, 346, 408, 411
learning error, 171–172, 178–179
least squares approximation, 280–281
least-angle regression, 18
least-square, 10, 18
least-square regression, 18, 24
least-squares optimization problem, 71
leave-one-out cross validation, 346
leave-one-out error, 25
length-scale parameter, 21
linear correction, 71
linear expansion, 306
linear interpolation, 63
linear mappings, 57
linear regression, 19
linear regression problem, 60
local sampling, 251, 257
local surrogate, 259–260, 263
log marginal likelihood, 21
loss function, 19

Index

low noise amplifier (LNA), 381, 386–389, 391–393, 397–400
low-fidelity model, 40–42, 44–45, 47–54, 56–59, 70–71, 262, 337, 340, 350
low-fidelity model development, 43
low-fidelity model selection, 49
low-fidelity modeling, 42, 44
low-fidelity simulations, 339

M

machine learning, 123
machine-learning assisted optimization of antennas, 126
machine learning techniques, 126
macromodeling, 385
manifold mapping (MM), 70–72
manufacturing tolerances, 296, 300
mapping, 61
mapping parameters, 54
material dispersion, 43
maximizing likelihood estimate (MLE), 345
maximum likelihood, 13
mean square error (MSE), 9, 224
measurements, 361–363, 365, 371
mechanical passive tuners, 362
medical imaging, 214
merit function, 50, 218–219
mesh coarseness, 44
mesh step, 47
metallization thickness, 43
metamodels, 2
method of moments (MoM), 339
microelectronics, 381
microstrip, 55
microstrip antennas, 214
microstrip bandpass filter, 58, 63
microstrip circuits, 2
microstrip filter, 54, 61, 63
microwave components, 246, 251
microwave couplers, 299, 305, 308
microwave engineering, 2
microwave filter, 53
millimeter-wave, 382

miniaturized rat-race coupler, 294, 307
minimax, 295, 298, 303
minimax problem, 299–300
minimax specifications, 299
minimum noise figure, 388
mismatch of poles, 90
mixed integer optimization, 346, 351
mixed integer-continuous problems, 325
mixed-integer optimization, 330
mixed-signal, 24
mixer, 332, 336
mixer design, 333–335, 337
MLP models, 256
MLP neural network, 257
model accuracy, 49
model complexity, 399–400
model correction, 49, 52, 54–56
model domain, 52, 58, 64, 213, 296
model fidelity, 48
model generalization, 25
model identification, 3, 28, 53
model optimization, 216
model order reduction techniques, 85, 296
model predictive power, 213, 217
model refinement, 87
model validation, 3, 42, 233
model verification, 55
model-order reduction (MOR), 3, 89
modeling error, 4, 28, 58
modeling process, 65, 218
modeling technique, 18, 70
modified candidate Search, 326
modified surrogate-based derivative-free TR optimization algorithm, 277
modified surrogate-based TR optimization algorithm, 277, 282
modified TR algorithm, 278, 281, 283
modified trust region algorithm, 271, 278
moment method solvers, 43
monopole antenna structure, 261
Monte Carlo, 273, 275, 295

Monte Carlo analysis, 300, 310–311, 314, 316, 319
Monte Carlo sampling (MCs), 245–248, 251, 253, 255–257, 409, 416
Monte Carlo simulation, 14, 294, 296
Monte Carlo yield estimate, 274
moving least squares (MLS), 20
MPVL algorithm, 114
multi-fidelity, 406
multi-fidelity local surrogate, 260, 266
multi-fidelity model, 339, 344
multi-fidelity modeling, 339
multi-fidelity simulators, 343
multi-fidelity surrogate, 405
multi-fidelity surrogate, 260
multi-fidelity surrogate model, 337
multi-fidelity surrogate-based optimization, 340, 343
multi-fidelity surrogate-based optimization with candidate search (MFSBO-CS), 337, 340–342, 351–354
multi-fidelity surrogates, 258
multi-harmonic tuner, 364, 371
multi-objective optimization, 325–326, 332, 334, 336
multi-objective RF circuit optimization, 325
multi-point OSM, 59
multi-point output space mapping, 59
multi-point response correction, 51–52
multi-point space mapping, 57
multi-point-corrected, 53
multilayer perception (MLP), 83, 256
multilayer perceptron (MLP) neural network, 255
multiphysics parametric modeling, 83
multiple-objective optimization, 355
multiplicative response correction, 50
multivariate polynomials, 15

N

narrow-band antenna, 65–66
Nelder-Mead, 186
nested kriging, 213, 216–220, 225–226, 228–229, 231, 234, 237
nested kriging framework, 213
nested kriging modeling framework, 222
nested kriging models, 229
nested kriging surrogate, 225, 232
network equivalents, 295
neural network, 3, 10, 19, 21–23, 162, 166, 169, 172, 204, 215, 276, 296
neural network model, 23
neuro-TF model, 86
neuro-transfer function, 84
noise figure, 381
nominal design, 299–301, 305–307, 310, 312–314, 316, 319
non-destructive testing, 405
nonlinear, 64 65, 67
nonlinear least-squares regression, 23
nonlinear regression model, 19, 58
nonlinear scaling, 54
nonlinearity, 65
nonlinearity of the system responses, 40
normalized root mean squared error (NRMSE), 415–416
novel compact quadruple band antenna for 2G/3G/4G/5G systems, 123, 148
novel compact quadruple-band indoor base station antenna for 2G/3G/4G/5G systems, 127, 134
numerical validation, 307

O

objective function, 69
objective space, 218, 220, 224, 227–228, 232, 238
on-chip, 386, 397–398
one-point response correction, 52
operating conditions, 213
optimal design, 405, 416

optimal design, 406, 411, 417–418
optimal system design, 272
optimization, 123, 384–386, 388–391, 393, 398, 401
optimization of antennas, 125
order-changing technique, 85
original neuro-TF model, 98
orthogonal array, 7
orthogonal arrays, 162–164
orthogonal matching pursuit (OMP), 24
orthogonal polynomials, 16
orthonormal polynomials, 15, 18
OSM, 53
OSM surrogate, 59
output probability distributions, 14
overfitting, 24, 384

P

parallel computing, 123, 126
parallel surrogate model-assisted, 123
parameter extraction (PE), 57, 283
parameter shift, 54
parameter space, 218, 220, 224, 320
parametric models, 82
parametric optimization, 1, 69, 213
parametric yield optimization, 24
passive microwave components, 22
passivity enforcement, 115
pattern classification, 22
PC-co-kriging, 405, 407, 409–410, 412–414, 421
PC-kriging, 296, 405, 412–413, 421
penalty coefficient, 219
penalty factor, 300
penalty function, 219
perceptrons, 162, 170
performance figures, 219–220, 299, 304
performance measures, 272
performance requirements, 295
performance specifications, 294, 300, 303
performance-driven, 213

performance-driven modeling, 213, 237, 297, 320
performance-driven surrogates, 305, 308, 321
Peripheral Component Interconnect Express (PCIe), 154, 156, 160, 172, 174, 194–195, 202
physical layer, 153–154
physics-based, 215
physics-based low-fidelity model, 50
physics-based model, 68
physics-based modeling, 42, 49
physics-based modeling techniques, 50
physics-based models, 39–41, 57, 65, 69
physics-based optimization methods, 215
physics-based surrogate, 39, 41–42, 44, 61
physics-based surrogates, 39–41, 50, 56, 69
pole–residue tracking, 94, 97
pole–residue-based transfer function, 93
pole-matching, 90
pole-zero-based transfer functions, 89
pole/zero-matching algorithm, 90
polynomial approximation, 3
polynomial-based surrogate modeling (PSM), 162, 165, 176–178, 180–182, 186–187
polynomial chaos expansion (PCE), 1, 10, 14–18, 295–296, 301, 407, 409, 412–413, 416, 421
polynomial regression, 9–10, 215, 246, 260
polynomial surrogate, 179
polynomials, 10
post-silicon validation, 153–156, 158, 162–163, 169, 172, 181, 187–189, 191, 196–197, 202–204
posterior distribution, 22
prediction capability, 25
predictive power, 2–3, 24, 224, 229, 238, 302, 311–312, 317, 320
preliminary training, 87, 99

principal component, 334
principal component analysis (PCA),
  296, 325, 326, 331–332, 334, 383
prior distribution, 22
probability density function (PDF),
  273–274, 277
probability distribution, 14, 275
problem-specific knowledge, 39–40,
  51, 70
process variations, 384–385
process, voltage, and temperature
  (PVT), 153, 158
projection method, 17
PSADEA method, 123, 127–129,
  131–135, 138–139, 141, 146–148
pseudo-random sampling, 6
pseudoinverse, 70

Q

quadratic model, 271, 277–279, 281
quadratic programming, 19
quadrature methods, 17–18
quality factor, 381, 386, 397–400
quasi-Monte Carlo sampling, 7

R

radial basis function (RBF), 1, 10–12,
  58–59, 215, 228–229, 234, 276, 301,
  327
radially symmetric functions, 11
radio frequency integrated circuits
  (RFICs), 382
random process, 13
rank-one Broyden, 304
rat-race coupler (RRC), 317–318
rational transfer function, 85–86
rectangular DRA, 68
reference design triangulation, 217
reference designs, 216, 218, 220,
  223–224, 227, 229, 231, 238,
  304–305, 309, 312–313
reference point, 57, 60
reflection characteristic, 66, 303
reflection coefficient, 45–47, 53,
  65–66, 68

reflection response, 67
region of interest, 57
regression function, 19
regression model, 11
regression surrogates, 20
relative error functions, 25, 27
relative RMS error, 310, 318
reliability, 384
residuals, 58
response correction, 50–51, 57, 215
response correction techniques, 57
response feature modeling, 65
response features, 65–67, 297
response scaling, 60
response surface (RS), 327, 338–339
response surface approximations, 296
response surface modeling, 24
response surfaces, 276
RF circuit designer, 396
RF circuit optimization, 325
RF power amplifiers, 22
RFIC, 382, 400
ring-slot antenna, 307–309
robust design, 296
root mean square (RMS), 255
root mean squared error (RMSE), 58,
  64, 68, 408, 410, 414–415, 418
root-relative squared error (RRSE),
  390, 393–394, 398

S

S_{11}, 392–393, 396–398
SADEA, 128–129
SADEA-II, 128–129
SAEAs, 126
sampling method, 265, 349
sampling techniques, 389
SAO-based algorithm, 321
SATA3, 192, 194
satellite communications, 214
scattering parameters, 49
second stage training, 88
second-generation SADEA, 128
semi-parametric, 21

sensitivity analysis, 83, 85, 383, 405–406, 410, 413–414
sensitivity information, 71
sensitivity-analysis artificial neural network, 110
sensitivity-analysis-based neuro-TF model, 98–99
sequential approximate optimization, 301, 317
sequential approximate optimization (SAO), 297, 301, 320
sequential design, 8, 385, 389, 392, 394, 398
sequential DoE, 7–8
sequential LHS, 8
sequential sample, 340, 349
sequential sampling methods, 5, 7–8, 215
sequential sampling strategies, 13
Serial Advanced Technology Attachment (SATA), 154, 156, 160, 172–174, 176, 178, 180–182, 187, 193, 196–198, 201
shape-preserving response prediction (SPRP), 61, 70
short-channel, 383, 385
signal variance, 21
simplex, 224–226
simplified physics, 43
simulation time, 44, 46, 49, 68, 71
simulation-driven design, 1, 24, 215
singular value decomposition (SVD), 70, 327
size reduction, 262
squared exponential (SE), 21
SM surrogates, 283, 271
SM surrogate model, 277–278
SM-fuzzy, 59
SM-Kriging, 59
SM-RBF, 59
SM-standard, 59
Sobol, 162–165, 174, 198
Sobol sampling, 349
Sobol sequence, 8, 349
Sobol sequence sampling, 349

Sobol's indices, 405–407, 410–411, 416–417
Sobol150, 175
Sobol50, 176–177, 179–183, 186–187
space mapping (SM), 54, 57–58, 70, 83, 153, 155, 196–197, 200, 203, 215, 261, 271–272, 276–278, 282, 289, 296, 326, 332, 338, 383–384
space-filling designs, 3, 5–6, 8
space-filling DoEs, 6–7
space-filling sampling method, 251
space-infill sampling method, 245, 247, 250
space-mapping model, 57
SPICE simulations, 338
SPICE/EM simulation, 327
split-sample method, 25
splitting pole, 96
square law, 383
standard SM model, 57–58
standard SM surrogate, 57
star distribution, 5, 41, 60
statistical analysis, 271, 275, 293–296
statistical approaches, 275
statistical design centering, 271, 275, 277–278, 283, 288
statistical figures of merit, 298
statistical moments, 14, 295–296, 298
statistical performance measures, 294
statistical surrogate model, 344
stochastic design, 294
stochastic moments, 15
stochastic variations, 14
stratified Monte Carlo, 275
structural risk minimization, 19
substrate height, 53
support vector machine, 162, 246
support vector regression (SVR), 1, 10, 18–19, 216
surface approximation, 295
surrogate, 295, 318, 406–409, 411–412, 415–416, 418–419, 421
surrogate-based optimization (SBO), 41, 50, 123, 126–128
surrogate model, 3, 6, 10, 19–20, 25, 50, 54, 56, 63, 65, 67–69, 222,

228–229, 232, 258–261, 263, 276–277, 297, 301–302, 306–307, 309, 312, 315, 318, 337, 340–341
surrogate model domain, 220–223, 238, 297, 302–305, 307–308, 310, 318
surrogate model-assisted evolutionary algorithms, 126
surrogate model-assisted global optimization for antenna design, 147
surrogate model-based optimization, 325
surrogate model-aware evolutionary search (SMAS) framework, 127–129
surrogate modeling, 1, 3–4, 28, 67, 126, 216, 245–247, 265–266, 297, 327, 396, 401
surrogate models, 64, 213, 215, 246, 276, 296, 309, 312, 315, 317
surrogate refinement, 260
surrogate-assisted algorithm, 69, 319
surrogate-assisted optimization, 8, 50, 69, 215, 245–247, 257–258, 261–262, 265
surrogate-assisted statistical analysis, 296
surrogate-assisted yield optimization, 293
surrogate-based model, 276
surrogate-based optimization, 41, 49, 123, 325, 337, 341
surrogate-based statistical analysis, 308
surrogate-based TR optimization algorithm, 278
surrogate-based yield optimization, 301–302
suspended DRA, 52–53
symmetrical bandstop microstrip filter, 284, 289
system margining, 158, 163, 175, 183, 187, 190
system margin validation (SMV), 158–159, 173–174, 197
systematic, 294

T

target operating frequencies, 298
testing, 155, 160–163, 170–172, 175–177, 179–182, 192, 204
tetrahedral mesh, 45, 68
time-domain analysis, 115
time-domain solvers, 43
time-to-market (TTM), 154, 156, 158, 203, 401
tolerance-aware design, 296, 298
tolerances, 293, 295
TR algorithm, 271, 278
TR optimization, 277, 280
TR sub-problem, 279
training, 162–163, 166, 169–173, 175, 178–180, 204
training cost reduced-SADEA (TR-SADEA), 128–129
training data, 2–4, 19, 21
training data acquisition, 3, 40, 222
training designs, 67
training error, 100
training set, 65
transfer function orders, 94
translation vectors, 62–63
transmission zeros, 59
trend function, 12
triangulation, 217, 224
triple-band dipole, 307
triple-band uniplanar dipole antenna, 313, 315
truncation, 18
truncation error, 18
truncation scheme, 17
trust region framework (TRF), 134
trust region (TR), 276–277, 279, 282, 288
trust region (TR) optimization algorithm, 271
trust-region gradient search, 304
tunable, 386–387, 398
tuner, 361–364, 366, 369, 371–373, 378

tuning, 153–158, 175, 183, 186–187, 196, 198, 201, 203–204
two-stage GPR, 216
two-stage training, 87

U

ultra-wideband, 71
ultra-wideband multiple-input–multiple-output antenna, 285
ultrasonic testing, 405, 413, 415–416
ultra-wideband (UWB), 71
uncertainties, 294, 296
uncertainty quantification, 1, 293, 295
uniform distribution, 301
uniform grid sampling, 6
uniform probability distributions, 309, 312, 314, 318
uniform sampling, 246
univariate polynomials, 15
USB3, 173, 182, 189–192, 194
USB3.1, 180, 186
Universal Serial Bus (USB), 154, 156, 160, 172, 174, 181, 183, 202
UWB MIMO antenna, 285, 286, 288–289
UWB monopole, 72
UWB monopole antenna, 263–265
UWB multiple–input-multiple-output (MIMO), 284

V

validation, 3, 24, 28, 410
validation set methods, 25
variability parameters, 414
variable thickness domain, 224
variable thickness model, 236
variable-accuracy models, 43–44
variable-fidelity models, 3, 216
variable-fidelity physics models, 43, 48
variable-fidelity simulations, 296
variable-fidelity training data, 13

variable gain amplifier (VGA), 158, 163, 198
variable-resolution, 43–44
variable-thickness domain, 213, 217, 222–223, 225, 227–229, 231, 234, 236, 238
variable-thickness models, 229
variable-thickness nested kriging model, 229, 234
variable-thickness nested kriging surrogate, 232, 236
variable-thickness surrogate model, 232, 236
variance reduction techniques, 275
VCO, 354–356
vector fitting, 93
vector-fitting approach, 86
vector-valued, 53–54, 57, 69
vector-valued models, 51
vector-valued outputs, 60, 64
velocity saturation, 383
Voronoi cell, 250
Voronoi diagram, 250–251
Voronoi tessellation, 8, 250

W

weight setting, 325
weighted least square, 20
weighting factors, 171–172, 185
wire-bonding, 392
wireless communications, 214
worst-case analysis, 24, 295

Y

yield, 274, 284–285, 287, 289, 293, 295–296, 298, 301–302, 304, 307
yield estimation, 301, 309, 311, 318, 320
yield function, 271–278
yield integral, 273–274
yield optimization, 288, 297–303, 305, 307–310, 312–315, 318, 320
yield optimization framework, 321
yield optimization problem, 298
yield values, 289

yield-driven design, 293, 296
yield-optimized design, 297, 312, 314–316, 319, 321
yield-optimum design, 302

Z

zero-matching, 92
zero-order consistency, 52, 69

CPSIA information can be obtained
at www.ICGtesting.com
Printed in the USA
BVHW051642200322
631697BV00002B/7